POLYMER BLENDS

Processing, Morphology, and Properties

POLYMER BLENDS

Processing, Morphology, and Properties

Edited by

Ezio Martuscelli
Rosario Palumbo

Institute of Research on Technology of
Polymers and Rheology, CNR
Naples, Italy

and

Marian Kryszewski

Center of Molecular and Macromolecular Studies
Polish Academy of Sciences
Lodz, Poland

PLENUM PRESS · NEW YORK AND LONDON

Library of Congress Cataloging in Publication Data

Joint Italian-Polish Seminar on Multicomponent
    Polymeric Systems, 1st, Capri, 1979.
    Polymer blends.

    Includes index.
    1. Polymers and polymerization—Congresses. I. Martuscelli, Ezio. II. Palumbo,
Rosario. III. Kryszewski, Marian. IV. Title.
QD380.J64 1979   547.8'4   80-22862
ISBN 0-306-40578-4

Proceedings of the First Joint Italian—Polish Seminar on
Multicomponent Polymeric Systems, held in Capri, Italy,
October 16–21, 1979.

PREFACE

Multicomponent polymeric systems, or polymeric blends, have
recently created considerable interest and they represent a new and
important challenge for research. These systems have already become
technologically important, but the prospects for their applications
have by no means been exhausted. For thermodynamic reasons polymer
blends do not usually form homogeneous mixtures but exhibit micro-
or macrophase separation. This incompatibility has some inherent
advantages as varying the composition and the processing conditions,
materials with different structures and morphologies can be obtained
whose properties may be superior to those of one of the components
e.g. high impact resistant plastics.

Investigations of multicomponent polymer systems constitute a
new branch of macromolecular science which now claims as much
interest as the behaviour of dilute solutions, crystallization,
statistics of chains, tacticity, and single crystal formation did
a few years ago. The complexity of the problems related to controlled preparation and properties studies of multicomponent polymer
systems is such that it is often more practicable to conduct them
on an international basis. The aim of the first Joint Italian-Polish
Seminar on Multicomponent Polymer Systems was to discuss recent
results obtained in that field in both countries. As the range of
topics to be covered was fairly wide, we thought it advisable to
invite scientists from other countries to share with us some aspects
of their own studies.

Considering the fact that no generally accepted definition of
polymer blend is as yet available we take a polymer blend, or a
polymer alloy, to be any combination of two or more polymers resulting from common processing steps, such as mechanical blending,
solution casting or, in some cases, from chemical synthesis. Since

block and graft copolymers have been dealt with at a number of
Symposia and in some recent publications, the discussion at the
First Italian-Polish Seminar concentrated on some aspects of polymer
blend preparation and studies of their properties in solid state
rather than in the molten state or in solution. That does not mean
that thermodynamic aspects of immiscibility, which is obviously the
basis for morphology and, by the same token, for the properties of
these systems, are felt to be unimportant.

The topics highlighted in the discussion included some new
aspects of preparation and properties studies of polymer blends,
such as their thermal stability, the effect of block and graft copolymer on polymer blends, as well as some problems related to the
formation of interpenetrating networks and their properties.

Polymer blends provide a large range of morphological structures. Aside from the discussion of the specific possibilities of
disperse phase systems by assignement of size and shape distributions
and geometrical arrangement of the particles, considerable attention
was paid to the formation and transitions of the complex structures
that arise in multicomponent systems containing crystalline polymers.

Properties, along with components and structure, constituted
one of the major topics discussed with special attention having been
devoted to the mechanical properties of partially crystalline and
amorphous blends in a wide temperature range. Also commented on
were the models proposed for the elucidation and prediction of the
properties of such blends. In view of the connection that exists
between polymer blends and filled materials in terms of phase
adhesion, we thought it fit to deal with this subject.

In deciding on the above topics we were fully aware that we
were leaving many gaps unfilled, a situation only to be expected
in a field developing as rapidly as the science and technology of
polymer blends. It seems, however, that the papers included in
this volume do indicate certain perspectives and will serve as a
source of inspiration for future work. The expected second volume
will contain the results to be presented at the Second Joint
Polish-Italian Seminar, and we hope that insights contained there
in will contribute to the further progress in the field.

The editors wish to express their gratitude to the authors who

contributed to this book and to the Polish Academy of Sciences and
Italian Research Council for their financial support to the reali-
zation of the First Italian/Polish Joint Seminar on Multicomponent
Polymeric Systems.

The editors are particularly grateful to Mrs.Cristina Carelli
for secretarial help.

Finally, they appreciate the help and cooperation of many
people of the Institute of Research on Polymer Technology and
Rheology of the Italian Research Council.

Istituto di Ricerche su Tecnologia   Ezio MARTUSCELLI
dei Polimeri e Reologia, C.N.R.,
Arco Felice, Napoli, Italy   Rosario PALUMBO

Centre of Molecular and   Marian KRYSZEWSKI
Macromolecular Studies
Polish Academy of Sciences
Lodz, Poland

CONTENTS

RECENT PROGRESS IN THE STUDIES ON THE PREPARATION

AND PROPERTIES OF POLYMER BLENDS

M. Kryszewski

Center of Molecular and Macromolecular Studies

Polish Academy of Sciences, Łódź, Poland

Owing to their potential technical importance polymer blends
currently command considerable scientific interest.

A polymer blend will be defined here as a combination of two
or more polymers resulting from common processing steps e.g. mixing
of two polymers in the molten state, casting from common solvent
etc. These preparation methods do not usually lead to chemical bonding between the components. This particular definition of polymer
blends or polymer alloys excludes graft copolymers, block-copolymers,
(partially) semiinterpenetrating networks and interpenetrating networks as well as crosslinked copolymers even though from the structural point of view these materials show many similarities with the
above mentioned blends eg. heterogeneity. It is for this reason
that such systems are often discussed together with typical polymer
blends.

For thermodynamic reasons which will not be discussed here in
detail, most polymer pairs are not miscible on molecular level.
Neverthless the degree of compatibility of two polymers may vary to
a high extent, and in discussing typical polymer blends with separated phases which may interact with each other it may be necessary
to pay some attention to the interfacial agents. These substances
are used to increase the miscibility of two polymers and are often
called compatibilizers. The proper use of compatibilizers leads to
marked improvement of many properties of blends eg. of mechanical
properties.

Two polymers can be expected to be miscible when a very close match in cohesive energy density or in specific interactions, which produce a favourable enthalpy of mixing, is involved. The ability to cocrystallize can provide an additional driving force for polymer miscibility. Important progress has recently been made in the study of these crystalline systems and some remarks on such miscible blends seem to be in order.

Polymer blends exhibit a wide range of morphological states from coarse to fine ones. In addition to the usual possibilities of obtaining separate phases with various size, shape and geometrical arrangements of inclusions, more complex structures are possible, especially in crystalline polymer blends. For these systems specific properties can be expected eg. for gradient polymers which contain crystalline elements. One of the major areas of studies on polymer blends is the dependence of the mechanical properties on composition. This is due to the fact that these complex systems exhibit a behaviour which does not simply follow the sum of the properties of the components.

It seems interesting to note that until quite recently, relatively little attention has been given to the interaction of the components in terms of the thermal stability of the composition. This aspect of polymer blends is of substantial importance for processing and as such it seems to be worth discussing.

BLENDS WITH INTERFACIAL AGENTS AND COMPATIBLE BLENDS

Polymers can be blended to form a wide variety of random or structured systems with desirable combination of properties but in practice, these theoretically expected properties are not achieved because of poor adhesion of the phases created. This sistuation can be changed using certain species referred to as compatibilizers[1] which alter the interfacial conditions between the different phases. The compatibility may be achieved even when, in thermodynamic sense, two polymers are not miscible. For this purpose one applies mostly block or graft copolymers with long segments or blocks. It has been shown that the blocks constituting these molecules at the interphase have to be miscible with phases A or B (the blocks or segments of interfacial agents are chemically identical or similar to A or B) to reduce the interfacial energy between the immiscible phases. They also ensure finer dispersion during mixing and provide a higher stability against separation[2,3].

It has been found that block copolymers are usually better interfacial agents than graft copolymers because, in the latter, multiple branches will restrict the penetration into similar homopolymer phases. For the same reason diblocks are more effective than triblocks. Block or graft copolymers have to segregate into two phases in order to localize at the blend interface. This specific behaviour of block – and graft copolymers as well as their immiscibility only in one of the homopolymer phase depends on the interactions between two segments and on their molecular weights. The amount of compatibilizer required depends on many factors of which conformation and molecular weight are the most important. It is possible to estimate this amount for a given molecular weight M of the compatibilizer to saturate all of the interfaces in a blend. In this calculations one takes into consideration the interfacial area per unit volume of the blend in correlation with the volume fraction Φ_A of polymer A which is dispersed in the form of spherical particles of radius R. When each of the compatibilizer molecule occupies an area a at the surface then the ratio of the mass m of block copolymer required per unit volume of the blend equals $\frac{3\Phi_A M}{aR \cdot N_o}$ where N_o is Avogadro's number. When the value of a is assumed to be 50 Å, one can calculate for $\Phi_A = 0.1$ and R = 1 μm that one needs about 20 % of the block copolymer by weight with $M = 10^5$ to fill up the interface. When the molecular weight of the compatibilizer is lower, eg. 10^4, then this amount drops to 2%. These arguments show the advantage of lower molecular weight compatibilizers but it should be mentioned that higher molecular weight compatibilizers are needed in order to penetrate deep enough into phase A and to be anchored firmly. On the other hand the conformation[4,5] of block copolymer molecules is usually such that some restriction exists at the interface and the amount of block copolymers as compatibilizer must be still lower depending on cohesive energy.

The morphology of blends of polymer A with B and copolymers AB has been investigated by many authors[2,6-13]. These studies have revealed the existence of a pronounced emulsifying effect in solutions of incompatible polymers, an effect on domain size and stability of dispersions. Examination of the morphology of polymer blends

containing block copolymers (mostly cast from solutions) by electron
microscopy (using appropriate stanning methods) has supported the
above suggestions concerning the emulsifying and stabilizing effect
of these compatibilizers against segregation. The extensive studies
of Riess on the optical properties (transparency) of blends of A and
B with added copolymer AB in terms of the ratio of molecular weight
of A and B to the molecular weight of blocks A and B in copolymer
AB show that the above mentioned suggestion is correct.

The understanding of the behaviour and role of added block copolymers to the blend of A and B is based on the theories of block
copolymers properties in solutions. Many conclusions reached from
the studies conducted by several authors in the early seventies are
relevant here, but we would like to mention only those which are related to the thickness of the interface; according to the interesting statistical mechanics approximation of Meir[14] this thickness
depends on molecular weights M and solubility parameter difference
$(\delta_A - \delta_B)^2$ of two segments of AB. When the molecular weights are
high and $(\delta_A - \delta_B)^2$ is large then the interface thickness is small.
When M_A and M_B are low and $\delta_A \approx \delta_B$ one has really only one phase,
the system is homogeneous (in the "solid state" only the T_g value
can be found whose value is intermediate between those expected for
both components). This behaviour was experimentally found for block
copolymers of α-methylstyrene and styrene. However, when the molecular weight increases the phase separation manifests itself in the
occurrence of two T_g's. These remarks show how sensitive the structure of block copolymers is to the molecular parameters of blocks
and point out that the proper choice of the chemical structure of
the compatibilizers is very important.

There are some other restrictions on the use of such compati-
bilizing agents. Solubilization usually occurs when the molecular
weight of the blocks is higher than that of homopolymers. This observation is important because it imposes new requirement on the
relation of molecular weights of the components of the added block
homopolymers to the molecular weight of both homopolymers. This
important relation is well established for blends obtained from common solvents but it is not clear whether it can be applied to blends
obtained by mixing two polymers in the molten state. More research
on this problem is obviously in order.

In the above discussion we have dealt with compatibilizers

which were added as separately formed chemical species. Similar
situations can be expected when the graft polymer is formed during
the polymerization of a monomer in the presence of another polymer,
or when block copolymers are formed by interchange reaction in the
course of processing condensation polymer blends. There is also a
possibility of graft formation through a chain scission mechanism
in such processing operations as mastification.

Blending of two polymers may enhance the bonding of two dissimilar polymers; e.g. it has been shown that polyethylene can be
firmly bonded to impact modified polystyrene with a tricomponent
blend of polyethylene of impact modified polystyrene and of a sty-
rene-butadiene block copolymer[15]. The role of polystyrene-butadie-
ne copolymer is not clear and much more work will have to be done
before his behaviour is elucidated. The role of compatibilizers
can be summarized in the following way. If the two polymers are fairly well miscible the addition of block copolymers of similar structure can result in the formation of one phase system, and blocks
of low molecular weight are sufficient. When the polymers are
not miscible and the difference of solubility parameters $(\delta_A - \delta_B)^2$
is high, then block copolymers can act as interfacial agents if
their molecular weights are high enough.

There is ample literature describing the effect of compatibilizers on the mechanical behaviour of dispersed blends and laminates. Very illuminating are the studies on blends of polystyrene
with polyethylene[16-18]. Without addition of interfacial agents
they show rather poor properties especially at high deformations.
It is generally agreed that poor adhesion between the phases plays
a significant role. The use of additives in the form of graft copolymers[16] (styrene grafted on polyethylene by radiation or Friedel-
Craft alkylation of polystyrene with low density polyethylene[17])
results in improvement of many mechanical properties, such as tensile strength and elongation at break. The responce to mechanical
stresses depends not only on the amount of the graft added but also on the way of preblending and, as might be expected, on the distribution of the graft copolymer particles on the dispersed phase
surface. It follows from these few remarks that the quantitative
influence of interfacial agents on the mechanical response of dispersed blends is complex and not very well understood. In the case
of laminates the layer structure of the system makes it possible
to study the interfacial activity of block and graft copolymers

(in some cases a homogeneous polymer may be a good adhesive to two
polymers). Interesting results have been published for the system
poly(vinyl chloride) natural rubber with graft copolymer of methyl
methacrylate on natural rubber[19] as well as for ethylene-propylene-
diene terpolymer (EPDM) and styrene butadiene rubber (SBR) with
graft copolymer consisting of grafted styrene and butadiene on
EPDM[20]. It has been shown that the presence of graft copolymers
promotes wetting and that interpenetration of segments between two
phases occurs. In some cases polymer blends are used as glue layers
between dissimilar, nonadhering layers. Their role is quite clear
when the glue-layer consists of two generally compatible polymers
but usually this is not the case and the behaviour of such a glue
is far from being understood. This problem is particularly difficult in the case of coextruded multilayer films for which the added
blend and/or copolymer may be an "extrusion aid" to adjust the viscosity. So far no quantitative theory is available that could predict
the interfacial adhesion in final two layer films but it seems quite
certain that the use of polymer blends will play an increasing role
in obtaining adhesion in coextruded products.

It has already been mentioned that there is only a small number
of polymeric systems which may be considered compatible. A particularly interesting polymer which shows miscibility with various
polymers is poly(ε-caprolactone) (PCL) whose synthesis was described by Brode and Koleske[21]. Block copolymers and graft copolymers
of ε-caprolactone are also known. PCL is a partially crystalline
polymer so, depending on its fraction in the blend and on the kind
of the second polymer in the system, one can obtain homogeneous or
heterogeneous compositions e.g. homogeneous mixtures with poly(vinyl
chloride) or phase separated structures (mechanical blends) with
poly(vinyl acetate). When PCL is mixed with other partially crystalline polymers interactions are mainly responsible for interesting mechanical properties of these blends. From that point of view
the crystalline blends of PCL can be treated as model systems
for analysing the properties-structure relations in other crystalline composites. An interesting review of the blends containing
poly(ε-caprolactone) has been recently published[22].

Amorphous polymers like PVC destroy the crystallinity of PCL
(the blends with PVC can be readly prepared by hot compounding process) and lead to soft compositions. When the fraction of PCL in
the blend increases (> 30%) these systems are more rigid due to

crystallization. It is interesting to note that the rigidity increases with time due to PCL crystallization. This ageing effect shows
that the compatibility depends on the time scale. Studies of the
crystallization kinetics of PCL in blends with PVC have shown that
the crystallization rate and the induction time for crystallization
is critically dependent on the concentration of the components.
There are two stages of crystallization; one is nucleation controlled and the other is diffusion controlled (the latter is slower).
Thus, at higher concentration of PCL many crystallites start to grow
simultaneously and they are responsible for faster overall change
of crystallinity degree.

Compatible blends were also obtained from PCL and poly(hydro-
xyether), cellulose acetate butyrate and with poly(epichloro hydrin).
The blends with polystyrene, poly(vinyl acetate) and other amorphous
polymers show two glass transition temperatures which leads to conclusion that the domain size of each of the polymers are greater
than 10-30 Å.

Crystalline interactions have been found to exist when PCL is
blended with polyethylene and polypropylene. It was found that the
α-relaxation in polyethylene was affected. Because this relaxation
is related to motion in polyethylene crystallites this effect could
be elucidated by assuming that the blend might be cocrystalline in
nature. Similarly the X-ray patterns of the two polymers can lead
to some difficulties in the interpretation of diffraction studies
thus the hypothesis of cocrystallization is not substantiated but
still shows that effects of this type may occur. Combinations of
polypropylene and PCL have been studied by X-ray diffraction. Pronounced alternations of the relative intensity of some X-ray diffra-
ctions suggest some unusual interactions thus one should pay more
attention in the detailed studies of polymer blends in which both
components are able to crystallize.

Interesting conclusions on the structure of semicrystalline
blends have been recently obtained by Stein et al.[23]. Blends of isotactic polystyrene with atatic polystyrene and of PCL with PVC were
examined by small angle X-ray scattering and differential scanning
calorimetry. In the range of concentrations used both blends
were crystalline and volume filled with spherulites ($\geqslant$ 70 weight
per cent in PS and $\geqslant$ 50 weight per cent in PVC). It was found that
in iPS/aPS blends the melting point of iPS and the lamellae thickness was not changed. This suggests that the segregation of the

atactic component occurs during crystallization inside of growing
spherulites. In the case of PCL/PVC blends the degree of crystallinity and melting temperature of PCL decrease but the lamellar thickness C and long period (C+A) increase. The indicates that the PVC
is included in the amorphous regions between the lamellae during the
crystallization process.

These examples show different behaviour of blends in which one
component can crystallize thus the mechanical properties of these
systems may be different depending on the rate of cooling (rate of
crystallization) and on composition. These aspects of polymer blend
studies are of importance and seem to be worth developing.

GRADIENT POLYMERS WITH CRYSTALLINE STRUCTURAL ELEMENTS

In the course of industrial processing of polymers, such as
e.g. molding, extrusion etc., the material has to undergo flow process in the molten state followed by rapid cooling. Due to the inherent microstructure of polymer blends an anisotropy can often be
achieved by such operations. These effects have been studied in
detail for inhomogeneous block copolymers eg. triblock copolymers
of styrene - butadiene - styrene. The existence of anisotropic
structure was confirmed by small angle x-ray diffraction. For such
systems that have undergone shearing at high temperatures the
stress-strain curves are quite different for samples cut normal or
parallel to the shearing direction[24]. Similar behaviour was found
for blends obtained by mechanical mixing eg. for blends of polystyrene with polyethylene extruded at high rates[25]. It was suggested,
in both cases, that melt shearing had deformed the spherical domains
or particles in the flow direction. The elongated particles in the
longitudinal direction can easily merge with each other leading to
an increase of continuity. The particles in the transverse direction will not have connectivities. These effects must influence
the mechanical properties of these systems in both directions.

Another interesting class of anisotropic polymer blends are
the gradient polymers. Gradient polymers are multicomponent systems
whose structure or composition is heterogeneous throughout the material. This means that there is a gradient of their structure or
composition. Blends of this type can be produced by diffusing a
guest monomer into a host polymer in order to establish a diffusio-

nal gradient profile which later can by fixed by polymerization.
Mechanical properties of such systems are different from those found
for interpenetrating networks of similar composition. Most of the
work on the preparation and properties of gradient polymers has been
devoted to amorphous systems eg. poly(methyl methacrylates) with methyl acrylates or with halogenated acrylic monomers as the second
component[26-28]. The diffusion rate of a monomer in a glassy polymer
is low, so when the polymer is removed from liquid monomer bath before reaching an equilibrium swelling, a concentration profile of
the monomer in the polymer will be established. This concentration
profile can be fixed by rapid polymerization of the guest monomer
in the host polymer. The resulting blend is a gradient polymer.
If the diffusion takes place for a long time, until reaching the
equilibrium conditions (equilibrium concentration of guest monomer
in the host polymer), an interpentrating network is formed. The
properties of the host polymer, of the interpenetrating network system and of the gradient polymers are different. A lot of work has
been devoted to these materials, but it will suffice here to present
only one example. Pure poly(methyl methacrylate) undergoes brittle
fracture at low strains. When a gradient polymer is obtained by
diffusing methyl acrylate into the poly(methyl methacrylate) the
fracture strain increases very much with the concentration of the
guest. This behaviour was not observed for interpenetrating networks
of the same composition as that of the gradient polymers.

We expected to obtain interesting new materials with unique
properties by preparing gradient polymers from partially crystalline polymers with specific supermolecular structures. The morphological structure of the host can be modified by appropriate thermal
treatment of the crystallizable component. Diffusion of the guest
monomer into prepared film and the fixation of the created gradient
by polymerization leads, in that case, to the formation of structures whose properties can be modified by the structure of the host
film and appropriate gradient profile of the guest[29]. This concept
was verified by preparing new gradient polymers from low density polyethylene (LDPE) with known degree of crystallinity and morphology using styrene as the second component (guest).

Before discussing the results we obtained using the above mentioned way of preparing gradient blends of a new type, it seems necessary to comment briefly on heterogeneous crystalline polymer
blends. In the case of polymeric blends, when one of the components

is crystallizable, its crystallization will play a role in determining the morphology of the system (in the case of amorphous blends the dominant factor in the formation of the morphology is the free energy of mixing of the components). Many studies have been devoted to the morphology of block copolymers with one crystallizable component eg. ethylene oxide isoprene - ethylenoxide. For pure poly(ethylene oxide) well formed spherulites can easily be seen in cross-polarized micrographs. When the fraction of the amorphous polyisoprene increases, the spherulitic structure is less perfect (polyisoprene domains are dispersed in spherulitic poly(ethylene oxide) matrix). The arising morphological structures can be discussed depending on the concentration of the amorphous component. It has been shown that, even at high concentration of polyisoprene, when poly(ethylene oxide) constitutes the dispersed phase in the form of small inclusions it is crystalline. The amorphous phase is rejected from the crystallites. The crystalline phase seems to be composed of folded chain crystals. An alternative model, fringed micelle of crystalline regions, could be accepted too. In the case of block copolymers with crystallizable segments the morphology of the system depends on the casting solvent too. Morphologies similar to the one mentioned above has been found for mechanical blends of two molten polymers in which one of the components was able to crystallize. In that case the structure depends on the composition, blending conditions and cooling rate. For both classes of these heterogeneous systems the structure is homogeneous throughout the sample (the dimensions of dispersions are, however, different) and the amorphous or crystalline phase is the continuous one. In the case of gradient crystalline polymers the amorphous phase is distributed in the material according to a given profile.

In order to obtain appropriate conditions for diffusing the guest monomer into crystalline polyethylene (host) several different brands of PE have been investigated. Preliminary experiments have shown that the most suitable is low density polyethylene (LDPE). Styrene (St) was chosen in this study as the guest because it can easily polymerized and because its concentration in the PE-films at different depth can be detected by UV absorption spectroscopy.

As the preparation of these gradients is discussed in some detail[29], only a brief description will be given here.

LDPE films 0.1 ÷ 0.5 mm thick were obtained by melt pressing

at about 400 K by pressure $3 \cdot 10^6$ N/m^2 between polished steel plates and by fast cooling in water at room temperature. After introducing the guest polymer these films were used for determination of
some mechanical properties. In order to determine the gradient of
St concentration specimens 4 mm thick were prepared under similar
conditions. The diffusion of liquid monomeric St, containing less
than 3 % of benzoyl peroxide was studied on thicker plates. From
these plates samples in the form of parallelepiped were cut out.
They were carefully shielded from all five sides, so diffusion of
liquid monomer could occur only from one side. Thin films and the
above mentioned blocks of LDPE (taken out of the styrene bath) were
exposed to a) UV-irradiation (mercury lamp) or b) glow discharge.
In order to avoid St losses in the course of polymerization in the
second method the samples were introduced into the glow discharge
polymerization equipment at liquid nitrogen temperature. The glow
discharge was initiated in argon at pressure 27 N/m^2 (0.2 Tr) the
glow current density being 1.4 nA/cm^2.

The gradient of PS concentration was established by cutting
thin 50 ÷ 100 μm slides. Following evaporation of small amount of
monomeric St which did not polymerize, these slides were analysed
by UV absorption spectroscopy. PS shows strong absorption bands at
x = 200 - 270 nm. PS concentration was determined from intensity
at λ = 269 nm. In the gradient polymers prepared by UV irradiation
the profile of PS depends on the time of diffusion. This effect is
less clear for gradient polymers prepared by glow discharge polymerization. These differences can be elucidated in the following way.
The amount of St imbibed by LDPE depends on the density and crystallinity of the host polymer. Generally the concentration of St does
not exceed 14% (at quasiequilibrium). The rather fast evaporation
of St at room temperature from the surface influences the initially
created gradient to a high extent. Polymerization at glow discharge conditions occurs at the surface and PSt is formed in thin layer
whose thickness does not exceed 150 μm. This is probably due to
limited diffusion rate of radicals formed and the concentration of
PSt does not depend very much on the time of PS diffusion in the
liquid state (the amount of St which was not polymerized was removed
from the shavings). The results of the absorption analyses of the
shavings taken from one side of LDPE blocks which were immersed in
St and then polymerized by UV irradiation are presented in Figure 1
as weight per cent of PSt in the function of L/L_0 (L_0 is the initial
sample thickness). The LDPE blocks were kept in St for various

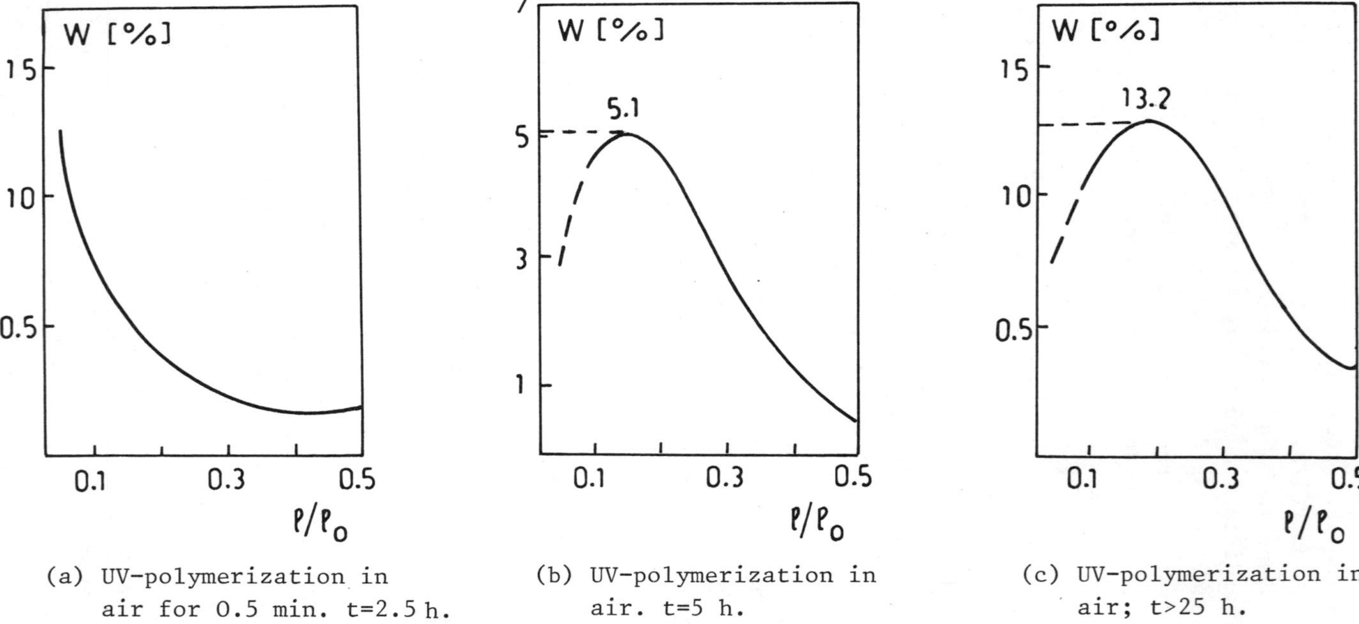

(a) UV-polymerization in air for 0.5 min. t=2.5 h.

(b) UV-polymerization in air. t=5 h.

(c) UV-polymerization in air; t>25 h.

Figure 1.   Gradient of PSt concentration (in weight percent) in LDPE films as function of $1/1_\circ$. Times t of immersion in liquid St and methods of polymerization indicated.

periods of time. It follows from Figure 1 that the gradient of PSt
concentration has a complex character and depends on the time of
immersion in monomeric St. The dotted lines denote the expected
gradients which could be achieved if the polymerization were istantaneous.

The differences of the PSt gradient formed show that in addition
to the time dependent swelling the changes in polymerization condition can lead to preparation of different structures.

To determine strain stress curves small oar-shaped samples
were cut out from thin films containing PSt and deformed with the
stretching rate of 50%/min or 100%/min using an Instron apparatus.
These studies led to the determination of the Young modulus (initial
slope of the stress-strain curve) of stress at break, elongation
at break and of the range of Hook's elasticity. The results of the
mechanical properties studies are collected in Table 1.

The stress-strain curves obtained for gradient polymers prepared in two above described ways do not show any substantial differences beside of some quantitative differences. These results and
observations of the surface of the deformed sample show that St do
not form a continuous phase in the matrix. This can be related to
internal stresses existing in the LDPE initial samples and in those
modyfied by the presence of PSt. This conclusion can be easily
reached from the twisting of thin films toward the side which was
modified by PSt.

A number of publications have been devoted to the preparation
and characterization of systems containing PSt inside PE (see eg.Ref.
30). Usually divinylbenzen (DVB) was added as a crosslinking agent.
Incorporation of the monomers which undergo interpolymerization
causes some changes in the structure of PE. The overall crystallinity degree does not change but the increasing amount of DVB leads
to a decrease of the average size of the crystallites and to an
increase in the amount of fibrillar crystallites. These structural
modifications have been correlated with the change of the PSt
dispersion inside PE. In absence of DVB, PSt is dispersed in the
form of small domains but the increased concentration of DVB results
at first in a decreased amount of isolated PSt domains and then in
network formation. This network has been considered responsible
for the internal mechanical stresses and the change in the crystal-

line structure of PE. These systems were prepared at higher temperatures than our gradient semicrystalline polymers and no direct
comparison of these two materials is possible, although some similarities in the LDPE structure are expected. The work that is now
being done will answer these questions and provide more information
on the influence of initial LDPE morphology on the properties of
these new gradient polymers.

THERMAL STABILITY OF POLYMER BLENDS

Polymer blends consisting of two polymers which show miscibility
usually exhibit one Tg and have mechanical properties which depend
in a nonlinear way on the composition e.g. blends of head-to-tail
polystyrene (HTPS) with poly(2,6-dimethyl-1,4 oxyphenylene) (PPO).
The high degree of miscibility found for these systems by many
authors results from specific interactions of the components. For
this reason one should also expect mutual influence of both components on their thermal stability. In order to verify this suggestion
studies on thermal degradation of HTPS blends with PPO were under-
taken using the thermogravimetric technique and an analysis of the
gaseous products evolved[31]. We have found that HTPS is stabilized
by the presence of PPO during thermal degradation at high temperature
range. In order to propose a possible mechanism of this stabilizing
effect investigations on thermal degradation of head-to-head polystyrene (HHPS) and of poly(α-methylstyrene) (PMS) blends with PPO
were performed[32]. Before discussing the principal results of these
investigations it seems necessary to characterize briefly some
results of thermal degradation studies of polymer blends.

Richards and Slater investigated thermal degradation of HTPS
with PMS in the temperature range 260-290°C in which pure HTPS is
thermally stable[33]. It was found that HTPS does not influence the
degradation of PMS (statistical chain scission and depolymerization
with evolution of α-methylstyrene). The degradation products of
PMS cause, however, a destabilization of HTPS due to diffusion of
small radicals from PMS domains into HTPS domains and due to their
reaction with HTPS chains. Polyethylene glycol in the blend with
HTPS influences the rate of thermal degradation of polystyrene
(decrease of thermal activation energy)[34]. In the blends of PVC
with HTPS the small stabilizing effect for both components occurs.
This effect was suggested to be related to reaction of Cl· radical
with HTPS and to decrease of intermolecular proton transfer because

of the heterogeneous structure of these blends[35]. Similar influence
on thermal degradation of HTPS exhibit poly(acrylo nitrile), poly
(vinylacetate) and poly(vinyliden chloride)[36]. Polypropylene and
polyethylene undergo a faster thermal degradation in the presence of
HTPS. In the system HTPS-PMMA no specific interactions between
components could be detected contrary to the blends of PVC and PMMA
for which a faster decomposition of PMMA and slower HCl evolution
from PVC was found. The detail studies of the thermal degradation
of this system lead to conclusion that the acceleration of decomposition or stabilization of PMMA depends on temperature range (stabilization of PMMA is connected with reactions of ester groups with
HCl and with anhydride formation.

From above mentioned papers one can conclude that:
a) during thermal degradation of polymer blends an important role
   is played by small radicals like $Cl^{\cdot}$;
b) thermal degradation of two component polymer blends does not
   generally result in the important changes of qualitative composition of produced compounds. The use of sensitive methods is
   necessary to detect new species (in comparison with products
   evolved during thermal degradation of pure components);
c) the change in the degree of heterogeneity of the blend does not
   usually influence the specific interactions, only leads to the
   modification of their extent.

Our investigations, in accordance to previous works, have shown
that mechanical blends of PPO with HTPS and of PPO with HTPS are
homogeneous (one Tg value) contrary to the blends of PPO with PMS
and to graft copolymers of PS onto PPO. Thermogravimetric investigations of PPO-HTPS blends with different amount of HTPS indicated
that at small concentration of HTPS in PPO the thermal stability
of the later decreases insignificantly while the same amount of PPO
in HTPS causes an increase of its stability. This effect is clearly
seen from differential thermogravimetric curves and from their comparison with those which were calculated on the basis of TG curves
obtained for pure components and on assumption of additivity of
weight loss of both components (Figure 2). The maxima of weight
loss for both components are well separated (calculated curve) while
the maximum for the blend is broad and shifted toward higher temperature. The products of thermal degradation of pure HTPS and of PPO
were studied by mass spectroscopy at different temperatures of ion
surce. The detected mass spectra of simple fragmentation products

TABLE 1

Some mechanical properties of gradient polymers LDPE - PS.

Sample	$E \times 10^{-6}$ N/m^2	$\left(\dfrac{\Delta l}{l}\right)_{\%}$ at break	$\sigma \times 10^6$ N/m^2 at break	Range of Hook's elasticity $\left(\dfrac{\Delta l}{l}\right)_{\%}$
		Stretching rate 50%/min		
LDPE	72	510	12,0	3,5
Gradient LDPE-PS, UV polymerization in air	74	490	12,1	3,6
Gradient LDPE-PS, plasma poly-merization	86	510	12,9	3,8
		Stretching rate 100%/min		
LDPE	82	380	9,1	1,7
Gradient LDPE-PS, UV-polymer	96	420	10,0	0,9
Gradient LDPE-PS, plasma poly-merization	110	380	9,9	1,2

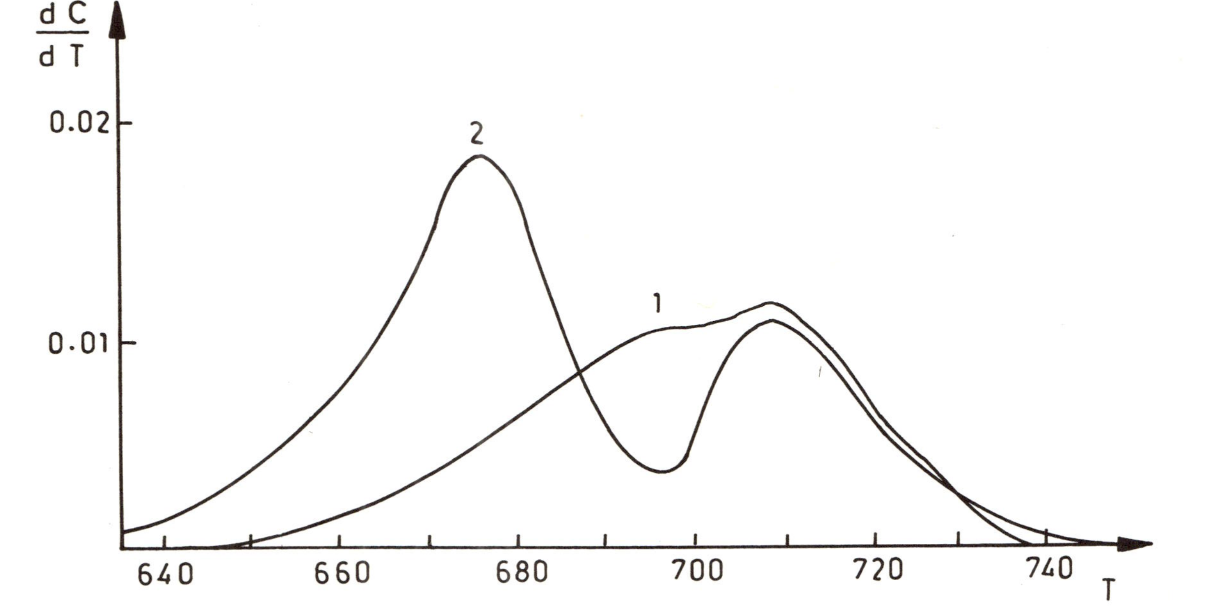

Figure 2. Differential thermogravimetric curves for the PPO and HTPS blend (PPO:HTPS = 60:40 weight percent).

1. Experimental curve
2. Theoretical curve assuming additivity of weight loss of components.

and of oligomers were compared with the mass spectra of model
compounds and species which could be isolated by gas chromatography.
Similar experiments were performed for the HTPS-PPO blends of different composition. The results of these analyses will not be presented
here in detail. It seems important to mention that in blends the
concentration of some degradation products of HTPS increases but the
composition of oligomers related to the presence of PPO is similar
to that of pure PPO. The yield of styrene and other principal degradation products has been measured in a quantitative way. These investigations has led to conclusion that the addition of PPO to HTPS
in the whole range of concentration did not influence the depropagation reaction because these products follow a linear relation describing the theoretical dependence of evolved species on HTPS fraction
in the blend. The presence of PPO, however, causes a change of the
yield of this product which is created by the intramolecular proton
transfer.

Some other information on the mechanism of HTPS degradation in
the blends with PPO were obtained from the studies of isothermal
thermogravimetric analysis. The experimental curves could be compared with the theoretical ones calculated on the basis of Boyd's works
(see ref. 37). The experimental results lead to a general conclusion
that the effect of PPO on the stabilization of HTPS in homogeneous
blend (the shift of the peak in DTG curves for HTPS) is related to
the change of initiation mechanism. In pure HTPS thermal degradation
starts by activation of chain ends but in the blends statistical
chain scission is more important. When cage effect is operating
(this effect was shown by the studies on dependence of thermal stability on molecular weight) then its influence on the position of
the effective chain scission should be affected by the presence of
PPO due to high concentration of protons which will desactivate radicals being formed. An additional argument for the change of the
degradation initiation mechanism is the established decrease of molecular weights. There is, however, another possibility to eluci-
date the HTPS stabilization in blends with PPO. This interpretation
follows from the analysis of degradation products (increase of the
concentration of 1,3-diphenylpropane and its analogues and the decrease of 2,4-diphenyl butan-1 and trimers 2,4,6-triphenylhexene-1
and their higher analogues). Reactions which are responsible for
the formation of these species may be considered as a depropagation
step because of generation of nonactive low molecular weight componds
and reproduction of a depropagation center. The second reaction is
the termination one because of formation of very mobile low molecular

weight radicals and desactivated chain ends. The contribution of
these two reactions to the thermal degradation of HTPS depends on
the enviroment. In pure HTPS the first reaction dominates with minor
importance of the second because of low probability of termination
of mobile radical. In the blends with PPO in which there are many
mobile protons the second reaction may be more important.

The thermal volatilization analysis as well as TGA studies and
the analysis of the products obtained from the blends of HTPS with
PPO (homogeneous) and of blends of PMS with PPO (heterogeneous) has
shown that these components do not interact in thermal degradation
conditions. The small shift of T_{max} of degradation of PMS observed
in TVA curves was not supported by TGA studies. This result is
consistent with previous discussion of thermal degradation mechanism
of HTPS-PPO blends. PPO does not influence the depropagation of
HTPS thus it should not change the course of this reaction in PMS.
Thermal degradation of PMS is not connected with inter-and intermolecular proton transfer. It is due to the statistical chain scission
thus these reactions should be independent on the PPO fraction.

The same conclusions are to the point in the case of HHPS-PPO
blends because HHPS undergoes thermal degradation only by statistical
chain scission. These remarks show that the thermal degradation
(or thermal stability) of polymer blends depends on the interactions
between the components which in turn are related to the phase structure of the system. In order to draw some conclusions on the thermal
stability of such complex systems like blends it is, however, necessary to carry out in detail the studies of thermal stability of
components both from mechanistic and kinetic points of view and then
to investigate the thermal degradation of blends of different composition.

At the end of these considerations it seems necessary to indicate that thermal stability of semicrystalline polymers depends on
the morphology. The same is valid for blends and, in fact, to a
higher extent because the presence of the second component (amorphous
or semicrystalline) may change very much the final structure of the
system. This problem is now discussed in many laboratories including ours.

In these considerations related to three selected but actual
areas of studies of polymer blends I have ommitted some important

problems and works which have contributed to the present understanding of behaviour of polymer blends but I purposely thought to give
some personal views on some questions which seem to be of interest.

It cannot be overemphasized that polymer blends are very complex
systems and that the results obtained cannot be a way easily and adequately elucidated. It means that further works are necessary in
order to reach informations which may help in the choice of models
for fundamental studies and works aiming in obtaining blends of
practical use and interesting properties.

REFERENCES

1. N. G. Gaylord in "Copolymers, Polyblends and Composites" ed.
    N. A. Platzer, Adv. in Chem. Ser. vol. 142 p. 76, Am. Chem. Soc.
    Washington D.C., 1975.
2. G. E. Molau in "Block Polymers" ed. S. L. Aggarwal, p. 79,
    Plenum, N. Y., 1970.
3. D. R. Paul, in "Polymer Blends" eds. D.R. Paul and S. Newman
    vol. 2, chap. 12, p. 35, Academic Press, N. Y., 1978.
4. D. J. Meier, J. Pol. Sci., part $\underline{C26}$, 81 (1969).
5. J. A. Manson and L. H. Sperling "Polymer Blends and Composites",
    Plenum, N. Y., 1976.
6. G. E. Molau, J. Pol. Sci., $\underline{A3}$, 1267, 4235 (1965).
7. G. E. Molau and W. M. Wittbrodt, Macromolecules, $\underline{1}$, 260 (1968).
8. G. E. Molau, Kolloid Z.Z. Polym., 238, 493 (1970).
9. G. Riess, J. Periard and Y. Jolivet, Angew. Chem. Int. Ed., $\underline{11}$,
    339 (1972).
10. G. Riess and Y. Jolivet in "Copolymers, Polyblends and Composi-
    tes" ed. N. A. Platzer, Adv. in Chem. Ser., vol. 142, p. 243,
    Am. Chem. Soc., Washington, D.C. (1975).
11. G. Riess, J. Periard and A. Banderet "Colloidal and Morphologi-
    cal Behaviour of Block and Graft Copolymers" ed. G. E. Molau,
    p. 173, Plenum, N. Y., 1971.
12. T. Inoue, T. Soen, T. Hashimoto and H. Kawai, Macromolecules,
    $\underline{3}$, 87 (1970).
13. M. Moritani, T. Inoue, M. Motegi and H. Kawai, Macromolecules,
    $\underline{3}$, 433 (1970).
14. D. J. Meier, Polym. Prepr. Am. Chem. Soc. Div. Polym. Chem.,
    $\underline{15}$, 171 (1974).

15. U. Koenig, German Patent 2.236, 903 (1974),
    Chem. Abstr., 81, 50711 (1974).
16. C. E. Locke and D. R. Paul, J. Appl. Phys. Sci., 17, 2597, 2791
    (1973).
17. W. M. Barentsen and D. Heikens, Polymer, 14, 579 (1973).
18. W. M. Barentsen, D. Heikens and P. Piet, Polymer, 15, 119 (1974).
19. T. D. Pendle in "Block and Graft Copolymerization" ed.
    R. J. Ceresa, vol. 1, p. 83, Wiley, N. Y., 1973.
20. C. F. Paddock, U.S. Patent 3.758, 435 (1973)
    Chem. Abstr., 80, 48968 (1974).
21. G. L. Brode and J. V. Koleske, J. Macromol. Sci. Chem. A6, 1109
    (1972).
22. J. V. Koleske in "Polymer Blends" ed. D. R. Paul and S. Newman,
    vol 2, p. 369, Academic Press, N. Y., 1978.
23. T. P. Russell, F. P. Warner and R. S. Stein, IUPAC 26 Interna-
    tional Symposium on Macromolecules, J. Luderwald and R. Weiss
    eds., vol. 2, p. 924, Mainz, Sept. 1979.
24. J. M. Charrier and R. J. Ranchoux, J. Polym. Eng. Sci., 11,
    381 (1971).
25. J. Grebowicz, T. Pakula, M. Kryszewski, umpublished results
26. M. Shen and M. B. Bever, J. Mat. Sci., 7, 741 (1972).
27. C. F. Jasso, S. D. Hong and M. Shen, J. Am. Chem. Soc. Polym.
    Prepr., 19 n. 1, p. 63 (1978).
28. G. Akovali, K. Biliyar and M. Shen, J. Appl. Polym. Sci., 20,
    2419 (1976).
29. M. Kryszewski, G. Czeremuszkin, Polym. Bull., in press.
30. H. Czarczyńska, W. Trochimczuk, J. Polym. Sci. Symp., 47, 111
    (1974).
31. J. Jachowicz and M. Kryszewski (in preparation).
32. J. Jachowicz, M. Kryszewski and O. Vogl (in preparation).
33. D. H. Richards, D. A. Salter, Polymer, 8, 127 (1967).
34. L. P. Blanchard, V. Hornoff, H. Lam and S. L. Malhotra, Europ.
    Polym. J., 10, 1057 (1974).
35. B. Dodson, I. G. Mc Neill, J. Polym. Sci. Chem. Ed., 14, 353
    (1976).
36. I. G. Mc Neill, D. Neil, Europ. Polym. J., 6, 569 (1970).
37. R. H. Boyd, J. Polym. Sci., 49, 51 (1961).

MORPHOLOGY, CRYSTALLIZATION PHENOMENA AND TRANSITIONS IN
CRYSTALLIZABLE POLYMER ALLOYS

E. Martuscelli

Istituto di Ricerche su Tecnologia dei Polimeri e
Reologia, C.N.R., Arco Felice (Napoli), Italy

INTRODUCTION

Binary alloys may be made of pairs of polymers where neither
one or both components are crystallizable.

Below the melting temperatures, blends with crystallizable
components are generally heterogeneous. In fact compatible blends
of one crystalline polymer with any other polymer should require
the formation of mixed crystals, but so far there seem to be no
established cases of cocrystallization phenomena. Thus miscibility
in polymer alloys is generally restricted to amorphous phases.

Until recently, there was relatively little information in the
literature on the behaviour and on the properties of alloy systems
with crystallizable components though many important and commercial
polymers crystallize.

Such systems were considered of no interest as the crystallization itself indicated immiscibility.

Recently, both for fundamental and practical reasons blends
with crystallizable components have received increasing attention
as it has been found that the two components may influence each
other giving rise to very interesting effects such as: depression
of the equilibrium melting temperatures; decrease or increase of
the crystallinity and of the rate of crystallization; drastic

change of some morphological quantities (lamellar and interlamellar thickness, and shape and structure of spherulites).

Heterogeneous blends can be organized into a variety of morphologies. Many properties and, subsequently, uses of such alloys will depend critically on the arrangement and relative dimensions of the phases and on the degree of adhesion between them.

The main goals of the present paper are:
a) to review the possible morphologies that can be ecountered in binary blends with crystallizable components in order to delucidate the influence of phase structure on the properties of such heterogeneous blends.
b) to review the crystallization and melting behaviour of crystallizable alloys in order to clarify the reciprocal influence of components on such phenomena.
c) to get information on the compatibility of the two polymers above the melting temperatures, in the melt and in the amorphous state, mainly by following the dependence of glass transition temperature on composition.

MORPHOLOGY AND PHASE STRUCTURE IN CRYSTALLIZABLE ALLOYS

In the case of alloys with one crystallizable component, under the assumption of a stationary process of crystallization, the following morphologies may be encountered:

Type I morphology: The spherulites of the crystallizable component grow in a matrix mainly consisting of the non crystallizable polymer.

Type II morphology: The non crystallizable component may be incorporated in the interlamellar regions of the spherulites of crystallizable polymer. The spherulites fill all the volume available.

Type III morphology: The non crystallizable component may be included within the spherulites of the crystallizable polymer forming domains having dimensions larger than the interlamellar spacing.

For blends having both crystallizable components the most probable morphologies are:

Type I'morphology: Crystals of the two components are dispersed

Type II' morphology: in an amorphous matrix.
                          One component crystallizes according to a
                          spherulite structure while other crystallizes
                          in a simpler structure.
Type III' morphology: Both components exhibit a separate spheruli-
                          tic structure.
Type IV' morphology: The two components crystallize giving rise
                          to the formation of mixed spherulites con-
                          taining lamellae of both polymers.

Structures of an order lower than spherulites, such as sheaves
or hedrites might be also encountered. The amorphous phase, may be
in turn homogeneous or heterogeneous as the two components may be,
in this state, compatible or not.

The morphology of blends with one crystallizable component has
been studied in detail by Ong et al[1] and by Kambatta et al[2] in the
case of poly(ε-caprolactone)/poly(vinylchloride) (PCL/PVC) alloys;
by Wenig et al[3] in the case of poly(2,6-dimethylphenylene oxide)
(PPO)/isotactic polystyrene (iPS) and by Warner et al and by
Martuscelli et al for atactic polystyrene (aPS)/isotactic polystyrene
(iPS) blends[4].

I. PCL/PVC and PPO/iPS blends

PCL/PVC blends are crystallizable (PCL is a highly crystalline
polymer) for PVC content not exceeding 60% by weight. No crystallinity is in fact observed in blends containing more than 60% PVC.
As found by Koleske and Lundberg[5] the PCL/PVC blends, over a wide
range of composition, show a single glass transition temperature
intermediate between that of the pure homopolymers.

The experimental data may be fitted by the following equation
derived by Fox[6]:

$$\frac{1}{Tg} = \frac{W_1}{Tg_1} + \frac{W_2}{Tg_2} \tag{1}$$

where W_1 and $Tg_1 = 202°K$, W_2 and $Tg_2 = 355°K$ are the weight fraction
and the glass transition temperature of PCL and PVC respectively.
This observation indicates that PCL and PVC are compatible in the
amorphous state. The state of compatibility for blends rich in PVC

showing no traces of crystallinity, has been also deduced by X ray
and light scattering by Stein et al[7].

From these studies it emerges that PCL is molecularly dispersed
in a PVC matrix suggesting compatibility at a segmental, as well as
molecular level.

The morphology of crystallizable PCL/PVC blends was extensively
studied using light and electron microscopy, small angle light scattering (SALS) and small and wide angle X-ray scattering (WAX and
SAXS).

The results of these investigations may be summarized as follow:
a) In the case of blends rich in PCL the films are almost completely
   filled with spherulites composed of lamellae radiating from the
   center and twisting regularly. The texture of the spherulites
   becomes increasingly coarse (the constituent fibers are of relatively large cross section) and open (the overall crystallinity
   is relatively small) with increasing PVC content.
b) The crystal thickness seems to be almost independent of composition whilst the repeat period or long spacing increases with
   increasing PVC content (see Figure 1). At the same time a
   decrease in the linear crystallinity, defined as the ratio
   between the crystal thickness and the long spacing, was observed.
   Furthermore an agreement between the values of bulk and linear
   crystallinity, was found.

Observations a) and b) may be accounted for assuming that
during crystallization from melt PVC molecules are incorporated
within the PCL spherulites filling interlamellar regions. This
would explain also why the amorphous layer thickness increases with
increasing PVC content.

By thermal annealing it is possible to obtain PPO/iPS blends
with iPS crystalline phase and an amorphous mixed PPO-iPS phase
(PPO cannot crystallize by thermal treatment alone).

A SAXS and WAXS study shows that the crystal lamellar thickness
decreases with increasing PPO content, while the amorphous layer
thickness increases, and the linear crystallinity agrees with bulk
crystallinity.

The mixed amorphous PPO-iPS phase is then concentrated between the lamellae of iPS spherulites.

PLC/PVC and PPO/iPS blends, thus, both crystallize according to a type II morphology.

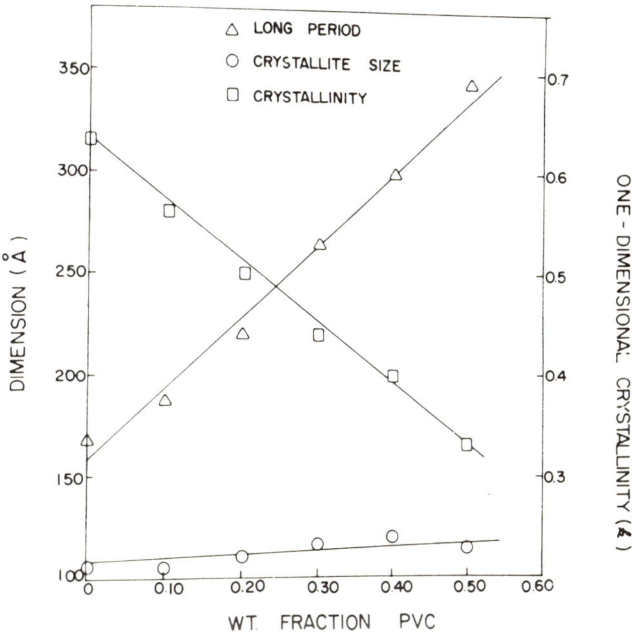

Figure 1. Variation of crystal lamellar thickness, repeat period and linear crystallinity with composition for crystallizable PCL/PVC blends (reference 7).

II. <u>aPS/iPS blends</u>

The crystal thickness, the amorphous layer thickness and the linear crystallinity result to be, for this system, independent of composition. The bulk crystallinity, on the contrary, decreases with increasing aPS content with values always lower than those of the linear crystallinity (see Figure 2). At the same time optical observations show that in blends films with low aPS content (< 30% by weight) iPS spherulites fill all the available volume. These observations lead to the conclusion that aPS is incorporated within iPS spherulites, not in interlamellar regions but forming larger

domains. It is likely that this blends crystallizes, at least for
low aPS content, according to a type III morphology. These results
are in agreement with a scanning electron microscopy study performed
by Martuscelli et al[4] on liquid nitrogen fractured films of aPS/iPS
blends. Such observations revealed that atactic rich blends, immersed in n-hexane at 40°C, developed a non intercommunicating network
of microvoids. Similar solvent treatment of isotactic rich blends
had little effect on the sample, infact there is no evidence of
voids development.

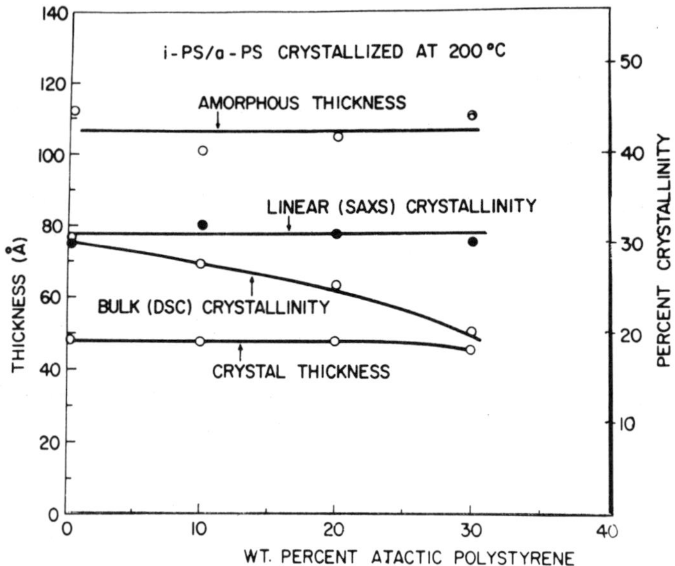

Figure 2. Variation of crystal lamellar thickness and amorphous
           interlamellar thickness with weight percent aPS in aPS/iPS
           blends (reference 7).

As suggested by Stein et al[7] blends with only one crystallizable
component will have a morphology of type I, II or III according to
the value of the parameter δ defined by Keith and Padden[8] as:

$$\delta = D/G \qquad\qquad (2)$$

In equation 2, G is the radial growth rate of spherulites and D
is the diffusion coefficient of non crystallizable component in the
crystallizing matrix. The parameter δ, having dimensions of length,
gives a measure of the distance that the amorphous component may
cover during crystallization. When δ is comparable with interlamellar distance then the non crystallizable component will be accomo-
dated in interlamellar regions (type II morphology). In the case
of a larger value of δ the amorphous component may reside in domains
within or quite outside spherulites (morphologies of type III and
I respectively).

The morphology of blends where both components can crystallize
has been studied only in few cases.

III. Poly(butylene terephtalate) (PBT)/Poly(ethylene terephtalate (PET) blends.

This system was studied in detail by Stein et al[7]. Highly
amorphous PBT/PET blends obtained by severe quenching from the melt,
show a single glass transition temperature intermediate between
those of homopolymers. The variation of Tg with composition is in
agreement with the Fox equation[6]. The existence of a single Tg
suggests that PBT and PET are compatible in the amorphous state.

PBT/PET blends with one component in excess, crystallize according to a type IV' morphology. The minor component is incorporated
within spherulites of the major component during crystallization.
Thus spherulites containing crystals of both components are formed.
When the relative amount of PBT and PET in the blends is comparable,
nonspherulitic morphologies are observed. The crystallized blends
consist, mainly of crystals or other smaller than-spherulitic
superstructures of the two polymers.

IV. Blends of crystalline PPO and crystalline iPS

By suitable thermal and solvent treatment PPO/iPS blends with
two crystalline phases and a mixed amorphous phase may be obtained.
A pseudo-phase diagram for crystallized blends of PPO and iPS, proposed by Neira-Lemos[9], is shown in Figure 3. Two crystalline phases
and a mixed glassy (T<Tg) phase exist for compositions between

20-50% of PPO. Blends with a weight fraction of PPO lower than 0.2
show only the crystalline iPS phase together with the mixed amorphous
phase. On the contrary, alloys with a weight fraction of PPO greater
than 0.5 consist of a crystalline PPO phase and a mixed amorphous
phase. No indications are so far reported regarding the morphology
of the crystalline phases.

V. Blends of isotactic polybutene-1 (iPB) and isotactic polypropylene (iPP)

The morphology of this system has been studied by optical microscopy by Siegman[10]. Both iPP and iPB homopolymers form spherulites when they crystallize from the melt.

The blends with 25% of iPP result in a solid comprised of well
developed spherulites resembling the iPP morphology, altough less
perfect with a coarser fine structure. In the 1:1 blends, no spherulitic morphology was observed. The blends consist of a mixture of
small crystalline aggregates of the two components. A very irregular or fragmental spherulitic structure is observed in blends with
high iPP content (iPB/iPP=1:3).

Thus the presence of molten iPB seems to affect the crystallization of iPP stronger than the presence of already crystalline iPP
in the case of iPB. Moreover in the intermediate composition range
a strong mutual disturbance seems to prevent the formation of spherulitic morphologies.

KINETICS OF CRYSTALLIZATION OF CRYSTALLIZABLE ALLOYS

The kinetics of crystallization of a polymer crystallizing
from a mixture containing another polymeric component may be strongly influenced by composition, particularly when the two polymers
have a certain degree of compatibility in the melt.

The effect of atactic polypropylene (aPP) and of atactic polystyrene on the morphology and on the kinetics of spherulitic crystallization of isotactic polypropylene (iPP) and isotactic polystyrene respectively was investigated by Keith and Padden[8,11].
It was found that the concentration and molecular weight of the

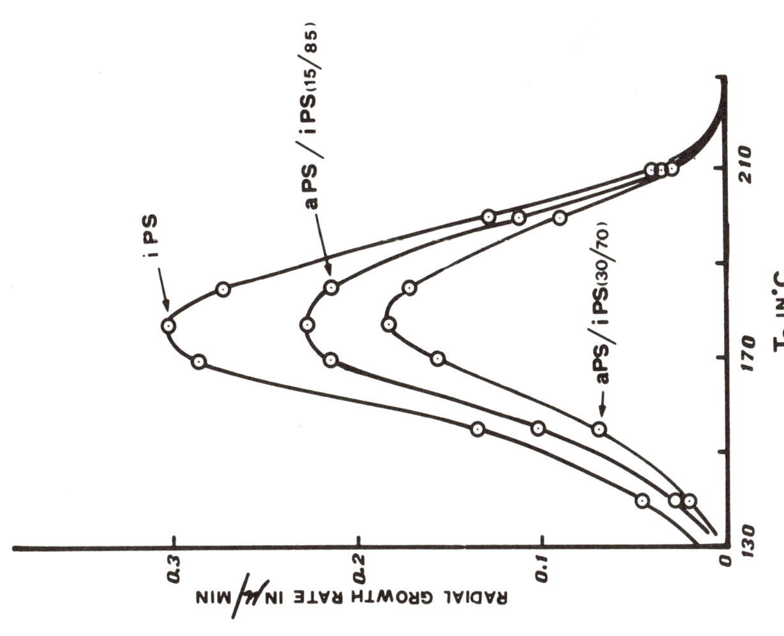

Figure 4. Radial growth rates of spherulites in iPS and iPS/aPS blends (reference 11).

Figure 3. Pseudo-Phase diagram for crystallizable PPO/iPS blends (reference 9).

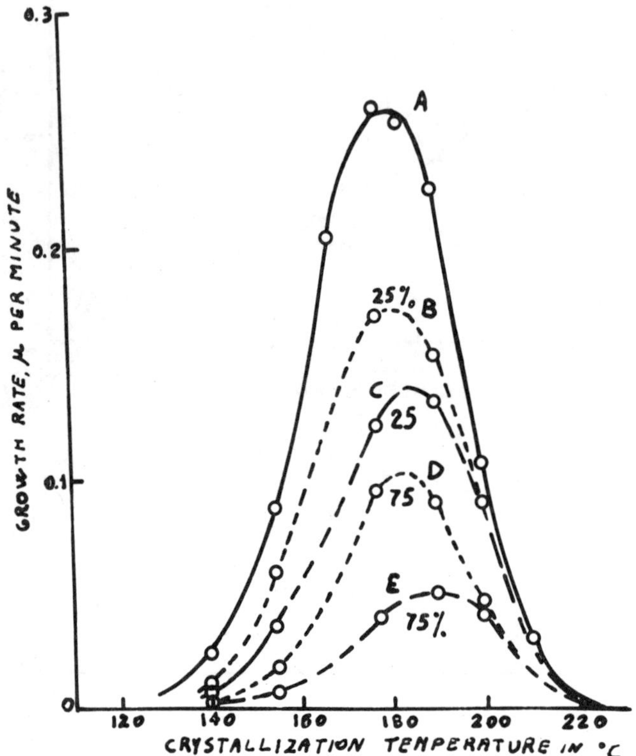

Figure 5.   Influence of aPS molecular weight on the growth rate of
iPS spherulites in iPS/aPS blends. A iPS; B iPS/aPS
(75/25) M.W. 41,700; C iPS/aPS (75/25) M.W. 247,000;
D iPS/aPS (25/75) M.W. 41,700; E iPS/aPS (25/75)
M.W. 247,000 (reference 11).

non-crystallizable polymers have considerable influence on the
texture and growth rates of spherulites. As shown in Table 1 and
in Figure 4 the radial growth rates of iPP and iPS are suppressed
by the addition of non crystallizable polymers. The effects are
greater the larger the concentration of atactic polymers in the
blends. Growth rates also depend upon the molecular weight of atactic polymers. As shown in Figure 5, in the case of iPS/aPS blends,
an increase in the molecular weight of aPS causes for the same composition and crystallization temperature a rather significant depression in the radial growth rates of iPS spherulites.

TABLE   1

Radial growth rates (μ/min) for various compositions in the case
of iPP/aPP blends (Data from Keith and Padden[11]).

Crystallization temperature °C	100% iPP	90% iPP	80% iPP	60% iPP	40% iPP
120	29.4	29.4	26.9	22.8	21.2
125	13.0	12.0	11.0	8.90	8.57
131	3.88	3.60	3.03	2.37	2.40
135	1.63	1.57	1.35	1.18	1.12
Melting temperature °C	171	169	167	165	162

More recently Yeh and Lambert[12] conducted a systematic study
of the effect of various molecular weight aPS (900 to 1,800,000)
on the crystallization kinetics, spherulitic growth rates as well
as the morphology and melting behaviour of iPS. In agreement with
Keith and Padden they found that for the same crystallization
temperature the growth rates are uniformly depressed with increasing
content of aPS. In addition, for the same crystallization temperature and composition, growth rates generally decrease with the molecular weight of aPS in the molecular weight range between 4800 and

19,800. Of particular importance is the sudden 30% increase in the
radial growth rate as the aPS molecular weight is increased from
19,000 to 51,000 (see Figure 6). According to Yeh and Lambert this
behaviour should be accounted for by a process of entrapment of non
crystallizable aPS molecules between the spherulite fibrils of iPS.
This would cause an increase in the effective interfacial concentration of crystallizable iPS molecules and then comparatively higher
growth rates. Morphological studies showed that the spherulites
at comparable crystallization temperatures become increasingly coarse
and open with increasing amount of aPS with a given molecular weight.
It is interesting to point out that for a given concentration and
crystallization temperature an increasing coarseness in the spherulite texture with molecular weight of the aPS was also observed.
Furthermore this effect was more pronounced just at an aPS molecular
weight of 51,000.

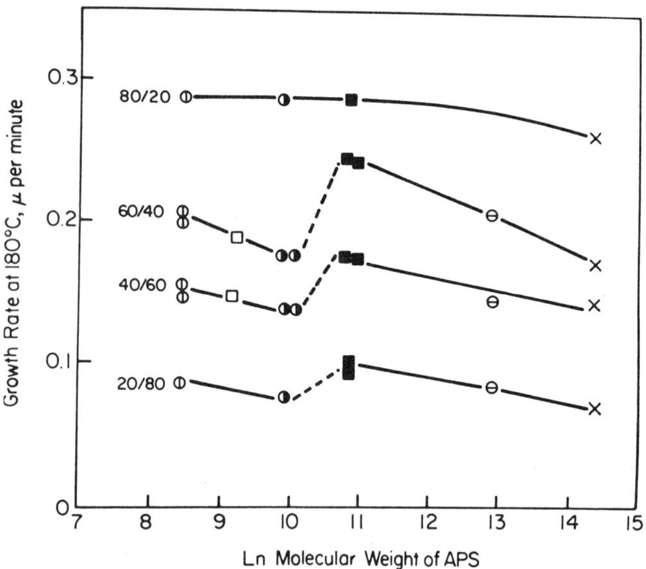

Figure 6. Growth rates at 180°C of iPS in iPS/aPS blends plotted
           as function of the ln of molecular weight of aPS (refe-
           rence 12).

The isothermal spherulitic growth rate and the overall crystallization kinetics of PCL from compatible PCL/PVC melt blends has been studied by Ong and Price[13]. As it can be seen in Figure 7, for the same crystallization temperature, diluition with PVC causes a depression of the spherulitic growth rate. At the same time is observed a tendency for the growth rate maximum to shift to higher temperatures. This shift may result from the change in the values of the glass transition temperature T_g and the melting point T_m of the blends.

The influence of non crystallizable polymers such as poly(methyl methacrylate) (PMMA) on the radial growth rate of poly(ethylene oxide) (PEO) spherulites has been studied by Martuscelli et al[14].

Beghmans et al[15] investigated the crystallization behaviour of iPS/PPO blends. Radial growth of spherulites in blends of poly(vinylidene floride) (PVF_2) and poly(methyl methacrylate) have been measured by Wang and Nishi[16].

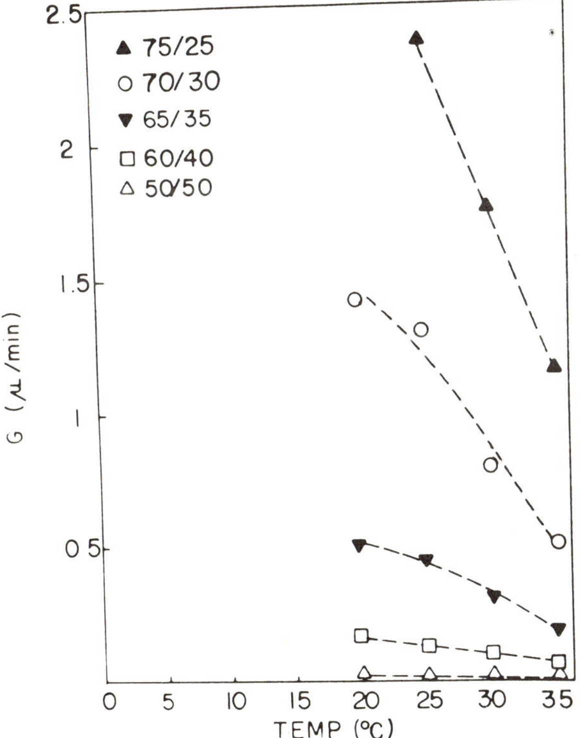

Figure 7. Plots of growth rate G against crystallization temperature for PCL/PVC blends (reference 13).

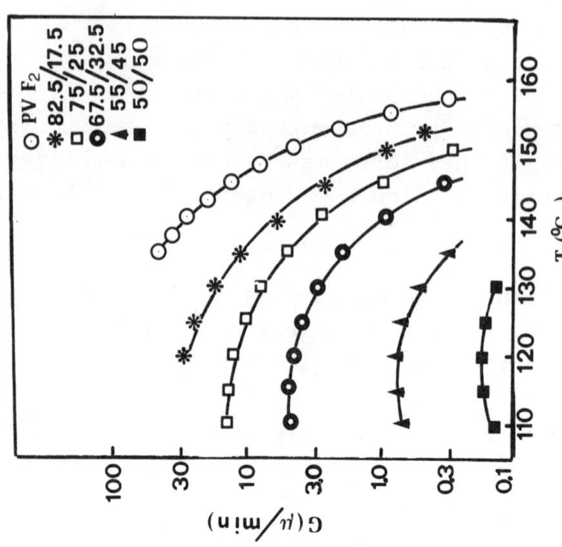

Figure 9. Radial growth rate G of spherulites in PVF$_2$/(PMMA) blends at various temperatures (reference 16).

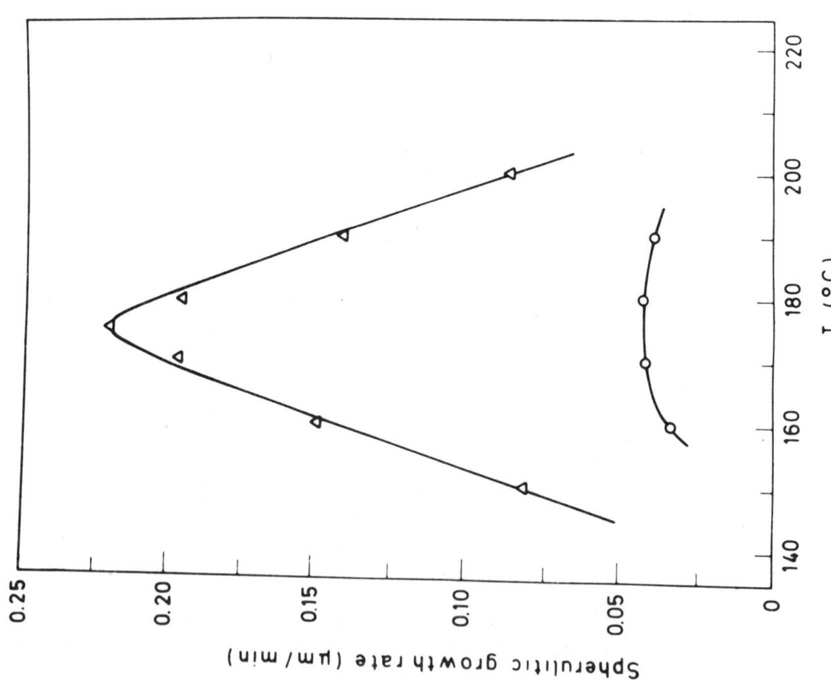

Figure 8. Temperature dependence of spherulite growth rate G: (o) iPS/PPO blend, W$_1$=0.80; (Δ) iPS (reference 15).

The presence of PMMA in PEO/PMMA and PVF_2/PMMA blends and of PPO in iPS/PPO reduces the rate of crystallization of crystallizable components (see Figures 8 and 9). No decrease of the growth rate with time was observed, regardless of concentration of non crystallizable component. This implies that the concentration of the amorphous polymer at the tips of radial lamellae remains constant throughout growth. At the same time no formation of a macroscopic separated phase of the non crystallizable component was observed neither prior to crystallization nor after. These observations indicate that during crystallization the radial diffusion of rejected non crystallizable component is outstripped by the more rapidly growing lamellae so that the molecules of these materials are trapped between the growing fibrils.

A case of non linear growth has been reported by Keith and Padden, for isotactic polypropylene containing an atactic diluent of very low molecular weight (540). As shown in Figure 10 growth curves, representing unimpeded nonlinear growth, confirm closely to parabolas of the form $r/t^{1/2}$ = constant. According to Keith and Padden[11], such behaviour is related to the fact that during crystallization the molecules of aPP can diffuse radially with a rate that surpasses the growth rate of lamellae. This leads to an increase in the concentration of non crystallizable material at the surface of a spherulite as, crystallization time increases. Then the radial growth rate will fall off.

The radial growth rates (G) of spherulites have generally been described by equation of the form:

$$G = G_O \ EXP \ (-\Delta F \ /RT) \ EXP \ (-\Delta\Phi \ /KT) \qquad (3)$$

Assuming that the non crystallizable component acts as diluent, then the work to form a nucleus of critical size $\Delta\Phi$ can be expressed as[17]

$$\Delta\Phi \ = \frac{4 \ b_o \ \sigma_u \sigma_e T_M}{\Delta H \quad (T_M - T_c)} - \frac{2\sigma_u \ T_c K(\ln\nu_2) T_M}{b_O \Delta H \ (T_M - T_c)} \qquad (4)$$

where σ_u and σ_e are the surface free energies parallel and perpendicular to the molecular chain direction; b_O is the distance of the two adjacent fold planes; ΔH is the heat of fusion per unit volume, T_m is the equilibrium melting temperature of the crystallizable polymer in the mixture with volume fraction v_2. Using for the free energy of activation related to the transport process of material from the liquid to the solid surface ΔF the empirical relation of William, Landel and Ferry[18]. G, can be represented by the following equation:

$$G = v_2 \; G_o \; \text{EXP}\left[- \frac{C_1}{R(C_2+T_c-T_g)} \right] \text{EXP}\left[- \frac{4b_o\sigma_u\sigma_eT_m}{K\Delta H \; T_c(T_m-T_c)} - \frac{2\sigma_u\ln(v_2)T_m}{b_o\Delta H(T_m-T_c)} \right] \quad (5)$$

The pre exponential factor G_o is multiplied by v_2, because the rate of nucleation is proportional to the concentration of crystallizable segments[19]. Using for σ_u the empirical equation[20]: $\sigma_u = 0.1 \; b_o\Delta H$ then equation 5 can be written as[13]:

$$\alpha = \log \; G_o - C_3 \frac{T_m}{T_c(T_m-T_c)} \quad (6)$$

where

$$\alpha = \log \; G - \log \; v_2 + \frac{C_1}{2.303 \; R(C_2+T_c-T_g)} - \frac{0.2 \; T_m(\log \; v_2)}{T_m-T_c} \quad (7)$$

$$C_3 = \frac{4b_o\sigma_e\sigma_u}{2.303 \; K\Delta H} \quad (8)$$

According to equation (6) a plot of α versus $T_m \; T_c \; \Delta T$ should give a straight line.

Ong and Price[13] found that growth rates for PCL/PVC blends can be described by equation 6 if a value of C_2 of 72°K is used (according to Williams, Landel and Ferry theory C_1 and C_2 should have values of 4120 cal/mole and 51.6°K respectively). From the intercept and the slope values of G_o and σ_e of 7.2×10^8 μm/min and of 27.0 erg/cm^2 were obtained respectively.

It is interesting to point out that the presence of PVC molecules contributes also to the reduction of the mobility of crystallizable PCL segments. The T_g of the system increases with PVC content. This according to the William Landel and Ferry equation, causes an increase in the value of the activation free energy ΔF related to the transport process at the liquid-solid interface and then a depression of the growth rate G.

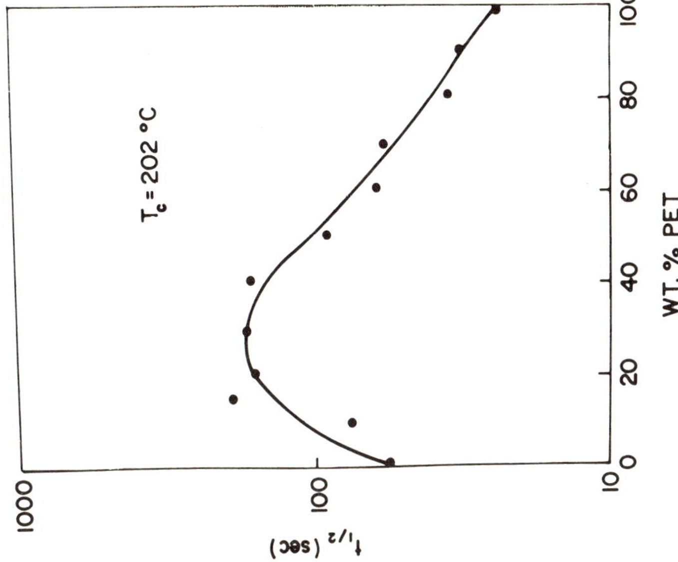

Figure 11. The variation of $\tau_{\frac{1}{2}}$ with wt% PET in a PBT/PET blend at 202°C (reference 7).

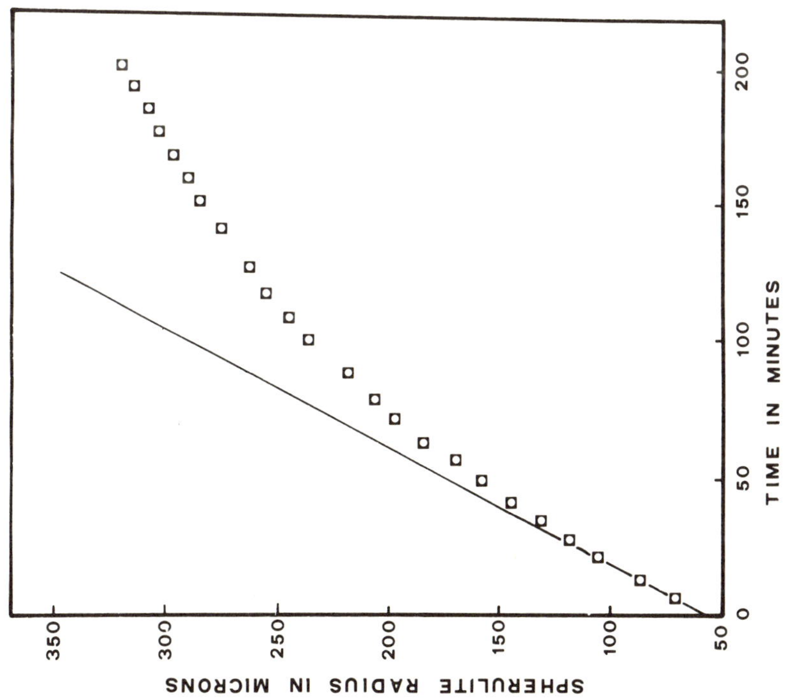

Figure 10. Radius as function of time for spherulites growing isothermally at 135°C in a blend of iPP and low molecular weight aPP (reference 11).

The overall rates of crystallization of PCL/PVC blends were obtained by density measurements by Ong and Price[13]. As expercted, for the same crystallization temperatures, the Avrami constant K decreases markedly with increasing PVC content. This phenomenon may be accounted for by:
a) the increase in the melt viscosity
b) the decrease in the number of nuclei as the PVC increases.
Effect b) results from the decrease in the value of supercooling, because of the depression of the melting point of PCL, that leads to a lower nucleation density. The Avrami index n has a value of 3 almost independent of PVC content. In the light of the above results Ong and Price[13] suggested the crystallization of PCL from PCL/PVC proceeds by heterogeneous nucleation, followed by a three dimensional growth.

The crystallization of blends, where both components can crystallize, has been studied by Martuscelli et al[21], in the case of high density polyethylene (HDPE)/isotactic polypropylene (iPP) blends and Stein et al[7] for poly(butylene terephtalate) (PBT)/poly(ethylene terephtalate) (PET) alloys. The investigation of Martuscelli et al led to the following results[21]:
a) For lower crystallization temperature it has been found that the kinetics of crystallization of HDPE may be markedly delayed by the addition of a small amount of iPP. In fact the HDPE (90%)/iPP(10% blends, shows an overall $\tau_{\frac{1}{2}}$ of order of 1260 sec. at $T_c=398°K$. This Figure is around three and four times larger than for pure HDPE and iPP respectively.
b) For crystallization temperature higher enough to prevent any polyethylene crystallization ($T_c>400$ K) the presence in the melt blends of liquid HDPE influences the crystallization of isotactic polypropylene. The half time of crystallization of iPP increases with the percentage of HDPE in the blend. The kinetics passes
a minimum at a well defined blend composition (around 60% in PP). Theoretical explanation of effect a) might involve, excluding co-crystallization phenomenon, an increase in the transport term following an increase in the melt viscosity of liquid polyethylene caused by the presence of a small fraction of polypropylene chains. For that, a certain degree of compatibility of the two polymers in the melt state must be probably invoked even if no diluent effect was observed. The equilibrium melting temperature of the two polymers was infact independent of composition. It is not clear at the moment why this process should present a maximum at a well defined composition. Observation b) may be accounted for by the minimum

found in the value of the surface free energy of folding of lamellar
crystals of isotactic polypropylene. The decrease in σ_e is likely
to be ascribed to crystals with a less regular fold surface i.e.
higher surface entropic content following a certain degree of solubilization of the shortest chains of HDPE with the formation of
intermolecular entanglements. This would also explain why at a
given T_c thinner crystals of isotactic polypropylene are obtained
with increasing the percentage of HDPE in the blend. It is likely
that the presence of intermolecular entanglements makes the growth
of larger lamellae more difficult. Also in this case however the
minima observed in the values of σ_e, in the lamellae thickness, and
in the overall rate of crystallization of iPP, for some blend compositions cannot be at the moment easily explained.

The crystallization of PBT/PET blends has been followed by
depolarized light intensity (DLI), DSC and IR techniques[7]. A typical
plot of $\tau_{\frac{1}{2}}$ versus wt% PET for PBT/PET blends crystallized at 202°C
is shown in Figure 11. The curve shows a well defined maximum for
a blend containing about 30 wt% PET.

In order to explain the anomalous minima observed in the values
of the overall rate of crystallization of some crystallizable blends
a correct analysis in terms of the effect of composition on nucleation, morphology and diffusion rate in the melt needs to be made.

MELTING BEHAVIOUR OF CRYSTALLIZABLE BLENDS

A substantial depression of the melting temperature of crystalline component has been often observed in crystallizable blends.
This effect is particularly relevant in blends where the two components are compatible in the amorphous state. The influence of
composition on the value of the equilibrium melting temperature of
isotactic polystyrene, poly(vinylidene floride) (PVF_2) and poly(e-
thylene oxide) (PEO) crystallized from iPS/aPS[4], PVF_2/PMMA[22] and
PEO/PMMA[14] blends respectively is shown in Table 2. This drop may
be also accompanied by an elevation in the glass transition temperature T_g of the system. As shown in Figure 12 for PCL/PVC blends[13]
the T_m-T_g interval may be markedly reduced with the result of severe
restriction on the crystallization process. The monotonic decrease
in the melting point of the crystalline component with added non
crystallizable polymer suggests that the non crystallizable component

may simply act as a compatible diluent for the crystallizable polymer. A quantitative analysis of this effect has been presented by
Nishi and Wang[22] and later by Imken, Paul and Barlow[23] based upon
assumed Flory-Huggins behaviour[24,25].

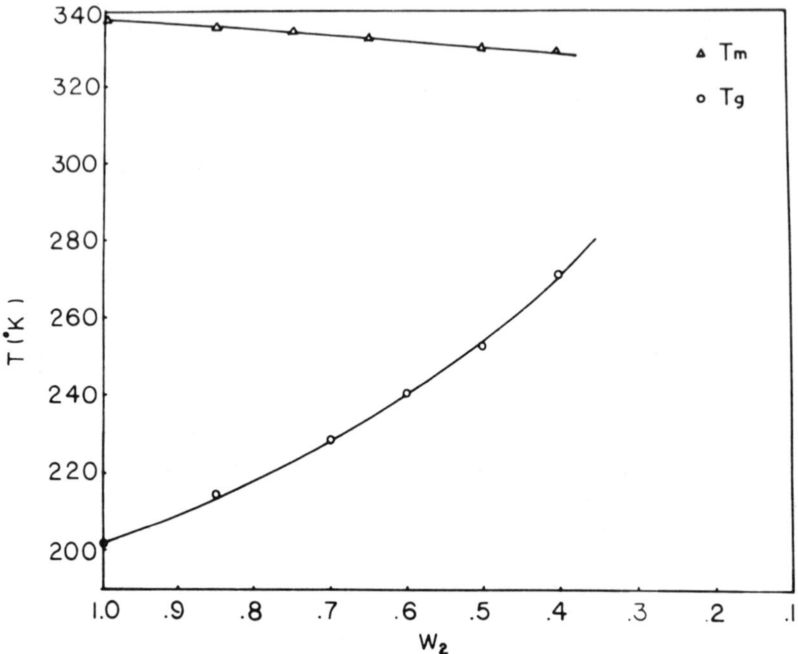

Figure 12. Plots of T_m and T_g against composition for PCL/PVC
blends (reference 7).

The treatment assumes that the chemical potential μ_{2u}^{1} per mole
of crystallizable polymer units in the melt blend relative to its
chemical potential μ_{2u}^{o} in the pure melt can be expressed as:

$$\mu_{2u}^{1} - \mu_{2u}^{o} = \frac{RT\,V_{2u}}{V_{1u}}\left[\frac{\ln v_2}{m_2} + \left(\frac{1}{m_2} - \frac{1}{m_1}\right)v_1\,\chi_{12}v_1^{2}\right] \qquad (9)$$

The subscript 2 and 1 denote the crystallizable and amorphous polymer respectively. V_u is the molar volume of repeating units, v is
the volume fraction in the melt, m is the degree of polymerization,
R is the gas constant, χ_{12} is the polymer-polymer interaction para-

meter and T is the absolute temperature. The difference between
the chemical potential μ_{2u}^{c} of crystallizable polymer in the crystals
and in the pure liquid μ_{2u}^{o} is given by the following equation:

$$\mu_{2u}^{c} - \mu_{2u}^{o} = - (\Delta H_{2u} - T \Delta S_{2u}) = -\Delta H_{2u}(1-T/T_{m}^{o}) \tag{10}$$

In equation 10 ΔH_{2u} and ΔS_{2u} are the enthalpy and entropy of fusion
per mole of repating unit. The ratio $\Delta H_{2u}/\Delta S_{2u}$ has been assumed to
be independent of temperature and equal to the equilibrium melting
temperature T_{m}^{o} of crystalline component. Under the condition that
at the equilibrium melting point of the blend T_{m} $\mu_{2u}^{c}=\mu_{2u}^{1}$ one obtains:

$$\frac{1}{T_{m}} - \frac{1}{T_{m}^{o}} = - \frac{RV_{2u}}{\Delta H_{2u}V_{1u}} \left[\frac{\ln v_{2}}{m_{2}} + (\frac{1}{m_{2}} - \frac{1}{m_{1}}) v_{1} + \chi_{12} v_{1}^{2} \right] \tag{11}$$

Since the degrees of polymerization m_{1} and m_{2} are usually high, the
first two terms in the brackets vanish in comparison with the third
term involving the interaction parameter. The algebric simplification is equivalent to describe the melting point depression entirely
by enthalpic effects rather than the entropic effects described by
the first two terms in brackets. This simplification leads to:

$$\frac{1}{T_{m}} - \frac{1}{T_{m}^{o}} = - \frac{R V_{2u}}{\Delta H_{2u}V_{1u}} \chi_{12}v_{1}^{2} \tag{12}$$

As shown by equation 12 the interaction parameter χ_{12} plays a decisive role on the melting behaviour of blends with one crystallizable
component. We should expect in fact a melting point depression
only if χ_{12} is negative; this implies a negative enthalpy of mixing.
In a more suitable form equation 12 may be written as:

$$\Delta T_{m} = T_{m}^{o} - T_{m} = - T_{m}^{o}(V_{2u}/\Delta H_{2u}) B v_{1}^{2} \tag{13}$$

$$\text{with} \qquad B = RT\chi_{12}/V_{1u} \tag{13^{1}}$$

According to Imken et al[23] a plot of ΔT_{m} versus v_{1}^{2} should thus be
linear with an intercept at the origin if there are no entropic
contributions to ΔT_{m}. The interaction parameter χ_{12} can be obtained
from the slope of such plots. The depression of the melting point

TABLE 2

Melting point depression in iPS/aPS, PVF_2/PMMA and PEO/PMMA blends.

Weight fraction of non crystallizable component	Equilibrium melting point °C	Melting point depression °C
iPS/aPS [*]		
0.0	228	0
0.10	224	4
0.25	222	6
0.50	220	8
PVF_2/PMMA [**]		
0.0	173.8	0
0.17	169.8	4.0
0.32	168.5	5.3
0.50	165.2	8.6
PEO/PMMA [***]		
0	76.1	0
0.05	72.0	4.1
0.10	71.2	4.9
0.15	70.2	5.9
0.20	68.6	7.5

[*] from reference 4

[**] from reference 20

[***] from reference 14

observed in blends such as PVF_2/PMMA[22]; PVF_2/poly(ethyl-methacrylate)
(PEMA)[23] is quite well described by equation 13. This is shown in
Figure 13 for PVF_2/PEMA blends. The straight line of Figure 13
however have a non-zero but very small intercept ($< 1°C$) which, according to Imken et al[23] may result from a small entropic effect.
From the slope a value of B of -3.18 cal/cm^3 was obtained. This
value compares quite well with the value of -2.98 cal/cm^3 found by
Nishi and Wang[22] for PVF_2/PMMA alloys. The interaction parameters
calculated from the values of B ($\chi_{12}=-0.295$ at $160°C$ for PVF_2/PMMA
blends) result to be resonable for compatible blends, that is, small
and negative.

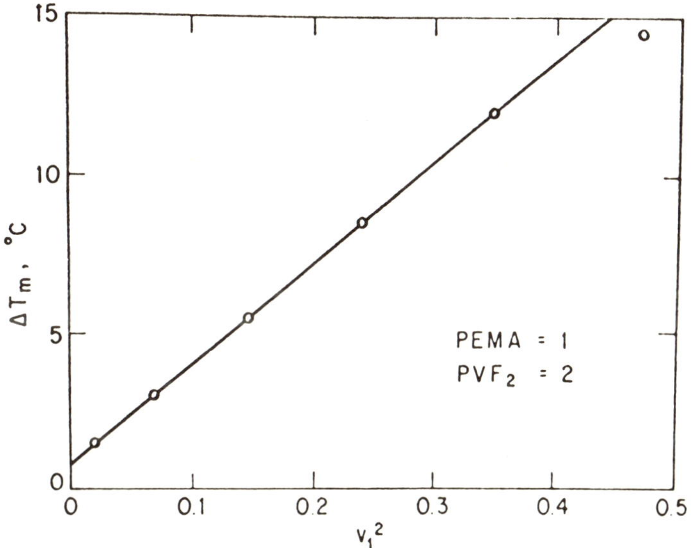

Figure 13. Melting depression ΔT_m as function of composition for
            PVF_2/PEMA blends (reference 23).

     This observation suggests that, at least for these systems,
the melting point depression seems to be explained by the diluent
effect of the non crystalline component on the chemical potential
of crystalline component in the melt blend.

In the case of iPS/aPS blends a non linear ΔT_m versus $v_1{}^2$ plot was obtained. This, according to Martuscelli et al[4], may result either from a significant variation of χ_{12} with $v_1{}^2$ or from non compliance with assumptions inherent in the construction of the Hoffman–Weeks plots[26] used to obtain the equilibrium melting temperatures of blend samples.

In some cases it has been found that the depression of the melting temperature may be dependent of molecular weight of non crystallizable component. Yeh and Lambert[12] found, in fact, that in the case of iPS/aPS blends the melting temperatures depression, at a given atactic concentration, increases with decreasing aPS molecular weight (below 19.000). The depression is particularly noticeable for atactic diluent of molecular weight 900.

It is interesting to point out that melting point depression has been observed also in systems which are not compatible in the melt state[27,28,29]. In such cases the T_m depression must be attributed to kinetic and morphological effects of the non crystallizable polymer producing a reduction in crystal size, lamellae thickness or crystal perfection. This morphological effects may play some important role also in the case of compatible blends. Only accurate and direct morphological studies may decide on the relative importance of thermodynamic versus morphological effects.

CONCLUDING REMARKS

Though the number of crystallizable alloys investigated in detail so far is not relevant results evident that the morphology, the kinetics of crystallization as well as the melting behaviour are dependent on factors such as compatibility of the components in the melt above the melting temperature and in the amorphous state below T_m; crystallizability of the two components and relative rate of diffusion and growing crystals. Finally, it has been found that the molecular weight of component and probably their distribution may play an important role on the final morphology and thermodynamic quantities of binary crystallizable blends.

REFERENCES

1. C. J. Ong, F. P. Price, J. Polym. Sci. Polym. Symp., 63, 45
    (1978).
2. F. B. Khambatta, R. Russell, F. Warner, R. S. Stein, J. Polym.
    Sci. Polym. Phys. Ed., 14, 1391 (1976).
3. W. Wenig, F. E. Karasz, W. J. Macknight, J. Appl. Phys., 46,
    4194 (1975).
4. E. Martuscelli, G. Demma, E. Drioli, L. Nicolais, S. Spina,
    H. B. Hopfenberg, V. T. Stannett, Polymer, 20, 571 (1979).

    F. P. Warner, W. J. Macknight, R. S. Stein, J. Polym. Sci.
    Polym. Phys. Ed., 15, 2113 (1977).
5. J. V. Koleske, R. D. Lundberg, J. Polym. Sci., A-2, 7, 795
    (1969).
6. T. G. Fox, Bull. Am. Phys. Soc., 2, 123 (1956).
7. R. S. Stein, F. B. Khambatta, F. P. Warner, T. Russell,
    A. Escala, E. Balizer, J. Polym. Sci., Polym. Symp. 63, 313
    (1978).
8. H. D. Keith, F. J. Padden, J. Appl. Phys., 35, 1270 (1964).
9. R. A. Neira-Lemos, Ph. D. Thesis, Univ. of Massachussetts (1974).
10. A. Siegman, J. Appl. Polym. Sci., 24, 2333 (1979).
11. H. D. Keith, F. J. Padden, J. Appl. Phys., 35, 1286 (1964).
12. G. S. Y. Yeh, S. L. Lambert, J. Polym. Sci., A2, 10, 1183 (1972).
13. C. J. Ong, F. P. Price, J. Polym. Sci. Polym. Symp., 63, 59
    (1978).
14. E. Martuscelli, G. Demma "Morphology and crystallization beha-
    viour of PEO/PMMA blends" (present book)
15. H. Berghmans, N. Overbergh, J. Polym. Sci. Polym. Phys. Ed.,
    15, 1757 (1977).
16. T. T. Wang, T. Nishi, Macromolecules, 10, 421 (1977).
17. J. Boon, J. M. Azcue, J. Polym. Sci., A2, 6, 885 (1968).
18. M. L. Williams, R. F. Landel, J. D. Ferry, J. Am. Chem. Soc.,
    77, 3701 (1955).
19. L. Mandelkern, "Crystallization of Polymers" p. 273,
    Mc Graw-Hill, N. Y. (1964).
20. D. G. Thomas, L. A. K. Staveley, J. Chem. Soc., 4569 (1962).

    J. D. Hoffmann, SPE Trans., 4, 315 (1964).

21. E. Martuscelli, M. Pracella, M. Avella, R. Greco, G. Ragosta, Makromol. Chem., <u>181</u>, 957 (1980); (present book).

22. T. Nishi, T. T. Wang, Macromolecules, <u>8</u>, 909 (1975).

23. R. L. Imken, D. R. Paul, J. W. Barlow, Polym. Eng. Sci., <u>16</u>, 593 (1976).

24. A. L. Scott, J. Chem. Phys., <u>17</u>, 279 (1949).

25. P. J. Flory "Principles of Polymer Chemistry" Cornell Univ. Press, Ithaca, N. Y. (1953).

26. J. D. Hoffman, J. J. Weeks, J. Research National Bur. Stand., Sec. A, <u>66</u>, 13 (1962).

27. M. Natov, L. Peeva, E. Djagarova, J. Polym. Sci. Polym. Symp. Ed., <u>16</u>, 4197 (1968).

28. R. Greco, H. B. Hopfenberg, E. Martuscelli, G. Ragosta, G. Demma, Polym. Eng. Sci., <u>18</u>, 654 (1978).

29. E. Martuscelli, G. Demma "Morphology and crystallization of PEO/aPS; PEO/aPP; PEO/EVA blends" in preparation.

PROPERTIES OF POLYETHYLENE-POLYPROPYLENE BLENDS: CRYSTALLIZATION BEHAVIOUR

E. Martuscelli, M. Pracella, M. Avella, R. Greco
and G. Ragosta

Istituto di Ricerche su Tecnologia dei Polimeri e
Reologia, Arco Felice (Napoli) Italy

INTRODUCTION

Most of theoretical and practical investigations on the properties of polymer blends have been concerned mainly with systems
containing amorphous components, while the properties of blends
with crystalline components have received less attention. Only few
examples of blends with crystallizable components have been investigated so far: poly(butyleneterephtalate)/poly(ethyleneterephtalate)
(PBT/PET)[1], high density polyethylene/low density polyethylene
(HDPE/LDPE)[2], nylon 6/nylon 11[3], isotactic polystyrene/poly(2,6-
dimethyl-1,4-phenylene oxide) (iPS/PPO)[4], isotactic polypropylene/
isotactic poly(1-butene) (iPP/iPB)[5], high density polyethylene/
isotactic polypropylene (HDPE/iPP)[6].

The most interesting aspects of these systems concern the influence of composition on the overall morphology, on mechanical
behaviour and on quantities such as the melting temperature, the
degree of crystallinity and the crystallization rate from melt of
components.

Even if these blends generally form a multiphase system, the
interphase adhesion may be sufficient to determine good properties.
Moreover some blends show in certain compositions better properties
than those of single components. In particular, studies on high
density polyethylene/isotactic polypropylene (HDPE/iPP) blends show

some interesting aspects both from the theoretical and practical
point of view. Deanin and Sansone[7], Greco et al.[8] found, for some
blend compositions, positive synergistic improvement in some mechanical and end use properties (such as modulus, ultimate tensile
strength, heat deflection temperature). The reasons for this
synergism have been qualitatively ascribed to interfacial effects
and partial miscibility of HDPE and iPP chains in the molten state.
The impossibility of cocrystallization phenomena, owing to the differences in the molecular structure of HDPE and iPP and the presence
of individual melting temperatures of the polymers in their blends
indicate the incompatibility of this system in the solid state.
In the liquid state some interactions between chains cannot be excluded owing to the small differences in polarity and free volume.
At the moment, we do not have sufficient information on the compatibility of these polymers in melt state.

In the present work, we report the results of an investigation
concerning the isothermal crystallization behaviour of HDPE/iPP
blends from melt. The objective of the investigation is to obtain
information on the influence that composition may have on some
thermodynamic and kinetic quantities related to the crystallization
process.

EXPERIMENTAL

a) <u>Materials</u>

  Melt blended specimens of high density polyethylene (HDPE)
($\bar{M}_w$=166,000; $\bar{M}_n$=10,200; M.F.I.=3.7 g/10 min; d=0.96 g/cc) and isotactic polypropylene (iPP) ($\bar{M}_w$=307,000; $\bar{M}_n$=15,600; M.F.I.=3.9
g/10 min; d=0.906 g/cc) with various compositions were prepared
using an C.S.I. mixing extruder at 180°C. The standard procedure
is reported in detail elsewhere[9].

b) <u>Calorimetric measurements</u>

The crystallization kinetic and thermal properties of the
homopolymers and their blends were studied by differential scanning
calorimetry. A Perkin Elmer DSC-2 apparatus was used with the following standard procedure: the samples (about 5-6 mg.) were heated
to a 20 degrees temperature above the complete melting and kept at

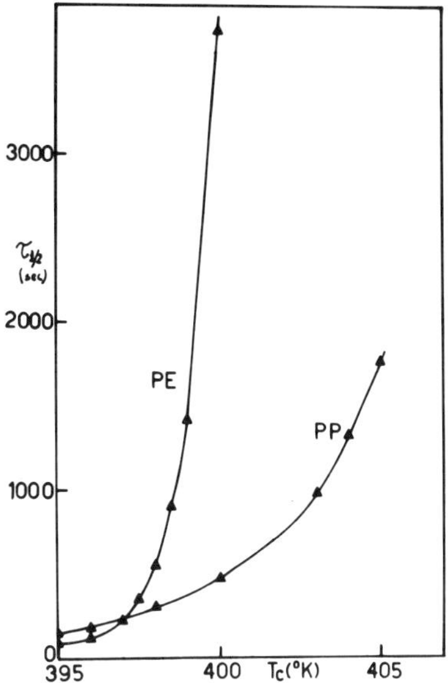

Figure 1.   Half time of crystallization, $\tau_{1/2}$ of polyethylene (HDPE) and isotactic polypropylene (iPP) as a funtion of the crystallization temperature, T_c. $\tau_{1/2}$ is given in s.

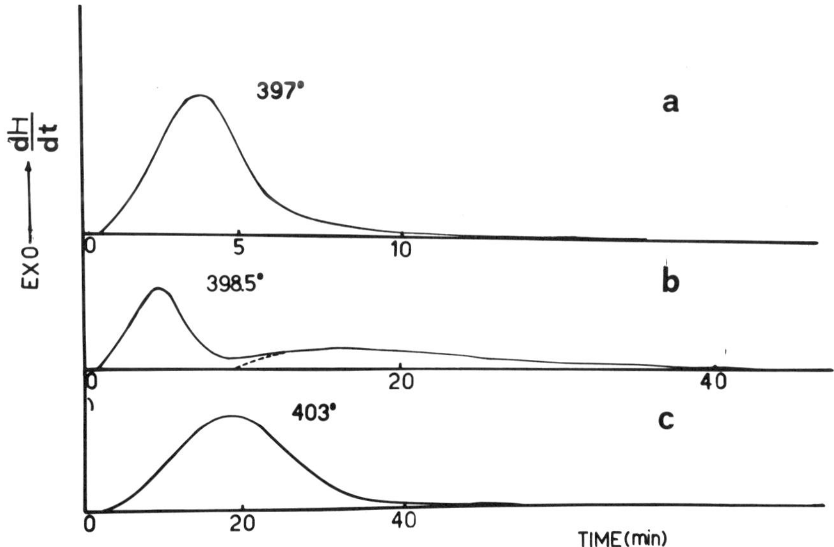

Figure 2.   Exotherms of crystallization of iPP/HDPE (50/50) blends at constant crystallization temperatures, T_c: a) T_c=397 K, b) T_c=398.5 K, c) T_c=403 K.

this temperature for 15 minutes. Thereafter the samples were rapidly
cooled to the crystallization temperature, T_c and the heat dH/dt
evolved during the isothermal crystallization at T_c was recorded
as a function of time. The weight fraction X_t of the material crystallized at time t was calculated from the ratio between the heat
Q_t generated at time t and the total heat Q_∞ corresponding to the
completion of the crystallization, according to the relation:

$$X_t = \frac{Q_t}{Q_\infty} = \int_0^t (\frac{dH}{dt})\ dt \Big/ \int_0^\infty (\frac{dH}{dt})\ dt$$

The observed melting temperatures T_m' and apparent enthalpies
of fusion ΔH_f^* of the isothermally crystallized samples were determined from the maxima and the area, respectively, of the fusion
peaks obtain d heating the samples directly from T_c to the melting
points of he two components at a scanning rate of 20°/min. The
compositions of the blends and the range of the crystallization
temperatures investigated are reported in Table 1.

RESULTS AND DISCUSSION

The half time of crystallization $\tau_{1/2}$ of polyethylene and
polypropylene as funtion of crystallization temperature T_c is
reported in Figure 1. According to data of this Figure the range
of T_c explored may be divided in three parts. For T_c lower than
398 K the $\tau_{1/2}$ of HDPE and iPP are of the same magnitude. As a
consequence both polymers in the blends will crystallize almost
simultaneously. The exotherms of crystallization will show practically a single peak (see Figure 2/a). When T_c ranges from about
398 K to about 400 K the $\tau_{1/2}$ of homopolymers has such values that
both iPP and HDPE are able to crystallize separately in experimental
times. The blends show, for these T_c, exotherms of crystallization
with double peaks more or less defined according to composition.
The two peaks are sharply separated only in the case of iPP/HDPE
(50/50) blend (see Figure 2/b). As shown in Figure 1, for T_c higher
than about 400 K, the $\tau_{1/2}$ of HDPE assumes values so high that
practically this polymer will be not able to crystallize in measurable times at these temperatures. Thus for $T_c > 400K$ the blends
will show only one peak as the crystallization of iPP phase alone
will contribute to the exotherm of crystallization (see Figure 2/c).

TABLE 1

Code, composition and range of crystallization temperatures explored.

Code	Composition: % iPP (W/W) in the blend	Crystallization temperature (K).
iPP	100	395; 396; 397; 398; 400; 403; 404; 405
iPP/HDPE (90/10)	90	396; 397; 398; 398.5; 400; 401; 403; 405
iPP/HDPE (70/30)	70	396; 397; 398; 398.5; 400; 401; 403; 404
iPP/HDPE (60/40)	60	398; 398.5; 399; 400
iPP/HDPE (50/50)	50	395; 396; 397; 397.5; 398; 398.5; 399; 400; 401; 403
iPP/HDPE (40/60)	40	398; 398.5; 399
iPP/HDPE (30/70)	30	395; 396; 397; 397.5; 398; 398.5; 399
iPP/HDPE (10/90)	10	395; 396; 397; 397.5; 398; 398.5; 399
HDPE	0	395; 396; 397; 397.5; 398; 398.5; 399; 400

This last observation enables us to study the influence of the melt polyethylene phase on the crystallization of isotactic polypropylene from iPP/HDPE blends.

Blends crystallized at T_c lower than 400 K show a maximum in the overall $\tau_{1/2}$ value for a composition around 90% in polyethylene (see Figure 3). The value of this maximum is temperature dependent (for T_c=398 K the $\tau_{1/2}$ of HDPE is 550s whilst that of iPP/HDPE (10/90) blend is 1260s). From the above results we may conclude that it is possible, for a given T_c, to reduce the rate of crystal-

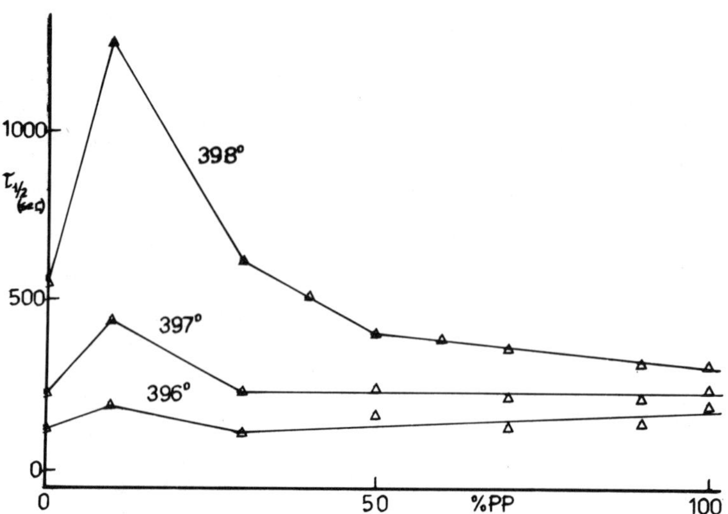

Figure 3.   Plots of half time of crystallization, $\tau_{1/2}$, versus blend
composition at constant T_c, as indicated. (The data refer
to crystallization temperatures lower than 399 K; at this
T_c both iPP and HDPE crystallize). $\tau_{1/2}$ is given in s.

lization of polyethylene simply by blending it with a small amount
of isotactic polypropylene.

The presence of a melt polyethylene phase seems to influence
the rate of crystallization of isotactic polypropylene. This is
shown in Figure 4 where the $\tau_{1/2}$ of the polymer is reported as
function of composition for values of T_c high enough to allow the
crystallization of iPP alone ($T_c > 400$ K). As shown in Figure 4 the
values of $\tau_{1/2}$ first increase monotonically with increasing the
percentage of HDPE in the blend, reach a maximum for a composition
around 60% in iPP, and thereafter start to decrease. Irrespective
of crystallization temperature and composition the crystallization
kinetic of iPP/HDPE follow the Avrami equation up to a high degree
of conversion. In fact, as shown in Figures 5, when the quantity
$\log\left[-\log(1-X_t)\right]$ is plotted against log t, a linear trend is observed.
The application of the Avrami equation, that may be written as,

$$\log \left[-\log(1-X_t)\right] = n \log t + \log \frac{K_n}{2.3} \tag{1}$$

to the straight lines of Figures 5 enables the calculation of the Avrami exponent n and of the overall kinetic rate constant K_n. Alternatively, K_n was also obtained by means of the relation:

$$K_n = \frac{\ln 2}{\tau_{1/2}{}^n} \tag{2}$$

The values of n, $\tau_{1/2}$ and K_n are reported for all the examined samples in Table 2. In the case of homopolymers the values of n are practically independent of T_c and in agreement with literature data[10],[11] (for iPP n=3, for HDPE n ranges from 2.2 to 2.4).

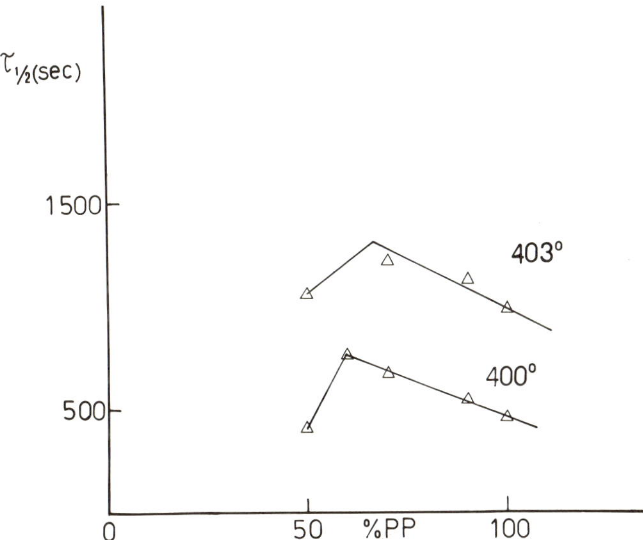

Figure 4.   Dependence of half time of crystallization, $\tau_{1/2}$, of isotactic polypropylene on the blend composition.(The curves refer to T_c high enough to prevent any HDPE crystallization).

Values of n larger than 4 are observed, for lower T_c, in the case of blends with an high iPP content, that give only one exothermic peak as the two polymers crystallize in comparable times. As expected, for lower crystallization temperatures (T_c lower than about

log[-log(1-X_t)]
+0.2
-0.4
-1.0
-1.6
396°
397°
398°
400°
403°
405°
2.0
2.5
3.0
3.5
log t (sec)

(a)

log[-log(1-X_t)]
+0.2
-0.4
-1.0
-1.6
396°
397°
398°
400°
403°
404°
2.0
2.5
3.0
3.5
log t (sec)

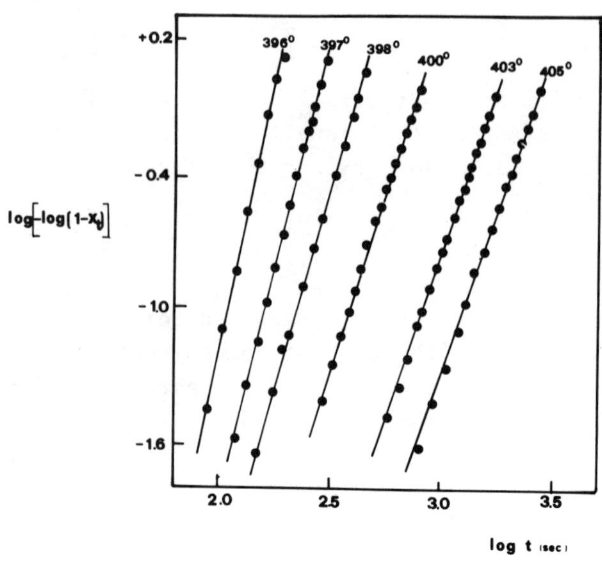

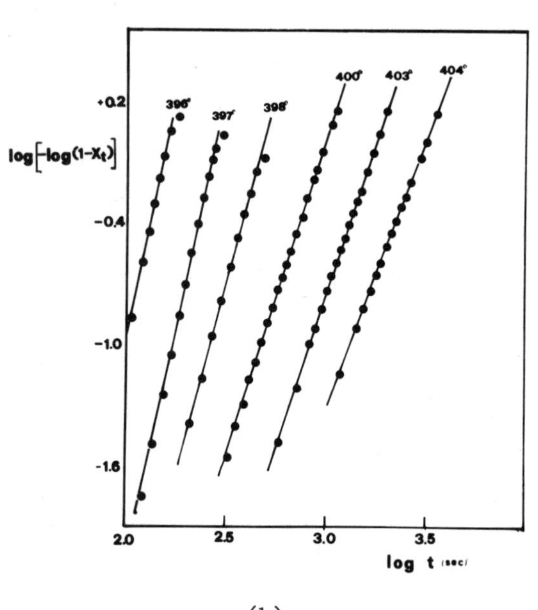

(b)

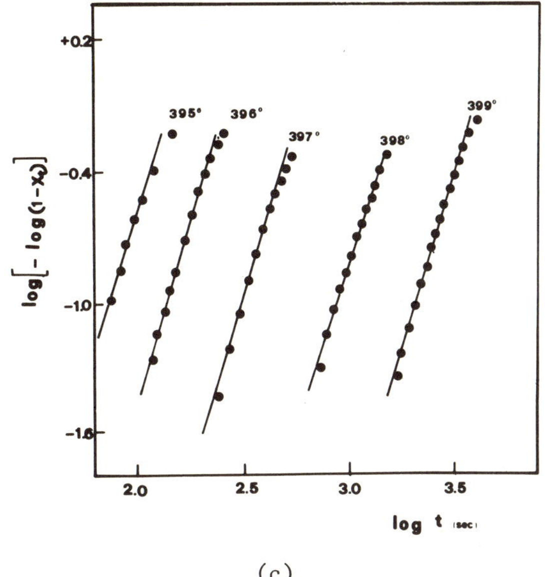

(c)

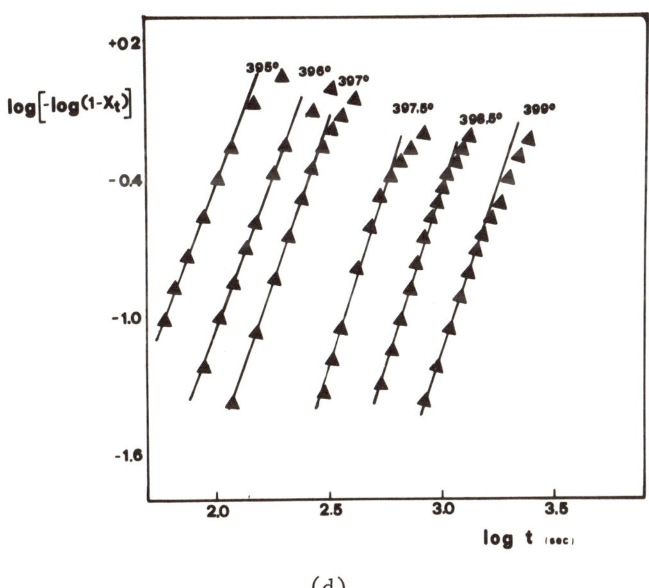

(d)

Figure 5. Avrami plots for iPP/HDPE blends: a) iPP/HDPE (90/10); b) iPP/HDPE (70/30); c) iPP/HDPE (30/70); d) iPP/HDPE (10/90). Crystallization temperature is given in K, time in s.

400 K) all blends show DSC curves with double fusion peaks relating
to melting of the polyethylene and polypropylene phases.

TABLE   2

Avrami exponent n, half time of crystallization $\tau_{1/2}$ and overall
kinetic constant K_n of homopolymers and blends as function of
composition.

Samples	T_c (K)	n	$\tau_{1/2}$ (s)	K_n (s^{-n})
iPP	395	3.0	152	1.96×10^{-7}
	396	3.0	187	1.05×10^{-7}
	397	3.0	237	5.18×10^{-8}
	398	3.0	305	2.43×10^{-8}
	400	3.0	456	7.28×10^{-9}
	403	3.0	981	7.34×10^{-10}
	404	3.0	1320	3.01×10^{-10}
	405	3.0	1763	1.06×10^{-10}
iPP/HDPE (90/10)	396	4.7	138	2.64×10^{-7}
	397	4.3	210	7.48×10^{-8}
	398	3.7	309	2.35×10^{-8}
	398.5	3.3	354	1.56×10^{-8}
	400	3.2	550	4.16×10^{-9}
	401	3.0	668	2.32×10^{-9}
	403	3.0	1140	4.68×10^{-10}
	405	3.0	1863	1.12×10^{-10}

Samples	$T_c(K)$	n	$\tau_{1/2}(s)$	$K_n(s^{-n})$
iPP/HDPE (70/30)	396	4.4	124	3.63×10^{-7}
	397	4.5	215	6.97×10^{-8}
	398	3.4	355	1.55×10^{-8}
	398.5	3.1	450	7.60×10^{-9}
	400	3.1	675	2.25×10^{-9}
	401	3.0	850	1.13×10^{-9}
	403	3.0	1210	3.91×10^{-10}
	404	2.9	2040	8.16×10^{-11}
iPP/HDPE (60/40)	398	3.1	420	9.35×10^{-9}
	398.5	2.8	566	3.82×10^{-9}
	399	2.5	647	2.56×10^{-9}
	400	2.5	763	1.56×10^{-9}
iPP/HDPE (50/50)	395	3.3	119	9.80×10^{-7}
	396	3.2	163	5.78×10^{-8}
	397	3.0	242	4.89×10^{-8}
	397.5	2.5	330	3.50×10^{-7}
	398	2.2	400	1.31×10^{-7}
	398.5	3.2	310	2.33×10^{-8}
(Only iPP phase)	399	3.2	324	2.04×10^{-8}
	400	3.1	408	6.76×10^{-9}
	401	3.0	572	3.70×10^{-9}
	403	3.0	1060	5.82×10^{-10}
iPP/HDPE (40/60)	398	3.1	558	3.99×10^{-9}
	398.5	2.8	847	1.14×10^{-9}
	399	2.7	1285	3.27×10^{-10}

Samples	$T_c(K)$	n	$\tau_{1/2}(s)$	$K_n(s^{-n})$
iPP/HDPE (30/70)	395	2.6	93	$5.28\ 10^{-6}$
	396	2.7	110	$2.13\ 10^{-6}$
	397	2.8	230	$1.69\ 10^{-7}$
	397.5	3.0	523	$9.06\ 10^{-8}$
	398	3.0	614	$2.99\ 10^{-9}$
	398.5	3.0	940	$8.34\ 10^{-10}$
	399	3.0	1777	$1.24\ 10^{-10}$
iPP/HDPE (10/90)	395	3.2	110	$2.03\ 10^{-7}$
	396	3.3	190	$2.09\ 10^{-8}$
	397	3.3	441	$1.30\ 10^{-9}$
	397.5	3.2	730	$4.76\ 10^{-10}$
	398	3.2	1260	$8.31\ 10^{-11}$
	398.5	3.1	1650	$7.35\ 10^{-11}$
	399	3.1	2880	$5.90\ 10^{-12}$
HDPE	395	2.2	91	$3.38\ 10^{-5}$
	396	2.3	120	$1.84\ 10^{-5}$
	397	2.3	223	$4.71\ 10^{-6}$
	397.5	2.3	354	$1.70\ 10^{-6}$
	398	2.3	548	$6.51\ 10^{-7}$
	398.5	2.3	900	$2.19\ 10^{-7}$
	399	2.3	1415	$8.08\ 10^{-8}$
	400	2.4	3750	$9.47\ 10^{-9}$

For higher T_c, on the contrary, only the endotherm of fusion of polypropylene phase is observed. The DSC traces were obtained by heating the samples directly from T_c to the melting without cooling to room temperature. In all blends and homopolymers the melting temperature of both HDPE and iPP phases increases linearly with T_c. The application to the experimental points of the following relation[12]:

$$T'_m = T_m \left(\frac{\gamma-1}{\gamma}\right) + \frac{T_c}{\gamma} \tag{3}$$

allowed the calculation of the equilibrium melting temperatures T_m. As shown in Table 3 the values of T_m seem to be slightly affected by blend composition.

TABLE 3

Values of the equilibrium melting temperature T_m of pure homopolymers and of homopolymers in the blends as function of blend composition.

Sample	T_m(iPP)	T_m(HDPE)
iPP	466.7	--
iPP/HDPE(90/10)	464.4	--
iPP/HDPE(70/30)	465.1	416.6
iPP/HDPE(50/50)	463.3	418.3
iPP/HDPE(30/70)	465.4	418.4
iPP/HDPE(10/90)	466.0	420.0
HDPE	--	416.7

In order to study the influence of a melt polyethylene phase on the thermodynamic and kinetic aspect of crystallization of isotactic polypropylene from iPP/HDPE blends, the crystallization rates, for low values of the undercooling ΔT ($\Delta T = T_m - T_c$), were analyzed using the secondary nucleation theory of Hoffman and

Lauritzen[12]. According to this theory the overall rate constant of crystallization K_n, may be written as:

$$\frac{1}{3} \log K_n + \frac{\Delta F}{2.3RT_c} = A_o - \frac{4b_o \sigma\sigma_e T_m}{2.3K\Delta H_F T_c \Delta T} \tag{4}$$

In equation 4, ΔF is the activation energy for the transport process at the liquid-solid interface, σ and σ_e are the free energies per unit area of the surfaces of the lamellae parallel and perpendicular to the chain direction, respectively; ΔH_F is the enthalpy of fusion and b_o is the distance of two adjacent fold planes. Under the assumption that the spherulite nucleus density is independent of time, ΔT, blend composition and T_m, then A_o may be considered constant. The transport term ΔF is usually calculated by means of the approximate relation[14]:

$$\Delta F = \Delta F_{WLF} = \frac{C_1 T_c}{C_2 + T_c - T_g} \tag{5}$$

where C_1 and C_2 are constants, generally assumed equal to 4.12 Kcal/mol (17.24 KJ/mol) and 51.6 K, respectively. T_g is the glass transition temperature (for our calculation we assume, irrespective of blend composition, as glass transition that of pure polypropylene, T_g=260 K). Plots of $\frac{1}{3} \log K_n + \Delta F/2.3RT_c$ against $T_m/T_c\Delta T$ are shown in Figures 6 for pure isotactic polypropylene and for blends iPP/HDPE (90/10), iPP/HDPE (70/30), and iPP/HDPE (50/50). Experimental values of K_n corresponding to values of T_c high enough to prevent any crystallization of polyethylene were used for all the blends. From the slopes of the straight lines of Figures 6 it was possible to obtain the quantity $4b_o\sigma\sigma_e/K\Delta H_F$ which turned out to be dependent on the blend composition. Using for b_o, σ and ΔH_F the following values of literature[15]: b_o=5.24 A, ΔH_F=50 cal/g (209 J/g) and σ=11 erg/cm^2 (11·10^{-7} J/cm^2) it is possible to calculate, from the known values of the quantity $4b_o\sigma\sigma_e/K\Delta H_F$ the free energy of folding σ_e of isotactic polypropylene lamellar crystals as function of blend composition. As shown in Figure 7, σ_e first decreases with increasing the percentage of polyethylene in the blend, reaches a minimum at about a composition of 70% in iPP and then start to increase.

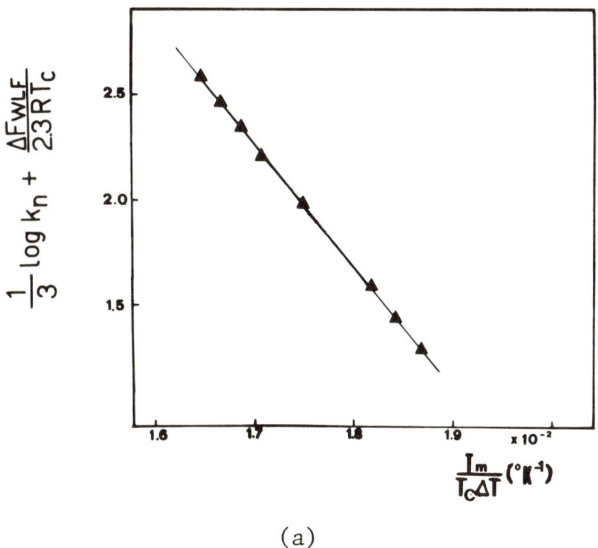

(a)

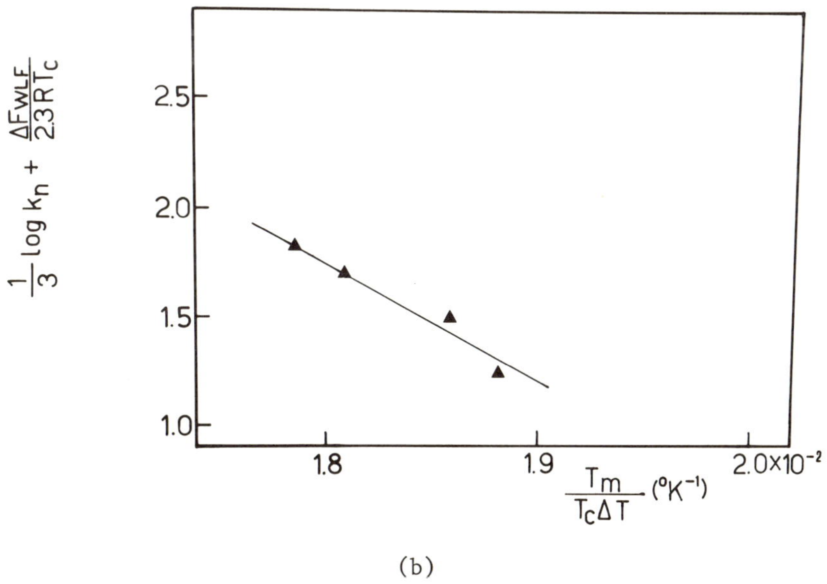

(b)

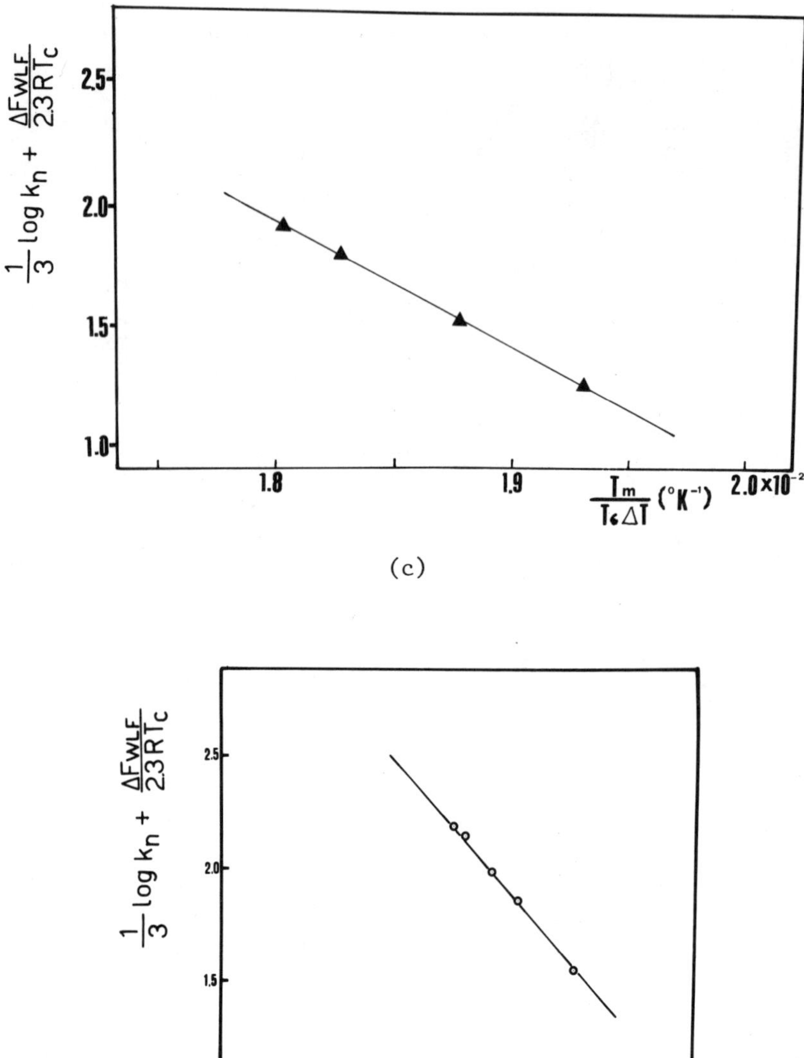

Figure 6.   Plots of equation 4 for pure isotactic polypropylene and
            for isotactic polypropylene crystallized from blends of
            different compositions: a) iPP; b) iPP/HDPE(90/10);
            c) iPP/HDPE(70/30); d) iPP/HDPE(50/50).

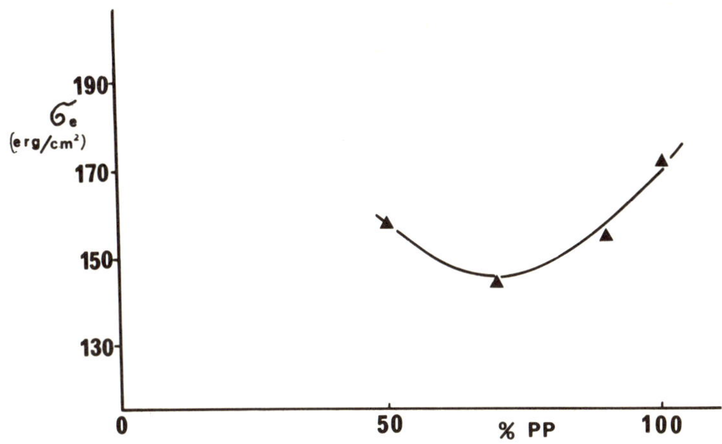

Figure 7. Variation of the free energy of folding, σ_e, of isotactic
polypropylene with the composition of blends.

CONCLUDING REMARKS

    The results of our investigation show that the crystallization
of iPP/HDPE blends presents some very complex aspects interesting
both from fundamental and practical point of view. It has been
shown in this study that:
a) The kinetic of crystallization of polyethylene is greatly delayed
   by the addition of a small amount of isotactic polypropylene
   ($\approx$10% by weight). In fact, for iPP/HDPE (10/90) blend, half crystallization times of the order of 1260 s were observed, at
   T_c=398 K. This figure is around three and four times larger than
   for pure polyethylene and polypropylene, respectively (see Figure
   3). At the same time a significant increase in the Avrami coefficient (from 2.3 for pure HDPE to about 3.2 for iPP/HDPE (10/90)
   blend) is observed. An increase in the Avrami exponent is usually
   attributed in literature to a change from istantaneous to sporadic
   nucleation.
b) For crystallization temperatures high enough to prevent any polyethylene crystallization (T_c>400 K) the presence in the melt blends

of liquid HDPE influences the crystallization of isotactic polypropylene. The half time of crystallization $\tau_{1/2}$ of iPP increases
with the percentage of HDPE in the blend. The kinetics passes a
minimum at a well defined blend composition (around 60% in iPP)
before the phase inversion. In order to explain points a) and b),
we must examine the thermodynamic and kinetic factors that
determine the overall rate of isothermal crystallization of polymers.

For a spherical geometric growth with istantaneous nucleation,
the kinetic constant K_n may be written[17]:

$$K_n = \frac{4 \pi \, d_c G^3 \, \overline{N}}{3 \, d_a (1-\lambda_\infty)} \tag{6}$$

where d_a and d_c are the densities of amorphous and crystalline phases;
G is the growth rate of spherulites, $\overline{N}$ is the number of potential
nuclei per volume unit, $(1-\lambda_\infty)$ is the polymer weight fraction that
is crystalline at time $t=\infty$. According to the kinetic theory, the
growth rate of spherulites G is given by

$$G = G_O \, \exp \left(- \frac{\Delta F}{RT_c}\right) \, \exp \left(- \frac{\Delta \Phi}{KT_c} \right) \tag{7}$$

where ΔF representing the activation energy connected with the
transport process of materials from a liquid to a solid phase, depends
on the melt viscosity and $\Delta \Phi$ representing the free energy formation
of a nucleus with critical dimensions, depends on the equilibrium
melting temperature on the surface free energies on the enthalpy of
fusion and on undercooling ΔT.

During the crystallization process of binary blends, it is
possible to find the following extreme cases:
a) the two components are incompatible in the melt;
b) the two components are compatible in the melt.
For what concerns case a), the only influence that a component may
have on the other, is on the value of nucleation density. This
effect should increase the number of nuclei per volume unit and
leads to increase K_n. In fact, a second component could act as
heterogeneous nucleant. When the two components are compatible in
the melt it is possible to make the following considerations:
1) the glass transition temperature of the blend, according to the

Fox's equation[18]

$$\frac{1}{T_g(\text{blend})} = \frac{W_{iPP}}{T_g(iPP)} + \frac{W_{HDPE}}{T_g(HDPE)} \tag{8}$$

is to be included among those of pure polymers. Particularly since $T_g(iPP) > T_g(HDPE)$ the result will be that:

$$T_g(iPP) > T_g(\text{blend}) > T_g(HDPE)$$

and so the transport energy, defined according to equation 5 will result, for a given blend composition and for a well defined crystallization temperature:

$$\Delta F_{WLF}(iPP) > \Delta F_{WLF}(\text{blend}) > \Delta F_{WLF}(HDPE)$$

2) The work necessary to form a nucleus of critical dimensions, may be expressed as[19]:

$$\Delta\Phi = \Delta\Phi_o - \frac{2\sigma K T_c T_m}{b_o \Delta H_F \Delta T} \ln\nu_2$$

where $\Delta\Phi_o$ refers to the pure crystallizable polymer and the second term is the result of the entropy contribution to free energy that represents the probability of choosing the required number of crystallizable polymeric sequences from a blend with volume fraction ν_2 of crystallizable component. Since $\ln \nu_2$ is negative, the value of $\Delta\Phi$ will be always higher than that of $\Delta\Phi_o$.

On the basis of the above mentioned considerations, the increase of $\tau_{1/2}$ observed for blend iPP/HDPE at a high HDPE content (>90%) may be interpreted only by admitting that the two polymers should be entirely, or partially compatible in the melt in the range of examined crystallization temperatures. In fact in this case, ΔF and $\Delta\Phi$ in equation 7 may increase and then determine a lowering of growth rate G and then of the kinetic rate constant K_n too in equation 6. But, it must be noted that in the presence of a diluent effect, a depression in the equilibrium melting temperature of HDPE, should also be present in opposition to our previous observations. This discrepancy cannot be explained on the basis of experimental data, that we actually have.

Even the increase of $\tau_{1/2}$ in iPP observed for blends iPP/HDPE at a high iPP content (>60%) at crystallization temperatures($T_c \geqslant 400K$) where HDPE does not crystallize can be interpreted by admitting a certain compatibility degree of two polymers in the melt. In this case, as previously mentioned, the term ΔF should result lower than that relative to pure iPP while the term $\Delta\Phi$, owing to the diluent effect, always will result higher. On the basis of experimental data, it is possible to conclude that the influence of thermodynamic factor on growth rate G, is higher than the transport factor, in the case of alloys with high iPP content that crystallize at high T_c.

The presence of two maxima in the plots of $\tau_{1/2}$ in function of the composition, observed both for blends rich in HDPE, crystallized at low T_c, and for blends rich in iPP, crystallized at high T_c, could be considered in connection with phenomena of reduction in the mutual solubility of the two polymers in the melt.

The linear trend observed in the graphs $1/3 \log K_n + \Delta F/2.3RTC$ versus $T_m/T_c\Delta T$ probably shows that, at least for the composition and temperature range of crystallization examined, the variations on T_g and on $\Delta\Phi$ value of iPP resulting from the diluent effect of HDPE, are very small and in any case would not be able to create a deviation from the linearity. Furthermore the resulting σ_e values may be considered very near to those which would result, considering the corrections to be made to T_g and $\Delta\Phi$ following the diluent effect.

The experimental results of this work show that, even indirectly, the two components (iPP/HDPE) the blends must have a certain compatibility degree in the melt.

REFERENCES

1. A. Escala, E. Balizen and R. S. Stein, Polym. Reprints, <u>15</u>,(1) 152 (1978).
2. Tetsuo Sato and Mikio Takahashi, J. Appl. Polym. Sci, <u>13</u>, 2665 (1969).
3. M. Inoue, J. Polym. Sci. A, <u>1</u>, 3427 (1963).
4. H. Berghmans and N. Overbergh, J. Polym. Sci. Polym. Phys. Ed., <u>15</u>, 1757 (1977).
5. A. Siegmann, J. Appl. Polym. Sci., <u>24</u>, 2333 (1979).

6. J. Grebowicz and T. Pakula, (present book).

7. R. D. Deanin and M. F. Sansone, Polym. Repr., $\underline{19}$, (1), 211 (1978).

8. R. Greco, G. Mucciariello, G. Ragosta and E. Martuscelli, J. Mat. Sci., $\underline{15}$, 845 (1980).

9. R. Greco, H. B. Hopfenberg, E. Martuscelli, G. Ragosta and G. Demma, Polym. Eng. Sci., $\underline{18}$, 654 (1978).

10. L. Amelino and E. Martuscelli, Polymer, $\underline{16}$, 864 (1975).

11. Yu. K. Godowsky and G. L. Slominsky, J. Polym. Sci. Polym. Phys. Ed., $\underline{12}$, 1053 (1974).

12. J. D. Hoffman, SPE Trans., $\underline{4}$, 315 (1964).

13. C. Mancarella and E. Martuscelli, Polymer, $\underline{18}$, 1240 (1977).

14. M. L. Williams, R. F. Landel and J. D. Ferry, J. Am. Chem. Soc., $\underline{77}$, 3701 (1965).

15. E. Martuscelli, M. Pracella and A. Zambelli, J. Polym. Sci., Polym. Phys. Ed., $\underline{18}$, 619 (1980).

16. M. Avrami, J. Chem. Phys., $\underline{7}$, 1103 (1939), $\underline{8}$, 212 (1940), $\underline{9}$, 177 (1941).

17. L. Mendelkern "Crystallization of Polymers", Mc Graw-Hill Inc., New York, 1964.

18. T. G. Fox, Bull. Am. Phys. Soc., $\underline{2}$, 123 (1956).

19. J. Boon and J. M. Azcue, J. Polym. Sci.A-2, $\underline{6}$, 885 (1968).

THE STRUCTURE AND MECHANICAL PROPERTIES OF BLENDS OF CIS
AND TRANS-POLYISOPRENE

A.J. Carter[*], C.K.L. Davies[*] and A.G. Thomas[+]

[*]Materials Department, Queen Mary College,
University of London, UK

[+]MRPRA, Hertford, UK

A series of blends of cis/trans-polyisoprene
(0-100%) show a single glass transition temperature and
are compatible at all compositions. The crystallinity
decreasing linearly with crystallizable trans content.
All the blends show impinged spherulites even at crystallinities of 1 to 2%. The spherulite growth rate decreases as the non-crystallizable content of the blends increases. TEM observations show that the crystals within
spherulites become more branched and dentrite like at
low crystallinities; fibrils of crystals are seen larger
non-crystalline spacings between the fibrils. The lamellar crystal thickness changes only slowly with crystallinity. The long period increases with decreasing crystallinity in the manner suggested by the TEM observations.
The tensile modulus decreases with crystallinity and
attempts have been made to describe the change with
simple reinforcement theory.

INTRODUCTION

Spherulitic crystallization in compatible blends of homo-poly-
mers has been studied by several investigators: Keith and Padden[1],
Yeh and Lambert[2], Robeson[3], Wang and Nishi[4], Wenig et al[5],
Khambatta et al[6] and Warner et al[7]. The non-crystallizable compo-

nent of the blend may be found to reside in interlamellar, interfibrillar or interspherulitic positions; the scale of segregation depending on the diffusion coefficient of the non-crystallizable component and the spherulitic radial growth rate[1]. In the case of isotactic polystyrene/poly(2,6-dimethyl phenylene oxide) (i-PS/PPO)[5], and polycaprolactone/poly-vinyl chloride (PCL/PVC)[6] the relatively slow diffusion of the non-crystallizable polymer results in interlamellar segregation. However, in the case of blends of isotactic/atactic polystyrene (i-PS/a-PS)[1,7] the relatively slower spherulite growth rates result in interfibrillar segregation. In both cases the evidence for the scale of segregation and hence of the detailed spherulitic morphology is derived from small angle X-ray scattering (SAXS) studies.

Spherulitic growth rates are found to decrease as the proportion of the non crystallizable component in the blends increases; the magnitude of the effect depending the relative properties of the non-crystallizable component. The effect is large if the glass transition temperatures of the two components is very different[4] and much smaller for blends of polymer isomers[1].

The present work on blends of cis and trans-polyisoprene was initiated to study the change in morphology and spherulite growth rate as the proportions of the crystallizable species (trans-polyisoprene) was decreased from 100% to 5%. This system was chosen for study as the glass transition temperatures of the two components are not very different and blends could be manufactured and crystallized with a very wide range of crystallinities. Furthermore, detailed electron microscope studies of crystallization of trans-polyisoprene[5,6] have provided information on the lamellar crystal morphology within spherulites. This electron microscope technique together with SAXS studies enabled the segregation of the non-crystallizable component in the blend to be studied in detail. The spherulitic morphology was also studied by optical microscopy and spherulite growth rates measured. The blends were used to study the variation of the initial tensile modulus with crystallinity and crystal morphology.

EXPERIMENTAL

The blends were prepared by milling each component separately

and then hot (60°C) mill rolling together quantities required for
each specific blend. The blends were then compression moulded at
100°C and slowly cooled to room temperature. Samples for a specific study were prepared from these sheets by remelting and recry-
stallization at a specific temperature. All the blends were manufactured from synthetic cis-polyisoprene (NATSYN 2200) which was
blended with naturally occurring trans-polyisoprene (Tjipiter).

Optical microscopy and spherulite growth rate studies of the
blends were carried out using a Mettler hot stage mounted on a Rei-
chart microscope. Spherulite structural studies were carried out
on a JEOL 100CX transmission electron microscope. Crystallization
was carried out in thin films, 100nm thick, which were stained prior
to examination with osmium tetroxide, using the technique reported
previously[5,6].

Glass transition temperatures were determined by measuring
tan δ over a range of temperatures (-100°C -20°C) at a limited range of frequencies (0.2 Hz to 15 Hz) using the technique of forced
non-resonance in tension or simple shear[7]. The dynamic, small
strain, tensile modulus was determined using the same apparatus.

Wide angle X-ray studies (WAXS) were carried out on a General
Electric Co. XRD-6 diffractometer with a step scan facility and a
diffracted beam monochromater. Small angle X-ray studies (SAXS)
were carried out using a Kratky camera set for infinite slit height.
In both cases Nickel filtered Cu Kα radiation was used. The SAXS
diffracted intensity versus angle curves were desmeared using a modified form of the computer programme FF SAXS 3 written by Vonk[11].
The diffracted intensities were also corrected for the Lorentz factor[12].

RESULTS AND DISCUSSION

The majority of the data presented in this paper will refer to
samples crystallized at 34°C which contain only low melting form
(LMF) crystals of trans-polyisoprene with an orthorhombic unit
cell[5,6,13]. Growth rate measurements have been carried out over a
range of temperatures for both the LMF crystals and the monoclinic
high melting form crystals (HMF)[5,6].

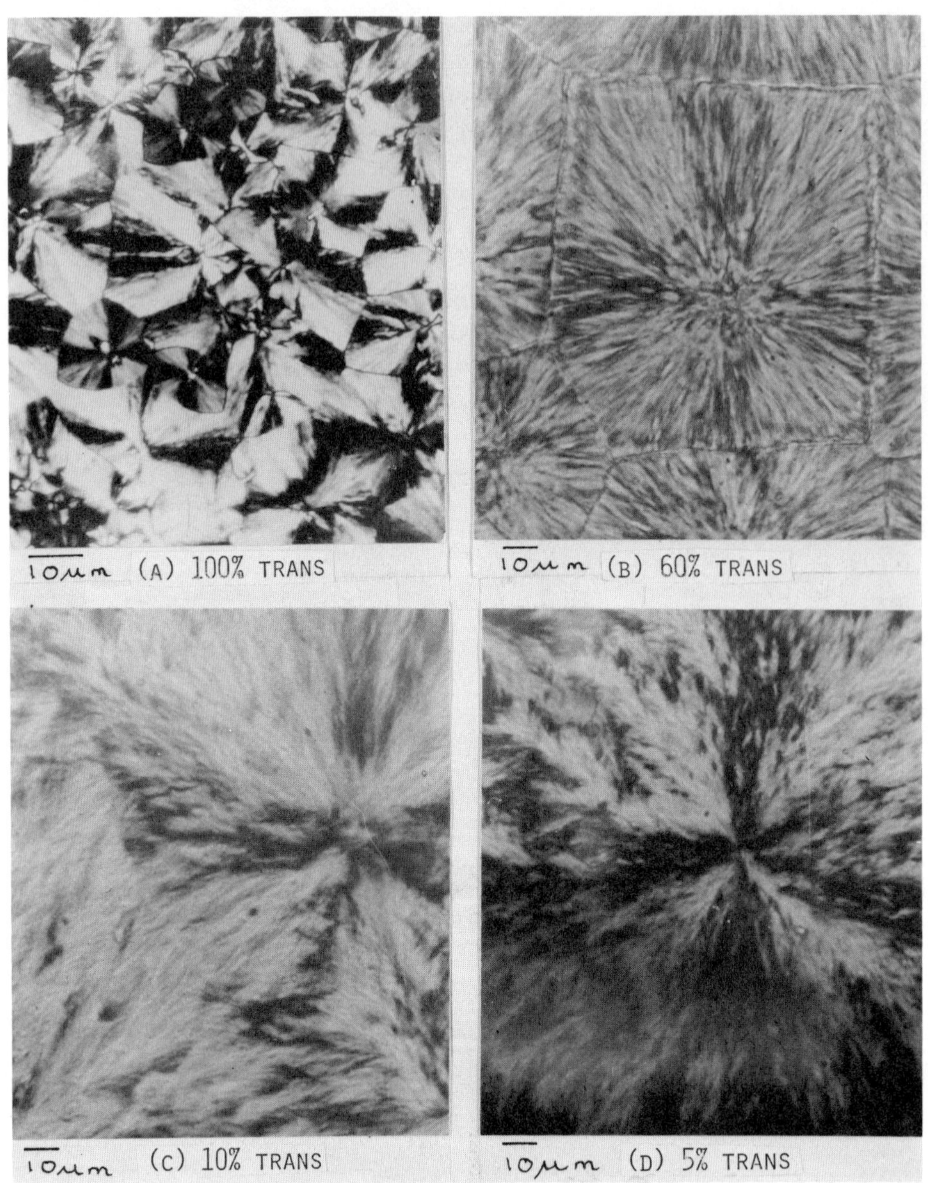

Figure 1. Optical micrographs of LMF spherulites in blends of cis/trans polyisoprene crystallized at 34°C.

Compatibility

In all the blends containing from 5% to 100% trans-polyisoprene spherulites are observed to nucleate and grow until complete impingement occurs (Figure 1). There is no evidence that the non-crystallizable component segregates at an interspherulitic level. The spherulites become increasingly coarse as the available crystallizable content of the blends decreases (Figure 1). It is clear that at least some interfibrillar segregation must occur. In all the blends a single maximum is observed in the plot of the dynamic loss factor (tan δ) versus temperature (Figure 2). This is good evidence of a single glass transition temperature and of compatibility of the components of the blends. Further evidence of compatibility is the fact that the trans-polyisoprene from the crystals remixes on crystal melting; the growth rate and crystal melting temperature being independent of the number of crystallization/melting operations.

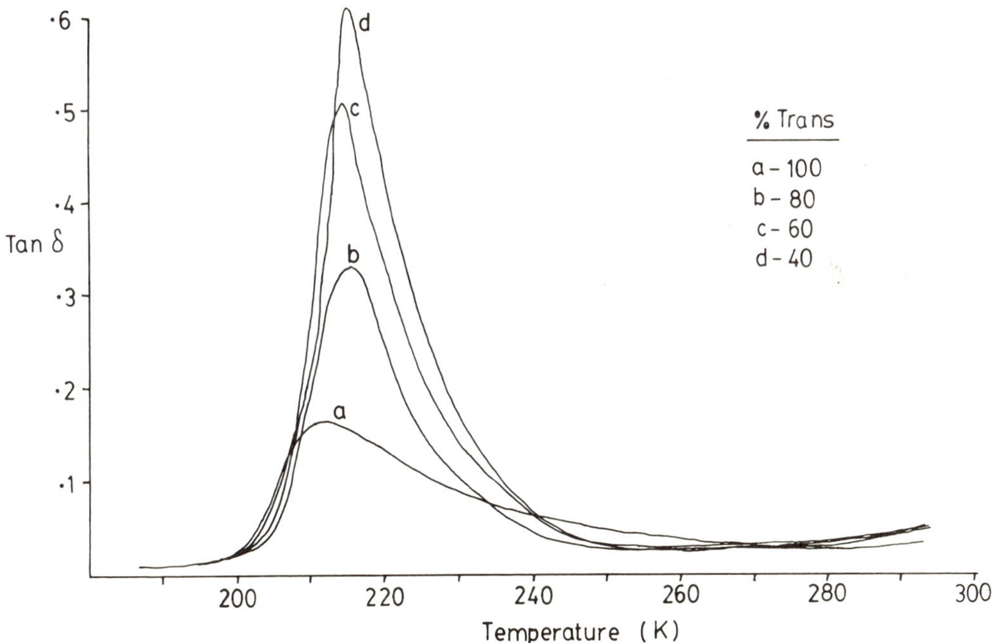

Figure 2. The dynamic loss factor (tan δ) as a function of temperature.

WAXS

The crystallinity of the blends was determined by WAXS using
simply the ratio of the total integrated scattered intensity due to
the crystals to the total real scattered intensity due to the sample
as a whole. The crystal structure of the trans-polyisoprene crystals is independent of blend content as can be seen from the costancy of the scattering angles (Figure 4). The scattering from
the non-crystalline regions however, increases as the proportion of
the non-crystallizable component in the blend increases (Figure 4).
The measured crystallinity varies almost linearly with trans content of the blends (Figure 5). It is interesting to note that
impinged spherulites are observed even in the blend containing only
5% trans-polyisoprene which must have a total crystallinity of only
1-2%.

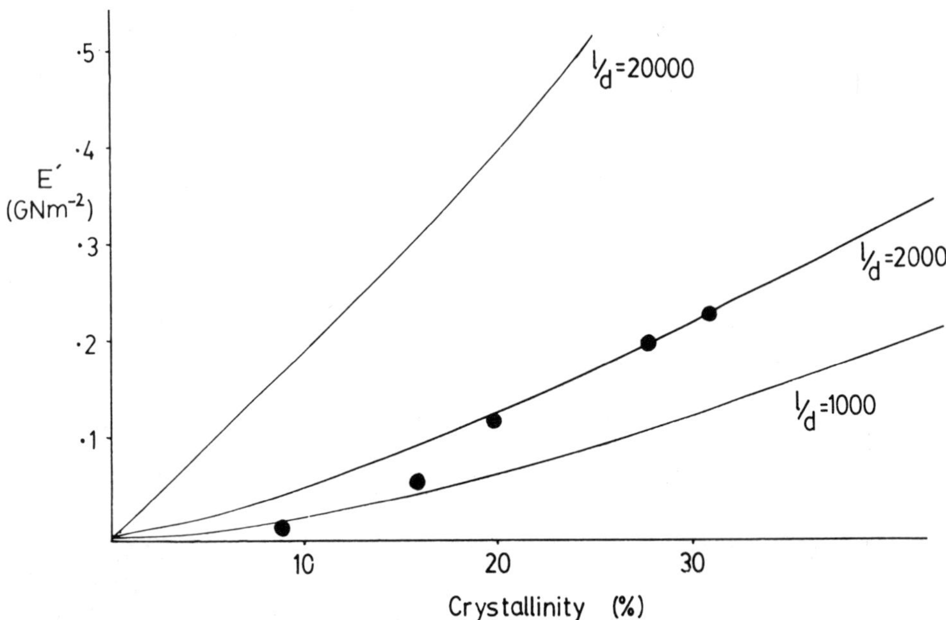

Figure 3.   The small strain dynamic modulus (E) as a function of
            crystallinity. The lines are calculated on the basis
            of the Halpin Tasi model for using various aspect ratio's
            (L/d).

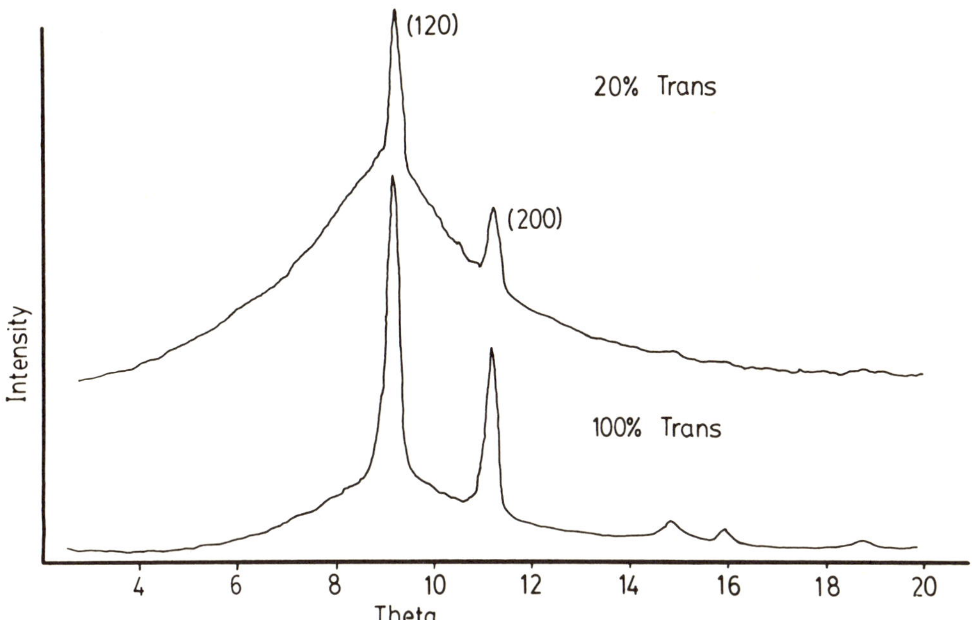

Figure 4. Wide angle X-ray scattering curves

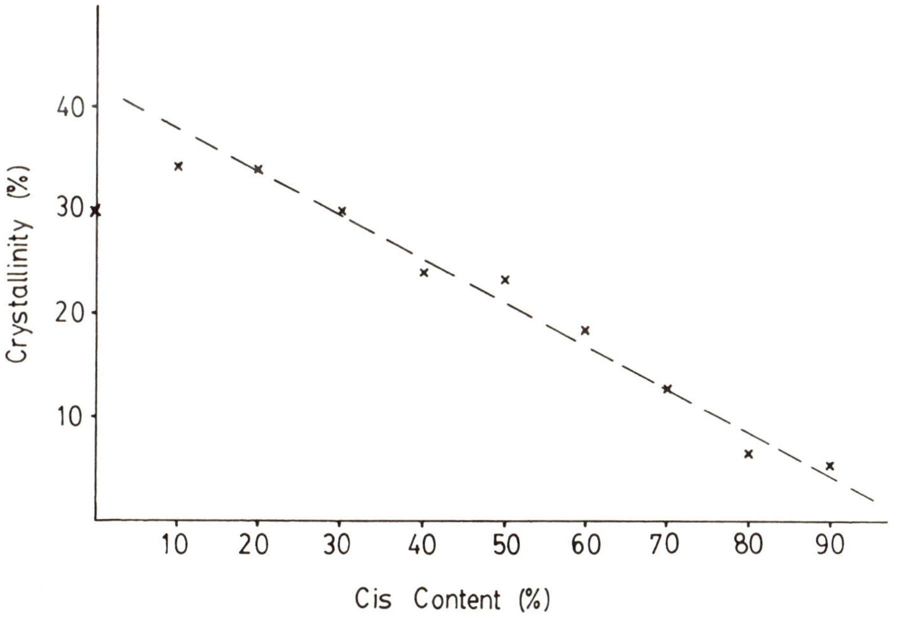

Figure 5. The X-ray crystallinity as a function of cis-polyisoprene content in the blends.

SAXS

The SAXS curves show intensity maxima which become less intense and broader as the crystallinity decreases (Figure 6). The position of the maxima shift to lower angles as the crystallinity of the blends decreases. This yields an increase in long period with decreasing crystallinity (Figure 7). If the non-crystallizable component is assumed to segregate at an inter-lamellar level, as in (i-PS/PPO)[5] or in (PCL/PVC)[6] the lamellar thickness can be calculated from the long period and the WAXS crystallinity assuming impinging regular stacks of equally spaced lamellar crystals (Figure 7). However, the lamellar thickness calculated using these assumptions becomes progressively more non sensible as the crystallinity decreases (Figure 7). This suggests that some material must segregate into interfibrillar regions as in (i-PS/a-PS) blends[7]. This suggestion is supported by the fact that there is little change in the electron microscope measured lamellar thickness (Figure 7) or in the melting temperature as the fraction of the non-crystallizable component in the blend decreases, at least for LMF crystals grown at 34°C.

The absolute intensities from the SAXS curves were used to calculate the correlation functions for comparison with the correlation functions derived from the simple two-phase lamellar model due to Vonk and Kortleve[14] as modified to allow for a finite stack size[15]. Good agreement was obtained for blends containing 40% trans-polyisoprene or more. However, at low crystallinities, poor agreement was obtained even using very small stack sizes. This is presumably due to the increasingly disordered stacks as the crystallinity decreases.

Transmission electron microscopy

The transmission electron microscopy studies yielded evidence of significant morphology changes as the crystallinity of the blends decreased. The crystals become much more heavily branched and dendritic in character (Figure 8). The interfibrillar regions increase in size, for a given crystallization temperature, as the crystallinity decreases. The scale of the segregation hence increases as the spherulite growth rate decreases. This is as predicted by Keith and Padden[1] when the diffusion coefficient is effectively

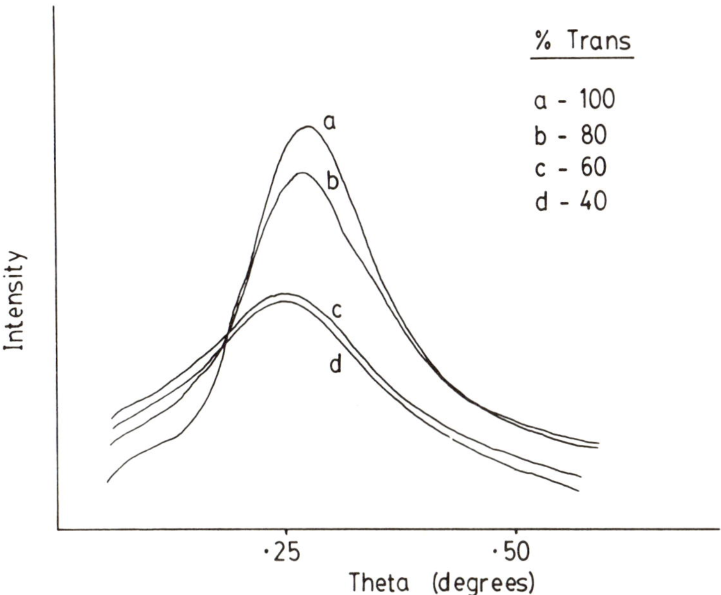

Figure 6. Small angle X-ray scattering curves

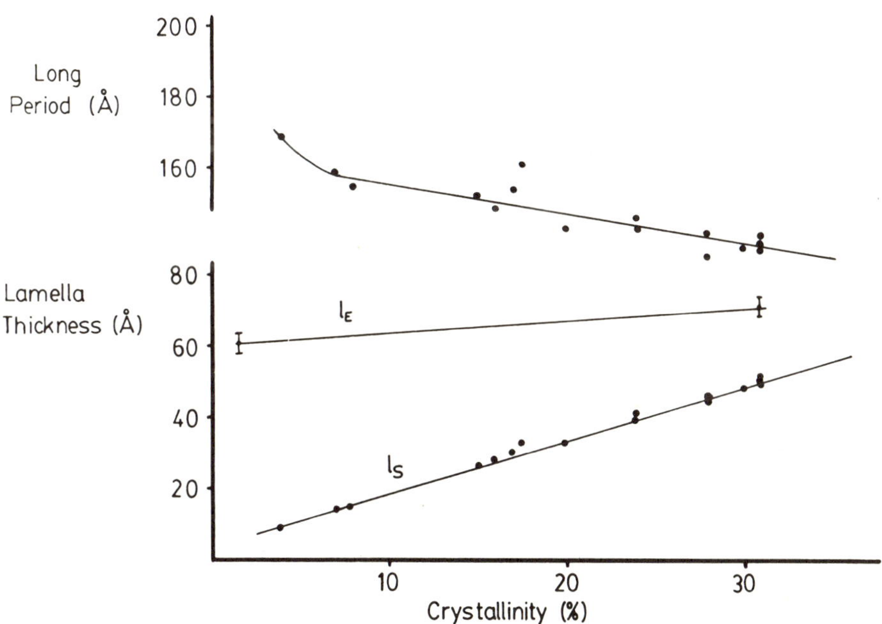

Figure 7. The long period (L), the SAXS lamellar thickness (L_S)
and the TEM lamellar thickness (L_E) as a function of
crystallinity.

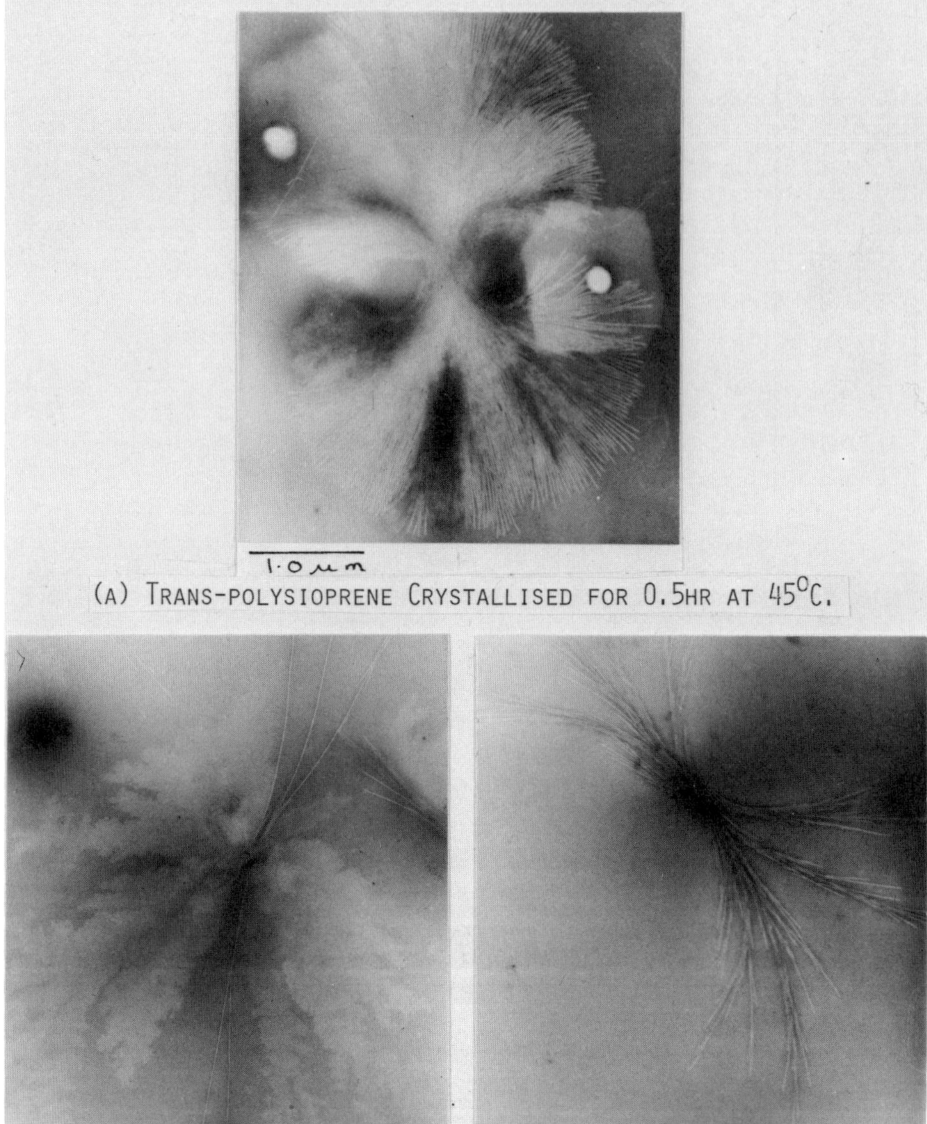

Figure 8.   Transmission electron micrographs of LMF spherulites in blends of cis/trans polyisoprene.

constant as crystallization takes place at a constant temperature
and the spherulite growth rate decreases.

Spherulite growth rates

    Both HMF and LMF spherulites grow at a constant rate at most
temperatures studied (Figure 9). However, at very low growth rates
there is some evidence of non-linear growth. Segregation would be
expected to occur on a spherulitic scale if the spherulitic growth
rate is low and diffusion of the non-crystallizable component is relatively rapid[1]. The growth rates of HMF and LMF spherulites decrease as the trans content of the blend increases (Figure 10). The magnitude of the effect is intermediate to that for atactic/isotactic
polystyrene[1] and for poly(vinylidene fluoride/poly(methylmetacryla-
te)[4]. The growth rate temperature curves are similar in shape for
all the blends as neither the glass transition temperature nor the
melting temperatures change significantly (Figure 11). The growth

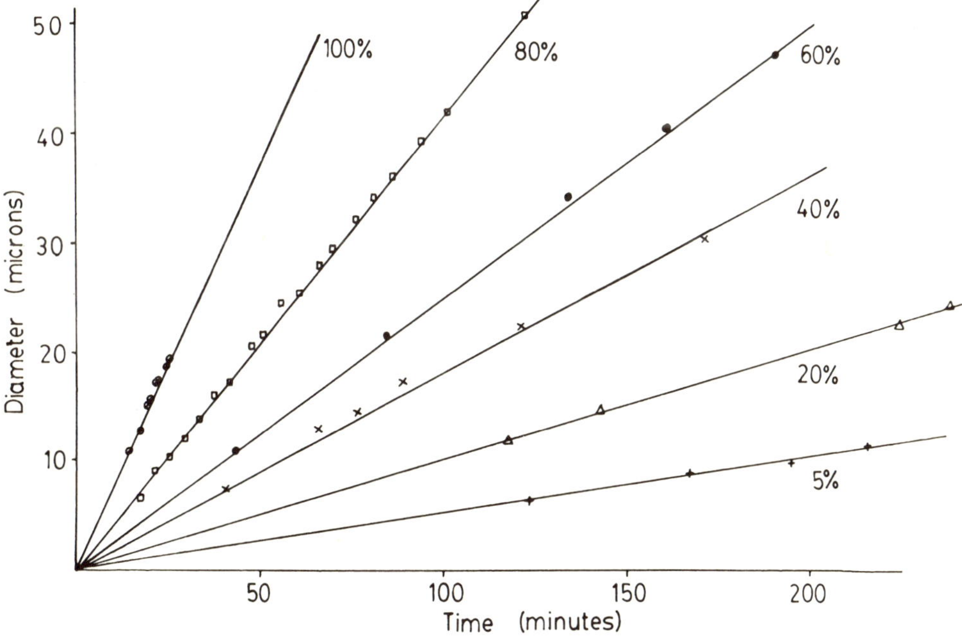

Figure 9.   The diameter of LMF spherulites as a function of time
            at 40°C.

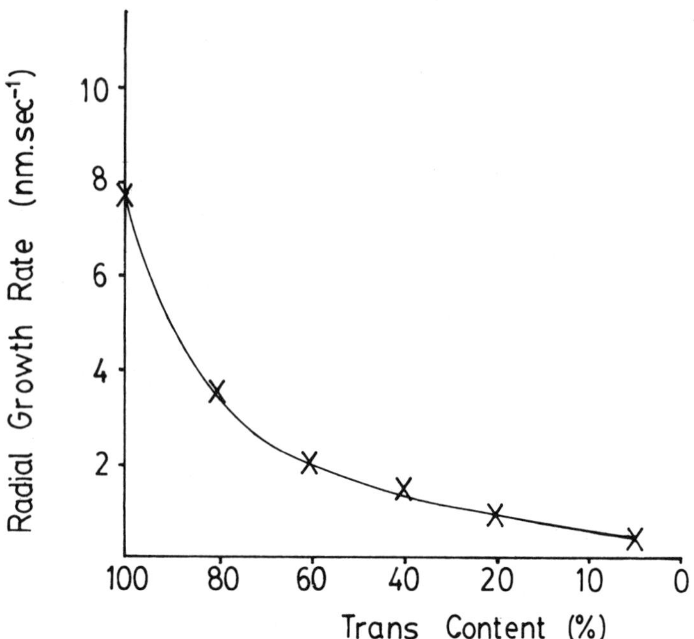

Figure 10. The growth rate of LMF spherulites as a function of trans content in the blends.

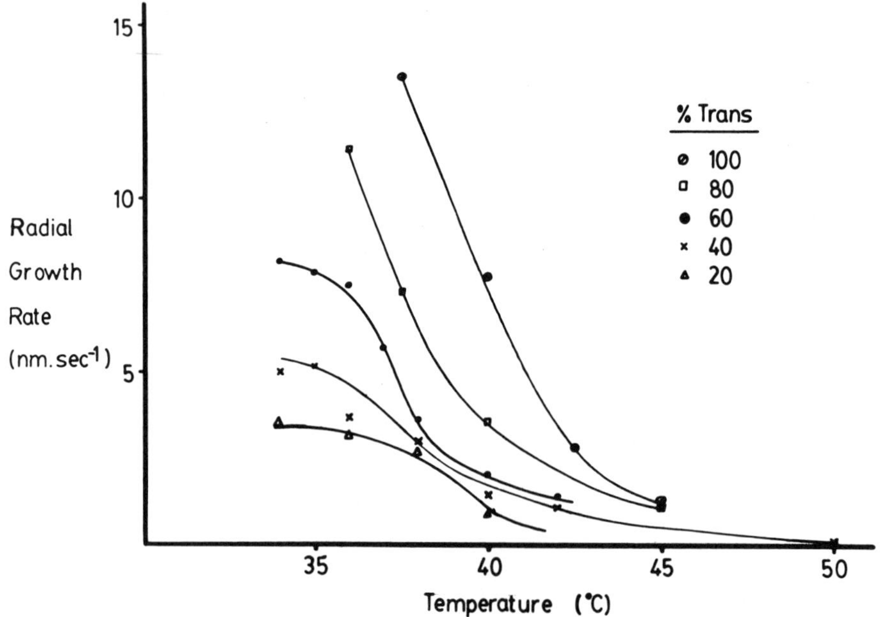

Figure 11. The growth rate versus temperature curves for LMF spherulites.

rate (G) data has been fitted to the Lauritzen and Hoffman[16] equation for homo polymers of the form,

$$G = G_O \exp\left[- U^* /R(T - T_\infty)\right] \exp\left[-Kg/T(\Delta T) \; f\right]$$

where G_O = prexponential factor, containing segmental jump frequencies

U^* = activation energy for transport of segments of molecules to site of crystallization

R = gas constant

T_∞ = temperature where all motion associated with viscous flow creases

K_g = nucleation costant

ΔT = $T_m^{\,\circ} - T_c$ = supercooling

f = $T_c/T_m^{\,\circ}$

$T_m^{\,\circ}$ = equilibrium melting temperature

with the value of the first exponential or mobility term approximated by the WLF relation[17]

$$\frac{U^*}{R(T-T_\infty)} = \frac{C_1}{R(C_2+T-T_g)}$$

where T_g is the glass transition temperature; C_1 and C_2 are constants taken to be 17.24 KJ/mol and 51.6°C respectively. The input data used is that for trans polyisoprene[18]. Both T_∞ and $T_m^{\,\circ}$ are assumed independent of blend content. T_∞ is very similar for cis and trans polyisoprene and will in any case not affect the fit greatly at small supercoolings. $T_m^{\,\circ}$ changes very little with cis content for LMF crystals but the assumption of costancy may be invalid for HMF crystals. The fit is shown graphically in Figure 12 and shows the major change to be a decrease in G_O, the intercept on the abscissa, as the fraction of crystallizable material in the blend decreases. This change can be attributed largely to the time taken to partition the melt as crystals with a similar thickness, melting temperature and melt diffusion coefficient grow at a given crystallization temperature independent of blend content.

Initial modulus

The dynamic, small strain, modulus decreases as the crystalli-

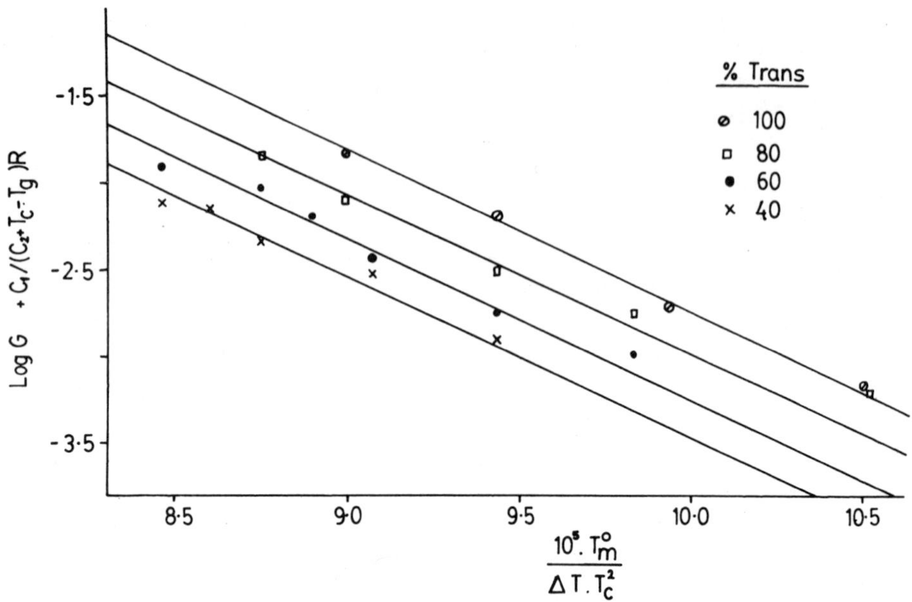

Figure 12. A plot of $\log G + \dfrac{C_1}{R(C_2+T_c-T_g)}$ versus $\dfrac{T_m^{\circ}}{(T_c)^2}$ for LMF spherulites with T_m°=351 and T_g=213 K. The intercept on the abscissa yields $\log G_o$.

nity decreases (Figure 3) by two orders of magnitude. The magnitude of the effect is similar to that observed by Flocke[19]. Attempts
have been made to fit this data to a semi empirical model of fibre
reinforcement due to Halpin and Kardos[20]. The aspect ratio was
taken as the ratio of the lamellar thickness to the spherulite diameter. The crystal chain modulus was approximated to the calculated value for cis-polyisoprene crystals as 7×10^9 Nm^{-2} as reported
by Avitabile et al[21] and the amorphous modulus as that for non-crystalline cis-polyisoprene. This approximation resulted in a prediction of the modulus change of the correct form but yielded moduli
which were too large (Figure 3). While this may be the result of
the very approximate calculation, a much better fit is obtained if
an aspect ratio on order of magnitude lower is used in the calcula-

tions (Figure 3). Some justification for this exists if the non-
branched length of a crystal is used in the determination of the
aspect ratio rather than the spherulite diameter.

SUMMARY

   The blends exhibit a single glass transition temperature and
are compatible at all compositions. Crystallization takes place by
the linear growth of spherulites which grow to impingement. The
final crystallinity of the blends decreases linearly as the proportion of non-crystallizable material increases. LMF crystals of the
same crystal structure with a similar lamellar thickness are formed
at a given crystallization temperature independent of blend composition. The non crystallizable cis-polyisoprene segregates increasingly to interfibrillar regions and the regions increase in size
as the cis content of the blend increases. However, some inter-lamellar segregation occurs resulting in an increase in the measured
long period. The crystals within the fibrils become more heavily
branched and dendritic in character. This combined with the increased crystal spacing, allowing crystals to wander geometrically, probably results in the disordering of crystal stacks at low crystallinities as seen by SAXS. The spherulite growth rate decreases as
the proportion of non-crystallizable material increases. The decrease in growth rate results largely from the time taken for the
segregation process.

   The initial, small strain, dynamic modulus of the blends decreases by two orders of magnitude as the crystallinity decreases.
This can be explained qualitatively, on the basis of simple fibre
reinforcement theory, if the dimensions of the reinforcing element
are taken to be the length and thickness of a non branched crystal.

ACKNOWLEDGEMENTS

   The authors would like to thank the Science Research Council
for financial support for this work. One of the authors A.J. Carter
would like to thank the SRC for the maintenance grant that enabled
this work to be carried out. Thanks are due to the Malaysian Rubber Producer's Research Association for producing the blends. The
authors would like to thank the National Physical Laboratory for

providing the facilities to measure tan δ and the dynamic moduli and particularly Dr. G.D. Dean. Thanks are due to Dr. C.G. Vonk for considerable help with the SAXS studies. Finally the authors would like to thank Departmental colleagues and particularly Prof. E.H. Andrews and Dr. D.A. Tod for helpful discussions.

REFERENCES

1. H. D. Keith and F. J. Padden, J. Appl. Phys., 35, 1270 and 1286 (1964).

2. G. S. Y. Yeh and S. L. Lambert, J. Polym. Sci.A-2, 10, 1183, (1972).

3. L. M. Robeson, J. Appl. Polym. Sci., 17, 3607 (1973).

4. T. T. Wang and T. Nishi, Macromolecules, 10, 421 (1977).

5. W. Wenig, F. E. Karasz and W. J. MacKnight, J. Appl. Phys., 46, 4194 (1975).

6. F. H. Khambatta, F. P. Warner, T. Russel and R. S. Stein, J. Polym. Sci.A-2, 14, 1391 (1976).

7. F. P. Warner, W. MacKnight and R. S. Stein, J. Polym. Sci., Pol. Phys., 15, 2113 (1977).

8. C. K. L. Davies and E. L. Ong, J. Mat. Sci., 12, 2165 (1977).

9. C. K. L. Davies and E. L. Ong, J. Mat. Sci., to be published.

10. B. E. Reed and G. D. Dean, Plastics and Rubber Materials Applications, 1 (1976).

11. C. G. Vonk, J. Appl. Cryst., 8, 340 (1975).

12. B. Crist and N. Morosoff, J. Polym. Sci. Pol. Phys.Ed. 11, 1023 (1973).

13. D. Fisher, Proc. Phys. Soc.,B66, 7 (1953).

14. G. Kortleve and C. G. Vonk, Kolloid Z.Z. Polym., 225 (1968).

15. C. G. Vonk, Private communication.

16. J. I. Lauritzen Jr. and J. D. Hoffman, J. Appl. Phys., 44, 4340 (1973).

17. M. L. Williams, R. F. Landel and J.D. Ferry, J. Am Chem. Soc., 77, 3701 (1955).

18. M. C. M. Cucarella and C. K. L. Davies, to be published in J. Mat. Sci.

19. H. A. Flocke, Kolloid Seit, 180, 118 (1962).

20. J. C. Halpin and J. L. Kardos, J. Appl. Phys., 43, 2235 (1972).

21. G. Avitabile, R. Napolitano and F. Riva, J. Polym. Sci. Phys., 16, 1983 (1978).

CRYSTALLIZATION AND MELTING OF COMPONENTS IN BLENDS
OF POLYETHYLENE AND POLYPROPYLENE

J. Grebowicz and T. Pakula

Centre of Molecular and Macromolecular Studies
Polish Academy of Sciences, 90-362 Łódź, Poland

INTRODUCTION

Blends of polymers are, in most cases, heterogeneous systems in
which components partially preserve their individual properties because of the incompatibility of polymers. In blends of semicrystalline polymers, the properties of components are strongly dependent
on crystallinity and crystalline morphology. If the crystallizing
polymer is dispersed in the blend, its properties are usually different from those in bulk. First of all this can be observed for
polymers which crystallize forming supermolecular structures like
the spherulites in which crystallization kinetics and resultant morphology are related to temperature dependent nucleation and growth
rates of spherulites. In such polymers in the bulk when the nucleation rate is not very high, one nucleus can involve the crystallization in a relatively large volume, i.e. the volume of the final
spherulite. In the blend in which the polymer is dispersed into
small particles, the range of the influence of an individual nucleus
is limited to the size of particles.

In some blends, partial molecular miscibility of components is
possible. In such a case the blend may consist of two phases, each
of them containing two molecularly mixed components in different
volume ratio[1]. Crystallization of components in such systems will
be influenced by the composition of individual phases and will involve demixing in the microscopical scale. Both the dispersion and

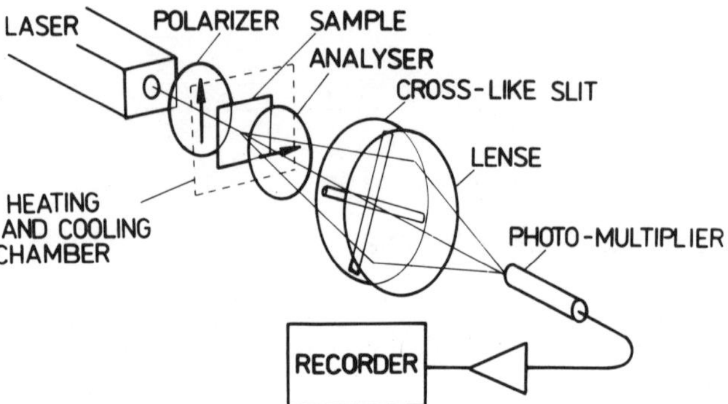

Figure 1.   Scheme of the SALS experiment.

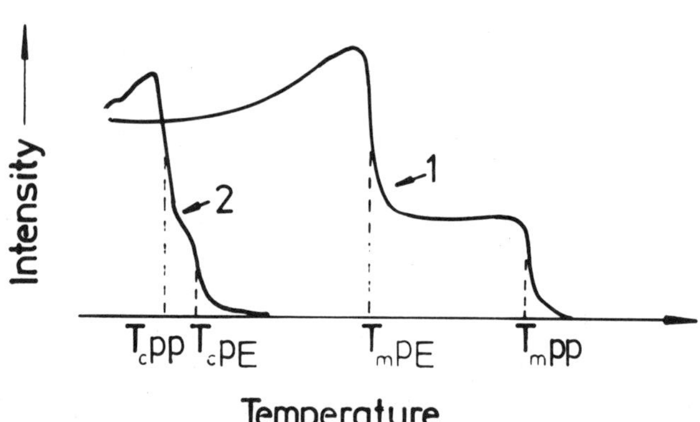

Figure 2.   Example of scattered light intensity changes recorded
during (1) melting and (2) crystallization of a blend.

the partial molecular miscibility of components in the blend can
involve altering of the crystallization kinetics and resultant morphology of polymers constituting the blend.

In these studies we have investigated the crystallization and
melting of polyethylene and polypropylene in blends using small angle
light scattering and comparative differential scanning calorimetry
techniques.

EXPERIMENTAL

Blends of linear polyethylene (PE) and isotactic polypropylene
(PP) were under investigation. The blends of various compositions
were prepared in a cyclic treatment consisting of press moulding of
components at 460°K, fast cooling in ice-water and granulating.
The cycle was repeated ten times for each sample. Finally the films
of 0.5 mm thickness were obtained. As it follows from our earlier
studies[2] such preparation procedures make it possible to obtain
particles of the dispersed component of sizes not exceeding 10 μm.

Melting and crystallization processes of all specimens were
under study. The investigations were conducted using two experimental techniques: small angle light scattering (SALS) and comparative
differential scanning calorimetry (DSC). Details of both techniques
were described elsewhere[3,4].

A version of SALS equipment used in this study to follow the
changes of scattered light intensity during crystallization and
melting of samples, is schematically shown in Figure 1. The sample
was placed between crossed polaroids and a flux of the light scattered at 45° angles with respect to polarization axes was selected
by a cross-like slit and was recorded during cooling or heating of
samples. The typical traces recorded in this way during crystallization and melting of the blends containing 65% PE and 35% PP are
shown in Figure 2. Both dependences show step-wise changes of the
light intensity vs. temperature indicating that crystallizazion and
melting of individual components in the blend can be distinguished.
The maxima of slopes at recorded traces were taken as the transition temperatures as indicated in Figure 2.

The comparative version of the DSC technique was elaborated earlier to investigate the partial melting of polyethylene[4]. When applied here for investigating blends it was based on mutual melting or crystallization of two samples: the first one, the sample of polymer blend and the second one (the reference), the system of two components without any contact between them. If both the sample and the reference have the same masses and contain constituting polymers in the same weight ratio, the recordered DSC trace (called comparative thermogram) contains direct information about differences in melting or crystallization behaviour of a given homopolymer in bulk (the reference) and its equivalent in the blend. Thus one can, on the basis of the comparative thermogram, estimate the effect of dispersion of the component in the blend on their properties in the range of transition. For example the effect of a shift of the melting endotherm along the temperature scale on the comparative thermogram is schematically illustrated in Figure 3. It shows a characteristic shape of the comparative thermogram in the case when the shift takes place without changes in the overall heat of fusion. In such a case the condition $S_1 = S_2$ should be satisfied (where S_1 and S_2 are the areas between the comparative thermogram and the base line as assigned in Figure 3) while the sum $S_1 + S_2$ can be treated as a measure of the shift. As a convention the sign of the sum $S_1 + S_2$ is assumed to be negative, when the transition in the sample appears at a temperature lower than in the reference, and positive in the opposite case. The sign of the sum is indicated on the comparative thermogram by the sequence of the maxima and minima in the transition region. For example, if with increasing temperature the maximum is followed by the minimum the sum $S_1 + S_2$ is negative (the case as in Figure 3).

In both described methods, melting of samples was observed during heating with the rate 10°C/min and crystallization during cooling with the rate 2°C/min. Finally thermal history of samples consisted of four stages: 1) fast cooling of the molten blend during sample preparation (fast cooled samples), 2) first melting with heating rate 10°C/min, 3) crystallization under cooling rate 2°C/min (slowly cooled samples), 4) second cooling with heating rate 10°C/min. The programme of temperature changes is schematically shown in Figure 4.

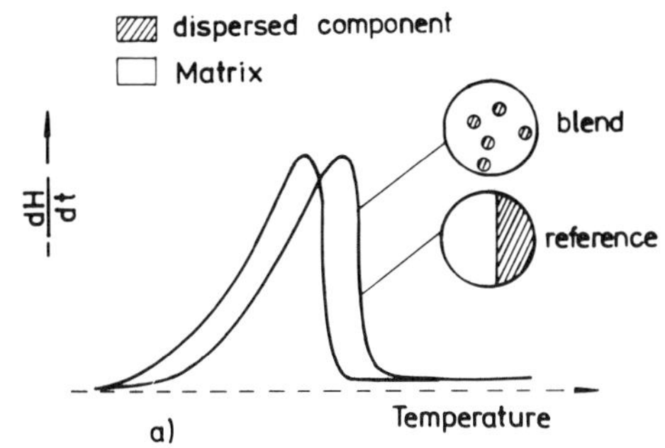

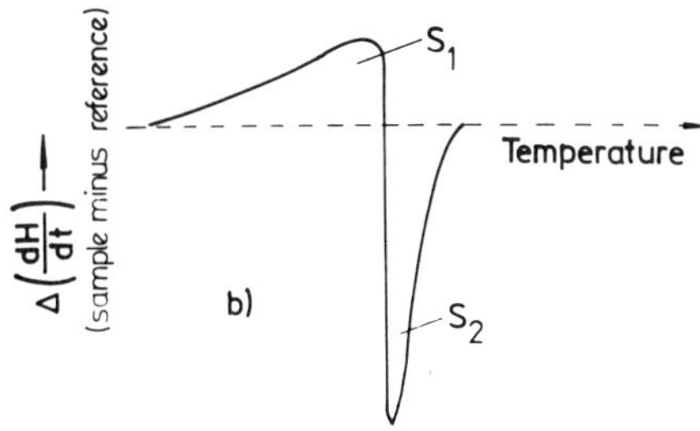

Figure 3. Illustration of the effect of a shift of melting endotherm along the temperature scale on the comparative
thermogram, (a) melting endotherms of the sample and
the reference, (b) the comparative thermogram.

RESULTS AND DISCUSSION

Crystallization

Nonisothermal crystallization of samples was studied during
slow cooling with the rate 2°C/min. In Figure 5 simple DSC thermograms recorded during crystallization of the blend containing 35%
PE and 65% PP is compared with the reference thermogram for unblended components. It follows from Figure 5a that in the blend the
components crystallize almost simultaneously and one mutual not
resolved maximum is observed on the thermogram. Similar observation

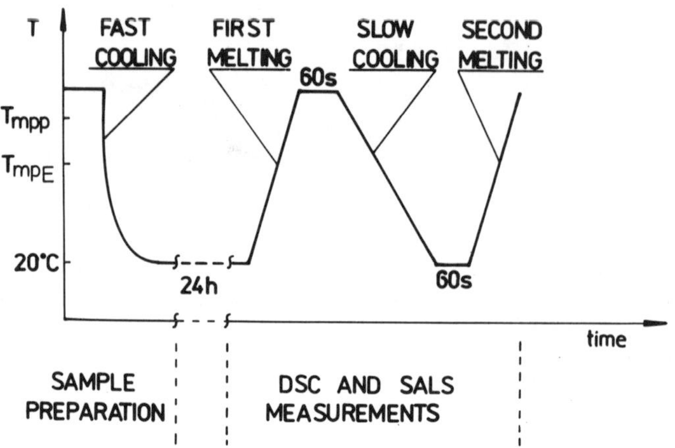

Figure 4.   Programme of the temperature changes in the cycle of
crystallization and melting of samples.

was made by other authors[5]. It is difficult to say, from such a
thermogram, which of the components crystallize earlier. The suggestion comes, however, from the thermogram for unblended components
(the reference sample) shown in Figure 5b. In this case two resol-
vable crystallization maxima are observed. It was established from
this thermogram that at this particular cooling conditions polyethylene crystallizes at higher temperatures than polypropylene.
This was also proved by recording comparative thermograms when the
PE sample was used as a reference for the PP sample during crystallization at various rates of cooling. Results are shown in Figure
6 from which it is clear that the sequence of crystallization of
PP and PE depends on the cooling rate. At a very slow cooling rate
(Figure 6a) PP crystallizes at higher temperatures while at high
cooling rate the opposite sequence is observed. For the cooling
rate 2°C/min (Figure 6b) only a very small shift of the crystallization temperature of PP against PE is observed while for the cooling
rate 50°C/min it becomes considerable (Figure 6c). The latter case
can be treated as close to the situation which takes place at fast
cooling of samples during their preparation.

The comparative thermogram recorded during simultaneous crystallization of the blend and its reference indicate that the compo-

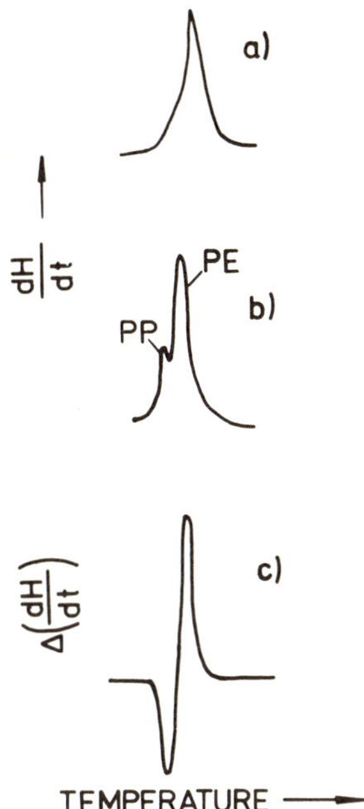

Figure 5. DSC thermograms recorded during crystallization of the
blend (a) and the reference (b), and the comparative
thermogram (c).

ments in the blend crystallize at slight different temperatures than
in the unblended samples. A shift towards higher temperatures is
observed for all compositions as shown in Figure 7 where the values
of the sum S_1+S_2 determined from comparative thermograms are plotted.
However, it is difficult to say, from the comparative thermograms,
what the overall shift means. More detailed information was obtained
from the SALS measurements. Temperatures of maximum rates of
increase of the scattered light intensity during crystallization of
components in the blend were determined during cooling of samples
with the rate 2°C/min. Results are shown in Figure 8. They show
that the shift of crystallization in the blend towards higher temperatures can mainly be attributed to crystallization of polypropylene.
It would happen when particles of PE, crystallized earlier, would
appear heterogeneous nuclei for polypropylene crystallization.
In such a case polypropylene in the blend could crystallize

earlier than polypropylene in bulk as it is observed. Presumably
this effect would be more clear during faster cooling as, for example,
during the initial preparation of samples.

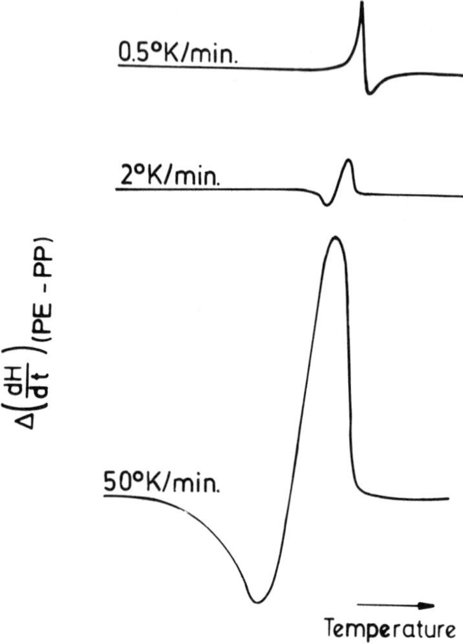

Figure 6.   Comparative thermograms recorded during crystallization
            of PE and PP with various rates of cooling.

Melting

     As it follows from the temperature programme shown in Figure 4
melting of two types of samples were studied: (1) those obtained
at fast cooling during initial preparation and (2) those slowly
cooled with the rate 2°C/min. For example in Figure 9 the simple

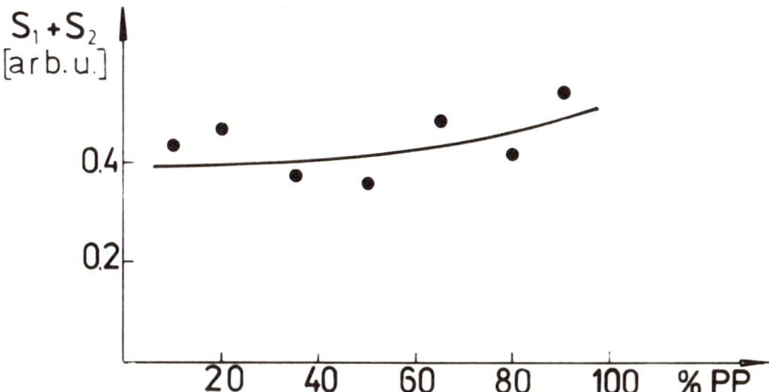

Figure 7. Indication of the shift of crystallization temperature
in blends with various compositions.

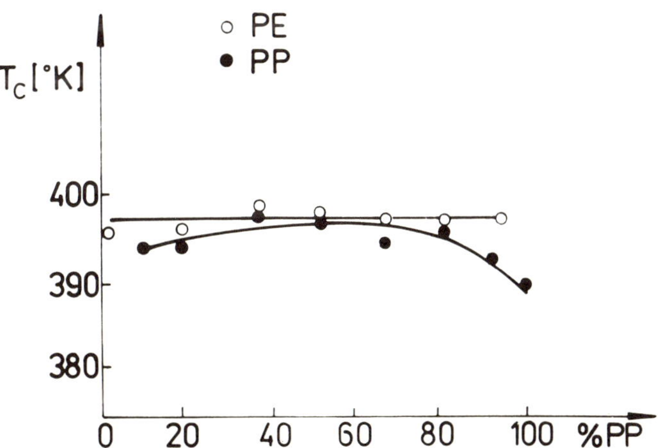

Figure 8. Crystallization temperatures of PE and PP in blends,
determined from SALS.

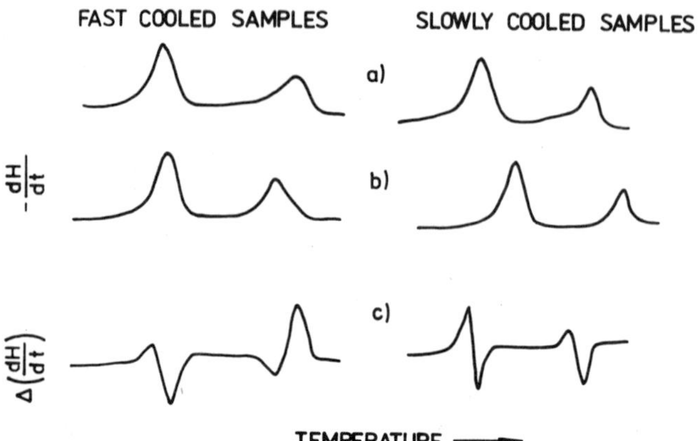

Figure 9.   Examples of melting endotherms of the blend (a) and
the reference (b), and the comparative thermograms
(c) for fast and slowly cooled samples.

melting endotherms of blended samples and the unblended references
with appropriate compositions are shown together with the comparative thermograms. The analysis of comparative thermograms for all
investigated blends showed that in fast cooled samples polyethylene
as a component of the blend melts at lower temperatures than in
bulk while polypropylene melts earlier in the reference sample than
in the blend. On the other hand, in slowly cooled samples both
components in the blend melt at lower temperatures than in the reference sample. These effects are shown in Figure 10 where the
values of the sum S_1+S_2 normalized by composition of a given polymer
are plotted as a function of blend composition. The negative or
positive values of $(S_1+S_2)/m_i$ indicate a shift of melting peak of
the component in the blend with respect to the melting of its equivalent in the reference sample to lower or to higher temperatures
respectively. According to this the increase of melting temperature
of the component when dispersed in the blend is observed only for
polypropylene in fast cooled samples.

Similar observations were also obtained from the analysis of
intensity changes of scattered light during melting of blends.

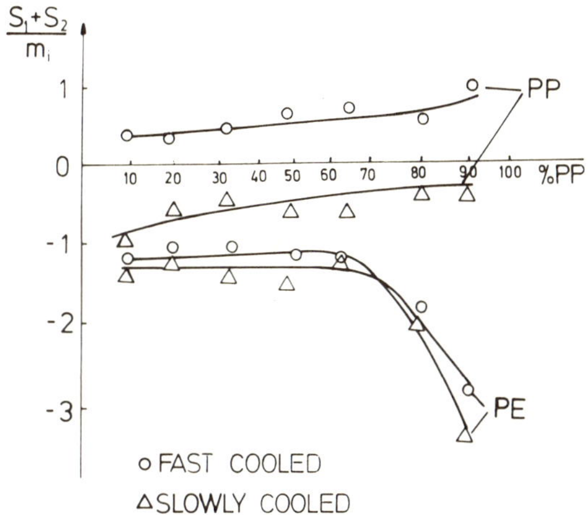

Figure 10.   Indication of the shift of melting temperatures of
components in fast and slowly cooled blends.

Melting temperatures obtained from these measurements are shown as
a function of blend composition in Figure 11. These results indicate also that in the case of fast cooled samples, the melting of
pure polypropylene takes place at lower temperatures than the melting of polypropylene in the blend.

The dependence of melting temperature of polypropylene on composition is not significant. However polyethylene, when dispersed in
polypropylene, melts at lower temperatures than when it constitutes
the matrix of the blend. This was observed for both fast and slowly
cooled samples (Figure 11). Two causes may be responsible for such
effect: (1) partial miscibility of components in the blend and

(2) the dispersion of polyethylene phase into very small particles.
Probably both of them contribute to melting behaviour of PE. Partial
molecular miscibility of components can involve lowering of melting
temperatures of PE in the whole range of composition because of
smaller perfection of PE lamellae which crystallize from the polyethylene melt in which certain amounts of polypropylene chains are
dissolved[6]. A smaller concentrations of PE in the blend the dispersion of the polymer into small particles can influence the crystallization kinetics and in this way lower also the melting temperature.

As suggested before, crystallization of polyethylene probably
causes the heterogeneous nucleation of polypropylene crystals which
grow in such a case at higher temperatures. This can result in
higher perfection and bigger sizes of PP lamellae and consequently
it can increase the melting temperature of PP as observed for fast
cooled blends.

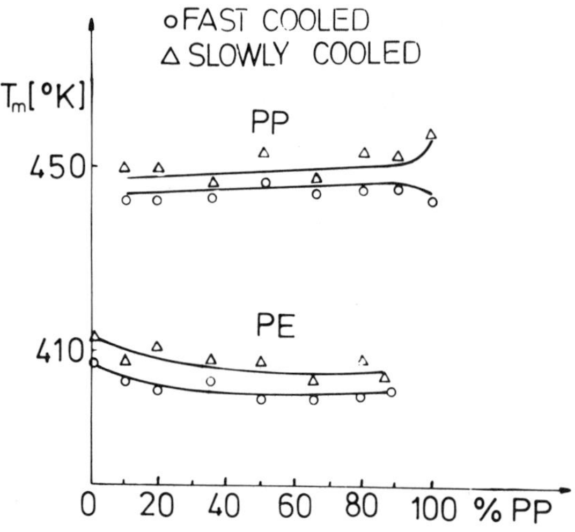

Figure 11. Melting temperatures of components in blends, determined
by SALS.

The results obtained from SALS measurements and those obtained
from DSC are only qualitatively similar but one must take into consideration that the changes of scattered light intensity are involved by changes in supermolecular structure of components while the
changes of heat supplied to samples in calorimeter are related to
changes in overall crystallinity. Considering this and the differences in melting behaviour of blends observed by these two different
methods one can suppose that the morphology of samples is not uniform. Crystals which are built up into spherulites are probably of
better perfection than the others which do not contribute to the
scattering of light because they do not have orientational or posi-
tional correlations. The latter can be formed for example at the
interfaces where, during crystallization separation of phases of
partially mixed components must take place to allow crystallization.

CONCLUSION

The results described in this paper show that in the investigated blends of crystalline polymers, the crystallization of components is not completely independent. The influence of one component
on the crystallization of the other one can both stimulate the nucleation of crystal growth and damp the nucleation and growth of
crystalline phase. This influence has been demonstrated in this
paper qualitatively. Studies of the kinetics of crystallization
and the resultant morphology should clear up these effects.

REFERENCES

1. M. Kryszewski, A. Galeski, T. Pakula and J. Grebowicz,
    J. Colloid Interface Sci., 44, 85 (1973).
2. T. Pakula, M. Kryszewski, J. Grebowicz and a Galeski,
    Polymer J., 6, 94 (1974).
3. T. Pakula and Z. Soukup, J. Polym. Sci. Phys. Ed., 12, 2437
    (1974).
4. T. Pakula and E. W. Fisher, in preparation.
5. M. Inoue, J. Polym. Sci., Part A, 1, 3427 (1963).
6. R. S. Stein, F. B. Khambatta, F. P. Warner, T. Russell,
    A. Escala and E. Balizer, J. Polym. Sci. Polym. Symp., 63,
    313 (1978).

7. M. Avella, R. Greco, E. Martuscelli, G. Mucciariello, M. Pracella
    and G. Ragosta, "MAKROMAINZ" Preprints of Short Comunications,
    vol. II, 958, September 1979.

MORPHOLOGY, CRYSTALLIZATION AND MELTING BEHAVIOUR OF POLY(ETHYLENE OXIDE)/POLY(METHYL METHACRYLATE) BLENDS

E. Martuscelli and G. B. Demma

Istituto di Ricerche su Tecnologia dei Polimeri e
Reologia, Arco Felice (Napoli) Italy

INTRODUCTION

The influence of uncrystallizable components on the kinetics
of crystallization and on the overall morphology of binary blends
with one crystallizable component has been studied in detail, only
in few cases.

The crystallization of isotactic polystyrene and isotactic
polypropylene blended with their atactic stereoisomers, has been
investigated by Keith and Padden[1], by Yeh and Lambert[2] and by
Martuscelli et al.[3]. It was reported that the concentration and
the molecular weight of the atactic component, strongly influence
the morphology and the growth rate of the spherulites.

More recently the spherulitic crystallization of poly(vinylidene fluoride) and poly(ε-caprolactone) blended with poly(methyl
methacrylate) and poly(vinyl chloride) respectively, has been studied
by Wang and Nishi[4] and by Ong and Price[5]. In both cases it was found
a reduction in growth rates for the addition of uncrystallizable
component to the crystalline polymers. Moreover the components were
found to be compatible in the molten and amorphous state. In fact,
the blends show a single glass transition temperature intermediate
between those of the respective components.

This work reports on the crystallization and thermal behaviour
of thin films of poly(ethylene oxide)/poly(methyl methacrylate)

blends (PEO/PMMA) obtained by solution casting from chloroform. The goal of this investigation is to understand the influence of a noncrystallizable component (PMMA) on the crystallization kinetics, on the spherulitic growth rates, as well as the morphology (shape and dimensions of spherulites) and melting temperature of PEO. The work is part of a more general research project for a better knowledge of the properties of polymeric binary blends with one crystallizable component.

EXPERIMENTAL

The samples of poly(ethylene oxide) PEO and poly(methyl methacrylate) PMMA used in this study were purchased from Fluka AG and BDH respectively. The characteristics of the two polymers are reported in Table 1.

TABLE   1

Characteristics of PEO and PMMA

	PEO	PMMA
Molecular weight	20,000	116,000 [a]
Melting temperature	65°C	
Glass transition temperature	-60°C	100-110°C
Melt flow index (10 Kg)		1.0
Melt viscosity (Shear rate 1120 sec^{-1}, 240°C)		3.25 K poise

[a] viscosity average (in chloroform) at 25°C .

The transparent thin films, approximately 10µ thick, were prepared by solution casting from chloroform on the glass plates placed on a hot plate set at 50°C then they were dried under vacuum at 70°C for 24 hrs. to ensure removal of the residual solvent.

The radial increase in size of the growing spherulites was

measured through a polarizing optical microscope equipped with a
proportional temperature hot stage. The overall temperature control
of the hot stage was around 0.2°C. Measurements of the spherulite
radial growth rates G=dR/dt (R=Radius of spherulite and t=time)
first involved melting of the films of PEO/PMMA blends at 80°C (about
10°C above the melting temperature of pure PEO) and then isothermally
crystallization of the films at the various desirable temperatures.
G was calculated by measuring the diameter of spherulites as funtion
of time at a given Tc. The measurements were carried out only on
blend films with PEO content above 65% and at Tc between 47 and 55°C.
The crystallization in fact is extremely low for blends with a PMMA
content higher than 35%. The melting temperature of the blends was
taken as the temperature at which the birefringence of spherulites,
observed under crossed polars, disappears totally.

The crystallization and thermal behaviour of PEO/PMMA blends
was also studied by DSC technique using a Perkin Elmer DSC-2 apparatus (scan rate of 20°C/min was used throughout). The melting
temperature was assumed as the maximum of the DSC endotherm of
fusion. The glass transition temperature in PMMA rich blends was
determined from DSC thermograms.

RESULTS AND DISCUSSION

a) <u>Morphology and growth rate of spherulites</u>

Thin films of PEO and PEO/PMMA blends crystallize according
to a spherulitic morphology. In fact when these films are observed
under the light microscope, with crossed polaroids, circular birefringent regions, truncated by impingement, are observed. The birefringent patterns displayed a Maltese cross whose arms are parallel
to the directions of the polarizer and analyzer. As shown in Figure
1 the radius of PEO spherulites, crystallized from PEO/PMMA blends,
increases linearly with time; no decrease of G with time is observed
over a long time. This observation indicates that during the growth,
the concentration of non crystallizing material at the tips of
radial lamellae is constant.

The examination of the morphological features of thin films
of PEO/PMMA blends shows that down to 70% PEO the sample is completely filled with spherulites (see Figure 2). No dark regions,

denoting higher concentration of non crystallizing material, are
observed.

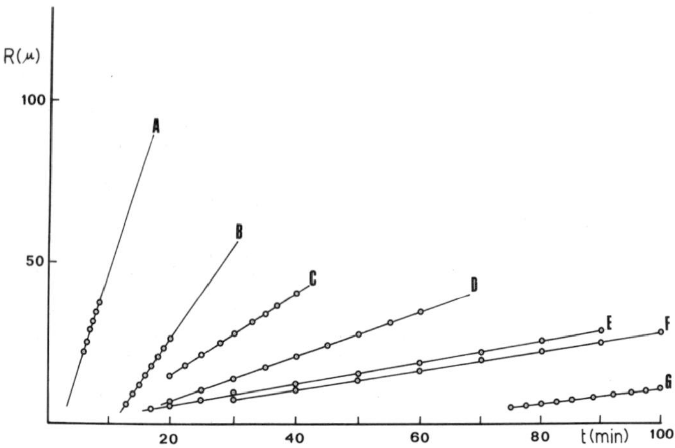

Figure 1.   Graph showing radius as a function of time for spherulites
            in PEO and its blends with PMMA at various temperatures.
            (A) PEO, T_c=50°C;   (B) PEO/PMMA (90/10), T_c=50°C;   (C)
            PEO/PMMA (80/20), T_c=50°C;   (D) PEO/PMMA (70/30), T_c=50°C;
            (E) PEO, T_c=55°C;   (F) PEO/PMMA (70/30), T_c=53°C;   (G)
            PEO/PMMA (90/10), T_c=55°C.

On the contrary a phase segregation fenomenon of the amorphous
components is clearly visible in the optical micrographs of thin
films of PEO blended with some others non crystallizable components
such as atactic polystyrene (aPS), atactic polypropylene (aPP) and
ethylene-vinyl acetate copolymers (EVA), shown in Figure 3.

These observations together with the costancy of G with time
suggest that PMMA molecules are probably incorporated, during crystallization, in interlamellar regions as the atactic polypropylene
and poly(vinyl chloride) in blends with isotactic polypropylene
and poly(ε-caprolactone) respectively[1-7].

The dilution of PEO with PMMA causes a depression of the spherulite growth rate. This depression is greater the larger the concentration of non crystallizing component and the lower the crystal-

(a)

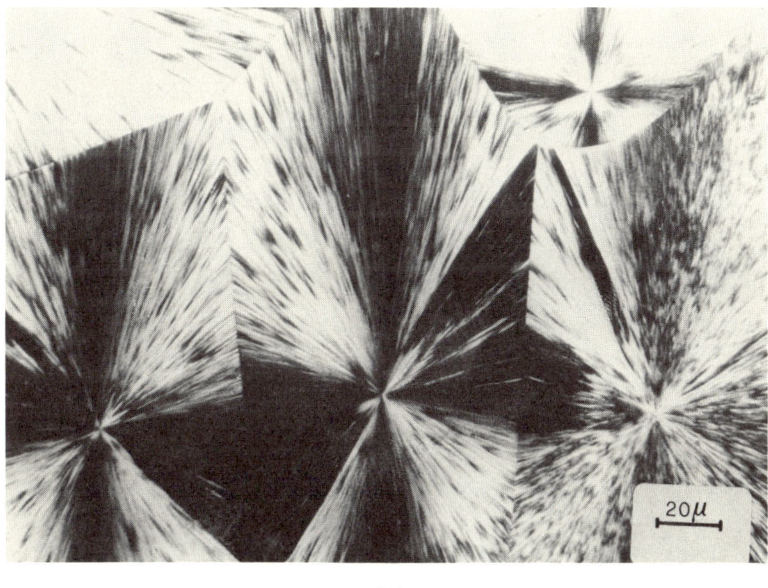

(b)

Figure 2. Optical micrographs (crossed polars) of thin films of:
           a) pure PEO; T_c=50°C.
           b) PEO/PMMA (70/30) blend; T_c=50°C.

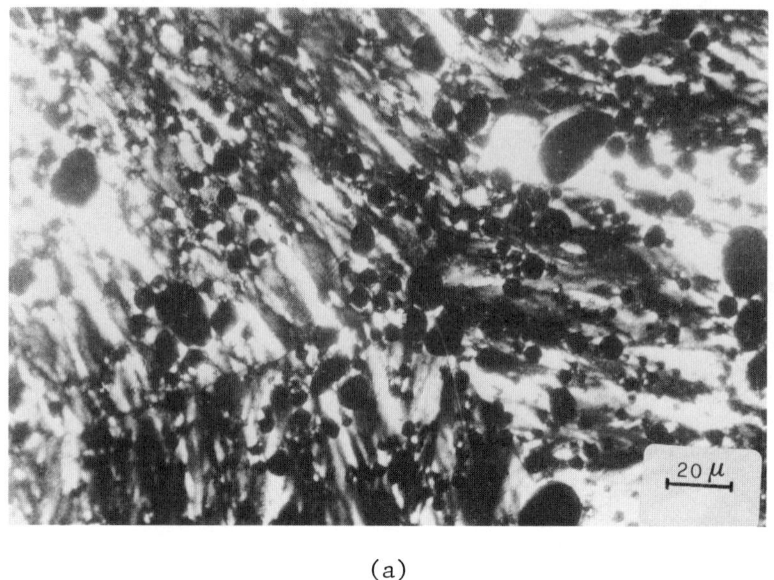

(a)

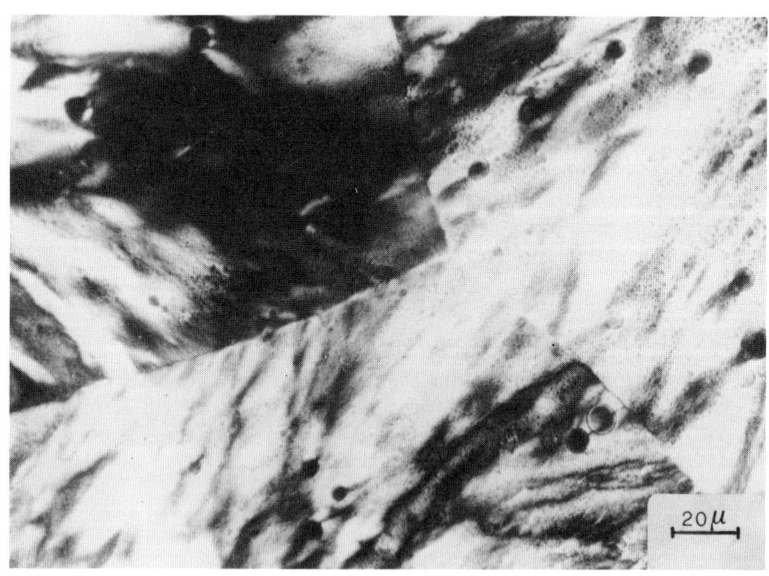

(b)

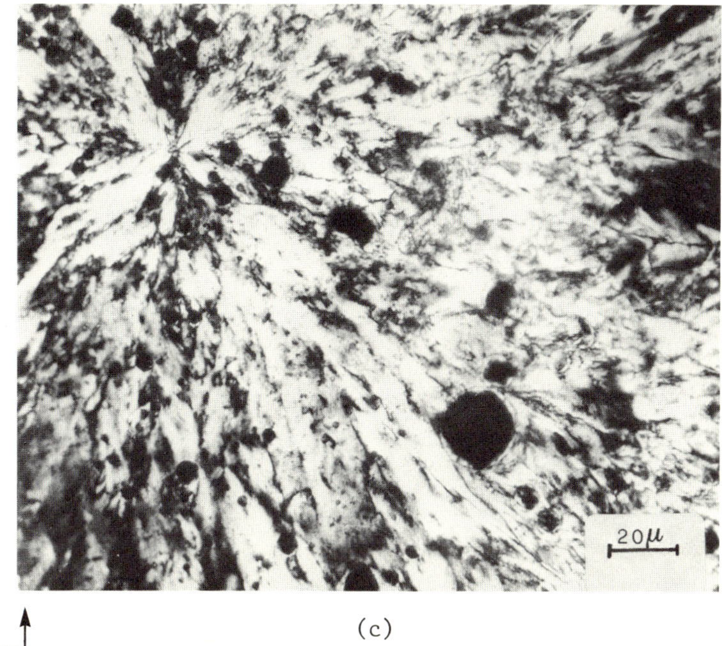

(c)

Figure 3. Optical micrographs (crossed polars) of thin film of:
            a) PEO/aPS (70/30) blend
            b) PEO/aPP (70/30) blend
            c) PEO/EVA (70/30) blend

lization temperature is (see Figure 4). For T_c=50°C values of G of 11.7×10^{-3} and of 0.7×10^{-3} cm/min are observed for PEO and for PEO/PMMA (70/30) blend respectively. On the contrary only a slight influence of aPS, EVA and aPP on the growth rate of PEO spherulites was observed.

b) <u>Melting behaviour</u>

     The melting temperature of pure PEO increases linearly with T_c. The extrapolation of the straight line to $T'_m = T_c$, according to the equation[8]:

$$T'_m = \frac{1}{\gamma} T_c + (1 - \frac{1}{\gamma}) T_m \qquad (1)$$

yields for the equilibrium melting point of PEO a value of 76.1°C.

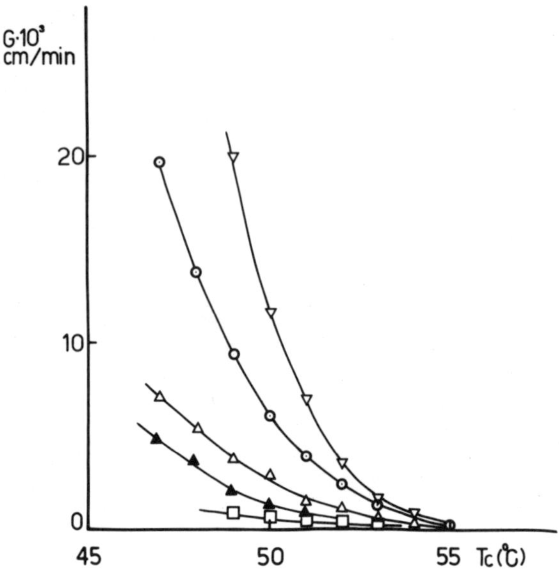

Figure 4.   Radial growth rate G (=dR/dt) of spherulites in PEO and
            its blends with PMMA at various crystallization tempe-
            ratures (∇ PEO; o PEO/PMMA (95/5); $\triangle$ PEO/PMMA (90/10);
            $\blacktriangle$ PEO/PMMA (80/20); $\square$ PEO/PMMA (70/30)).

The variation with T_c of the melting temperature T_m' of PEO/PMMA
blends is much more complicated (see Figure 5).   The following considerations may be made when the trend of the curve T_m' - T_c are
examined in detail:

i)   generally at the same T_c the melting temperature of PEO/PMMA
      blends is lower than that of pure PEO. This effect is more
      pronounced at lower T_c, whilst it is negligible at low under-
      cooling.

ii) T_m' increases linearly with T_c for high values of undercooling.
      At a well defined T_c, an abrupt increase in the slope is ob-
      served becoming the trend non linear and the melting point
      depression decreasing rapidly.

The experimental data of the linear regions may be fitted by equation 1. The values of the slopes are independent of blend composition that is $\frac{1}{\gamma}$ is constant ($1/\gamma=0.2$ for all blends and 0.3 for pure PEO). The rectilinear part of the curves $T'_m - T_c$ extrapolate to a different melting point T_m (T_m = 72.0, 71.2, 70.2, and 68.6° for PEO/PMMA (95/5), (90/10), (85/15) and (80/20) respectively).

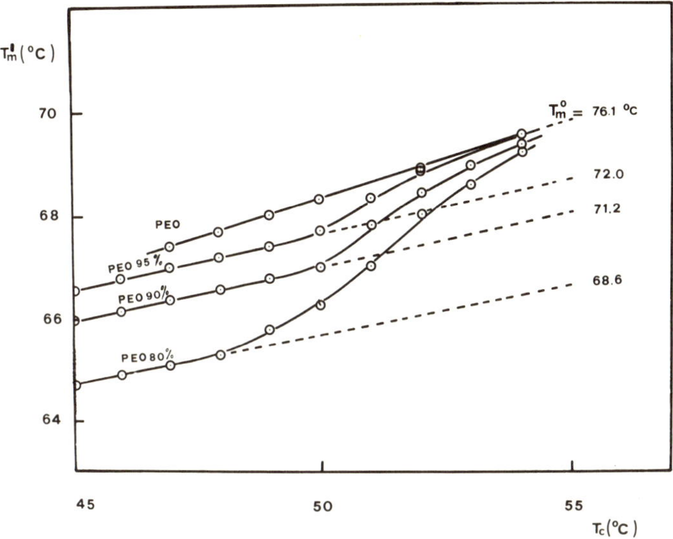

Figure 5. Variation with the crystallization temperature T_c of the observed melting temperature of PEO and its blends with PMMA.

According to Nishi and Wang[9] the finding that the morphological and stability parameter $1/\gamma$ is constant and independent of blend composition and that the lines $T'_m - T_c$ extrapolate to different equilibrium melting points indicates that the melting point depression is primarily ascribed to the diluent effect of the non crystallizable polymer on the chemical potential of the crystallizable component as the two components are compatible in the melt. If the depression is mainly due to morphological effects i.e. reduction in crystal size or lamellar thickness then $1/\gamma$ would not be independent of composition and the plots of T'_m versus T_c would have different slopes extrapolating to a single equilibrium melting temperature T_m.

The melting point depression observed at lower T_c in the case

of PEO/PMMA blends, according to the above considerations, must be
probably ascribed to the diluent effect of PMMA. The two component
are thus at these T_c compatible in the melt. Then the extrapolation
of the linear part of the $T'_m - T_c$ curves should give the equilibrium
melting temperature of PEO crystallized from a one phase melt blends.
The non linear variation of T'_m and the tendency to vanish of the
observed melting point depression at lower values of undercooling
may be exaplained by a decrease in mutual solubility of PEO and PMMA
in the melt as the temperature increases. From this, it emerges
that PEO/PMMA systems must exhibit a phase separation in the melt
at higher temperature with a lower critical solution temperature
(LCST).

The lower solution temperature behaviour has been reported for
many systems, viz. polystyrene/poly(vinyl-methyl ether)[10], poly
caprolactone/poly(styrene-co-acrylonitrile)[11] and poly(methyl methacrylate)/poly(styrene-co-acrylonitrile)[12]. A phase separation at
higher temperature in polymer-polymer blends, which are miscible
at lower temperatures, is also predicted by more recent mixing
theories such as the "equation of state" approach of Flory[13] and
by the "lattice model" of Sanchez[14]. Thus it may be concluded that
LCST behaviour should be rather common for compatible blends.

From the above considerations then, in first approximation,
the temperature corresponding to the deviation from linearity of
the plots $T'_m \rightarrow T_c$ may be assumed as equal to the lower critical
solution temperature of PEO/PMMA blends. These temperatures together
with the corresponding observed melting points are shown in Figure 6
as function of composition. We can conclude that in PEO/PMMA melt
two or only one phase exist when the isothermal crystallization
temperature considered is above or below the LCST line respectively.

The observations that the growth rate depression vanishes at
lower undercooling (see Figure 4) is also explained by the decrease
in mutual solubility of PEO and PMMA in the melt at higher T_c.

It is interesting to point out that a linear trend of T'_m against
T_c is observed in the case of PEO/aPS and PEO/aPP blends for all
the range of crystallization temperature explored (45-55°C). The
extrapolation to the line of equation $T'_m = T_c$ gives for these systems
values of the equilibrium melting temperatures that turn to be
dependent on blends composition and lower than that of pure PEO
(see Table 2).

TABLE   2

Equilibrium melting temperature (°C) and melting point depression ΔT_m for PEO in its blends with PMMA, aPP and aPS as function of composition.

PEO/PMMA	T_m^*	ΔT_m	PEO/aPP	T_m	ΔT_m	PEO/aPS	T_m	ΔT_m
100/0	76.1	/	100/0	76.1	/	100/0	76.1	/
95/5	72.0	4.1	95/5	74.1	2.0	95/5	75.9	0.2
90/10	71.2	4.9	90/10	72.5	3.6	90/10	75.7	0.4
85/15	70.2	5.9	70/30	72.0	4.1	85/15	75.6	0.5
80/20	68.6	7.5	/	/	/	/	/	/

*These values are obtained by extrapolation of the linear part of the curves $T_m' - T_c$ (see text).

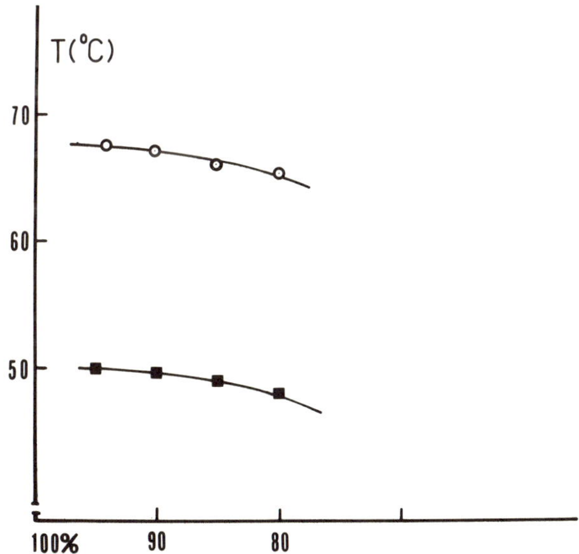

Figure 6.   Plots of the temperature corresponding to the deviation from the linearity of the plots $T_m' \rightarrow T_c$ (■) and the corresponding observed melting temperature (o) versus percentage of PEO in the blends.

A quantitative analysis of the melting point depression in binary polymer blends with components compatible in the melt has been presented by Nishi and Wang[9] and later by Imken et al.[15] based upon earlier treatment offered by Scott[16]. The results of this analysis lead to the conclusion that a plot of the melting point depression ($\Delta T_m = T_m^\circ - T_m$) versus the square of the volume fraction of non crystallizable component should be linear with an intercept at the origin if there are no entropic contribution to ΔT_m. The following equation for ΔT_m was in fact derived:

$$\Delta T_m = - T_m^\circ \left(\frac{V_{2u}}{\Delta H_{2u}}\right) B v_1^2$$

$$\text{with } B = \frac{RTx_{12}}{V_{1u}}$$

(2)

In equation 2 the ratio $\Delta H_{2u}/V_{2u}$ gives the latent heat of fusion of 100 percent crystalline component per unit volume; V_{1u} is the molar volume of non crystallizable component and x_{12} is the Flory-Huggins interaction parameter.

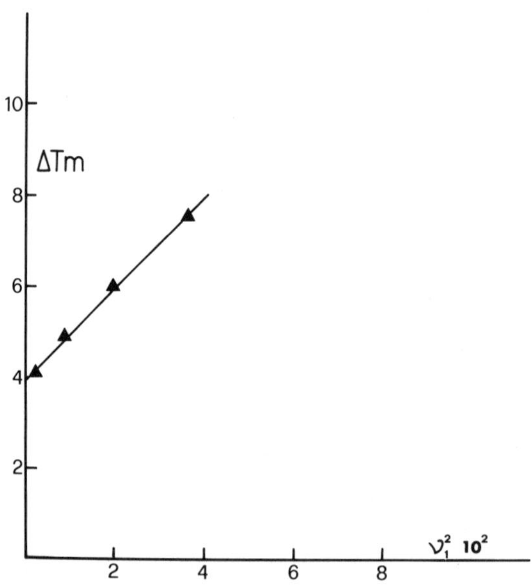

Figure 7.   Composition dependence of the melting point depression of PEO/PMMA blends.

As shown in Figure 7 the melting point depression, observed in PEO/PMMA blends, plotted against the square of the volume fraction of PMMA gives a straight line of equation:

$$\Delta T_m = 99.65 \ v_1^2 + 3.95$$

published density[17] data were used to convert weight fractions into volume fractions at 76°C. By using equation 2, from the slope of the line in Figure 7, we deduce a value of B = -15.6 cal/cm^3 from which a value of -1.93 at 76°C was calculated for x_{12}. In Table 3 these quantities are compared with those obtained in the case of poly(vinylidene fluoride)/poly(methylmethacrylate) (PVF$_2$/PMMA)[9] and poly(vinylidene fluoride)/poly(ethyl methacrylate) (PVF$_2$/PEMA)[15,18] blends, whose components are found to be compatible in the melt.

TABLE 3

Values the interaction energy density B and of the interaction parameter x_{12} for binary blends with components compatible in the melt.

System	B (cal/cm^3)	x_{12}
PVF$_2$/PEMA		
(Imken et al.)	− 3.18	
(Kwei et al.)	− 2.86	− 0.34 (at 160°C)
PVF$_2$/PMMA		
(Nishi et al.	− 2.98	− 0.295 (at 160°C)
PEO/PMMA		
(Present paper)	−15.6	−1.93 (at 76°C)

It must be underlined that for systems such as PVF_2/PMMA the intercept of the straight line $\Delta T_m \rightarrow \nu_1^2$ is very close to the origin ($\simeq 1°C$) and the value of x_{12} obtained from the slope of the linear plot were reasonable, for compatible blends. This finding suggests the adequacy of the thermodynamic treatment used to describe the experimental data.

In the case of PEO/PMMA blends the intercept assumes, on the contrary, a value of 3.97°C and the absolute value of x_{12} seems to be rather large in comparison with literature data on compatible binary blends. These discrepancies may be accounted for by:

i)   non negligible entropic effects that occur during mixing of the two polymers

ii)   non compliance with the assumptions inherent in the extrapo-lation of T'_m by using the Hoffman-weeks plots

iii)   the inadequacies of the Flory-Huggins theory to describe the melting behaviour of all polymer-polymer systems.

c) <u>Glass transition temperature</u>

The T_g of PMMA is between 100 and 110°C and can be quite easily detected from DSC thermograms. We were unable to detect, by DSC, the T_g of PEO, probably because of the very high crystallinity of this material, and the T_g of PEO/PMMA blends with high PEO content. Then the dependence of glass transition temperature on composition was investigated by DSC only in the case of PEO/PMMA blends with high PMMA content (>60%). As shown in Table 4 T_g decreases with increasing PEO content. The experimental values are in agreement with values calculated according to the Fox equation[19]:

$$\frac{1}{T_{g_{BLEND}}} = \frac{W_{PEO}}{T_{g_{PEO}}} + \frac{W_{PMMA}}{T_{g_{PMMA}}} \tag{3}$$

In this equation W_{PEO} and W_{PMMA} are the weight fractions of PEO and PMMA, respectively. The literature value of $-60°C$ was assumed for $T_{g_{PEO}}$[20]. This last result indicates that the two components, at least in the region of the composition explored, are compatible in the amorphous.

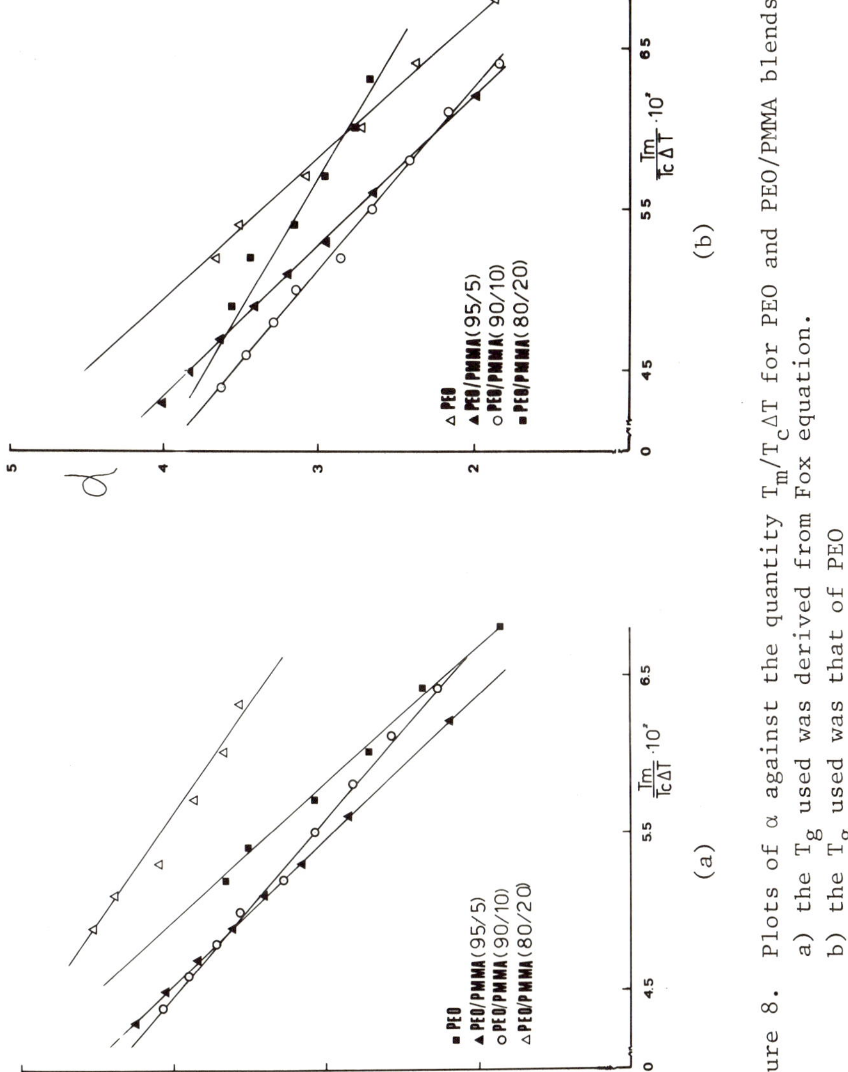

Figure 8. Plots of α against the quantity $\dfrac{T_m}{T_c\Delta T}$ for PEO and PEO/PMMA blends
a) the T_g used was derived from Fox equation.
b) the T_g used was that of PEO

TABLE 4

Values of T_g, detected by DSC, for PEO/PMMA blends and comparison
with values calculated by using the Fox equation.

PEO/PMMA Blend	T_g (Experimental) (°C)	T_g (from Fox Equation) (°C)
0/100	112	/
100/0	-60 *	/
2.5/97.5	100	97
5/95	87	89
10/90	79	77
20/80	54	53
30/70	36	34

*From literature data.

d) <u>Temperature dependence of growth rate of spherulites</u>

The steady state growth rate of spherulites may be described
by the modified Turnbull-Fisher equation:

$$G = G_o \, \text{EXP} \, - \Delta F \, / K T_c \quad \text{EXP} \, - \Delta \Phi \, / K T_c \tag{4}$$

where $\underline{K}$ is the Boltzmann's constant, $\underline{\Delta F}$ is the activation energy
for transport of crystallizing units across the liquid-solid inter-
face, $\underline{\Delta \Phi}$ is the free energy required to form a nucleus of critical
size and G_o is a constant.

According to Boon and Azcue[21] for a polymer-diluent system,
$\underline{\Delta \Phi}$ can be expressed as:

$$\Delta \Phi = \frac{4 b_o \sigma_u \sigma_e T_m}{\Delta H_{2u} (T_m - T_c)} - \frac{2 \sigma_u K T_c T_m (\ln \nu_2)}{b_o (T_m - T_c) \, \Delta H_2 \mu} \tag{5}$$

In equation 5 σ_u and σ_e are the interfacial free energies for unit
area parallel and perpendicular respectively to molecular chain
axis; b_o is the distance between two adjacent fold planes; ΔH_{2u}
is the enthalpy of fusion per unit volume; T_m is the equilibrium
melting point of the crystalline phase in the blend; v_2 is the volume
fraction in the melt of crystallizable component. The term containing $(\ln v_2)$, results from entropic contributions to $\Delta\Phi$ and is
related to the probability of selecting the required number of
crystalline polymer sequences in the blends. If ΔF is described
by the following relation of William Landel and Ferry[22]

$$\Delta F \quad = \quad \frac{C_1 T_c}{C_2 + T_c - T_g} \tag{6}$$

(where the constants C_1 and C_2 assume the values of 4120 cal/mole
and 51.6°K respectively) and σ_u is calculated as $\sigma_u = 0.1\ b_o \Delta H_{2u}$,
then the following expression may be written for $\log G$:

$$\log G - \log v_2 + \frac{4120}{2.303 R (51.6 + T_c - T_g)} - \frac{0.2(\log v_2) T_m}{(T_m - T_c)} = \alpha =$$

$$= \log G_o - \frac{0.4\ b_o^2 \sigma_e T_m}{2.303 K T_c\ (T_m - T_c)} \tag{7}$$

Equation 7 was applied to PEO/PMMA growth data with the following
assumptions:
i)   The T_m of blends are coincident with the values obtained by
     extrapolation of the rectilinear part of $T_m' \rightarrow T_c$ plots (see
     Table 2).
ii) The T_g of mixtures follow the Fox equation also in regions of
     high PEO content.
In order to check the influence of T_g on the applicability of
equation 7 α was also calculated using for all blends the T_g of pure
PEO ($T_g = -60°C$).

Strainght lines, whose equations are reported in Table 5, are
obtained plotting α against $T_m / T_c \Delta T$ as shown in Figure 8. From the
slopes of these lines σ_e can be calculated using the value 2.1×10^9
erg/cm^3 for ΔH_{2u} and 4.65×10^{-8} cm for b_o[23].

As shown in Figure 9 the surface free energy of folding σ_e

TABLE 5

Equation of straight line describing the dependence of the term α
in equation 7 on $T_m/T_c\Delta T$.

PEO	$\alpha = -115.32\ T_m/T_c\Delta T + 9.70$
PEO/PMMA 95/5	$\alpha = -109.11\ T_m/T_c\Delta T + 8.95$ [a] $\alpha = -107.62\ T_m/T_c\Delta T + 8.67$ [b]
PEO/PMMA 90/10	$\alpha = -89.92\ T_m/T_c\Delta T + 8.02$ [a] $\alpha = -88.21\ T_m/T_c\Delta T + 7.51$ [b]
PEO/PMMA 80/20	$\alpha = -73.49\ T_m/T_c\Delta T + 8.4$ [a] $\alpha = -68.37\ T_m/T_c\Delta T + 6.90$ [b]

[a] T_g (Fox)

[b] T_g (PEO)

decreases linearly, increasing PMMA content in the blends ($\sigma_e=42$
and 27 erg/cm^2 for PEO and PEO/PMMA (80/20) blend respectively).
As can be seen from the data of Figure 8 no appreciable differences
were observed when the T_g of PEO was used in the calculation of α
term of blends.

Considering Figure 8 again it should be noted that extrapolation
of the lines to the intersection with ordinate axis gives log G_o
for PEO and PEO/PMMA blends. As shown in Figure 10 the addition
of PMMA to PEO causes a drastic reduction of G_o ($G_o=50\times10^8$ and
2.5×10^8 (cm/min) for PEO and PEO/PMMA (80/20) blend respectively).

Ong and Price[5] have demonstrated that growth rate data for
poly(ε-caprolactone)/poly(vinyl chloride) (PCL/PVC) crystallizable

blends could be described by equation 7 only if C_2=51.6°K is replaced by C_2=72°K. The values of α calculated for pure PCL and PCL/PVC blends are fitted by the same strainght line. For this system then, quantities such as the surface free energy of folding σ_e and G_o result not to be dependent of composition.

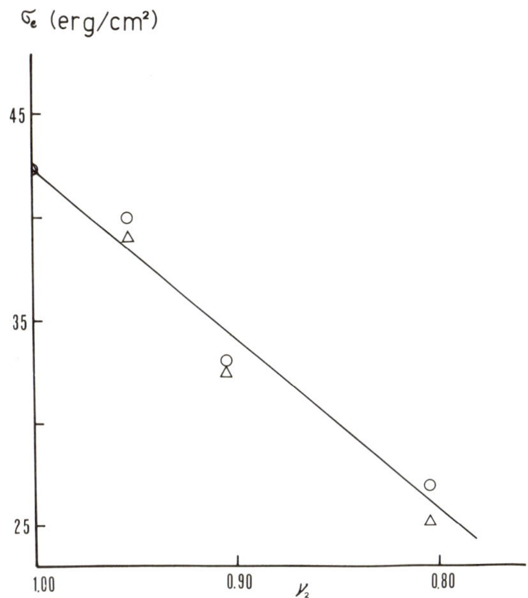

Figure 9. Plot of σ_e against ν_2 (o T_g(Fox); $\triangle$ T_g(PEO)).

Wang and Nishi[4] analyzed growth rate data for PVF_2 blends by incorporating changes in T_g and T_m^o into a rate theory developed by Lauritzen and Hoffman for homopolymers. Changes in viscosity of the melt and in T_g were assumed as responsible for the observed composition dependence. From this analysis it emerges that σ_e is almost independent of composition while the preexponential factor G_o, as in the case of PEO/PMMA system, falls rapidly with the decrease of ν_2.

The dependence of σ_e on composition found for PEO/PMMA blends may be partly accounted for an effect of concentration. In fact, according to Wang and Nishi[4] the concentration dependent part of the surface free energy is only a few ergs per square centimeter. It is likely then some PMMA molecules trapped in interlamellar regions may induce some perturbation on the surface of PEO lamellar crystals.

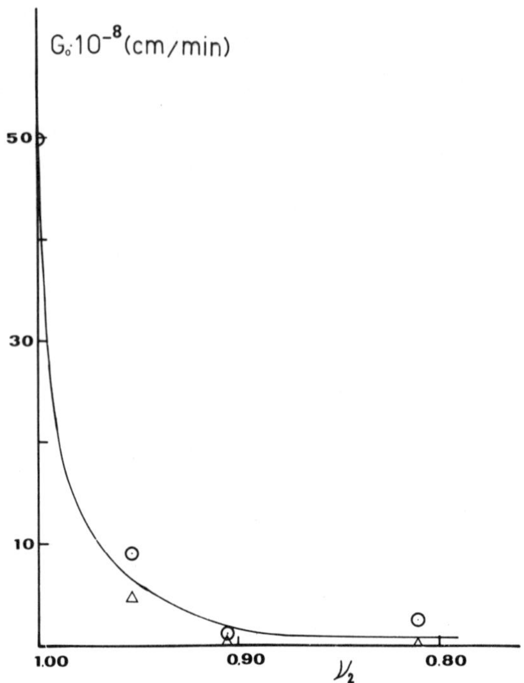

Figure 10. Plot of G_O versus ν_2 $\left[\circ\ T_g(Fox);\ \Delta\ T_g(PEO) \right]$.

On the contrary, the influence of composition on G_O is not easy to explain as this quantity, according to the latest kinetic theory on polymer crystallization, is strongly dependent on a number of factor[24] which are not valuable easily on the basis of the experimental data shown in the present paper.

REFERENCES

1. H. D. Keith and F. J. Padden, J. Appl. Phys., 35, 1286 (1964).
2. C. S. Y. Yeh and S. L. Lambert, J. Polym. Sci. A-2, 10, 1183 (1972).
3. E. Martuscelli, G. Demma, E. Drioli, L. Nicolais, S. Spina, H. B. Hopfenberg and V. T. Stannett, Polymer, 20, 571 (1979).
4. T. T. Wang and T. Nishi, Macromol. 10, 421 (1977).
5. C. J. Ong and F. P. Price, J. Polym. Sci. Polym. Symp., 63, 59 (1978).
6. H. J. Cantow, G. Meyerhoff, Z. Elektrochem., 56, 904 (1952).

7. C. J. Ong and F. P. Price, J. Polym. Sci. Polym. Symp., $\underline{63}$, 45 (1978).

8. J. D. Hoffman, SPE Trans., $\underline{4}$, 315 (1964).

9. T. Nishi and T. T. Wang, Macromolecules, $\underline{8}$, 909 (1975).

10. M. Bank, J. Leffingwell and C. Thies, J. Polym. Sci. A-2, $\underline{10}$, 1097 (1972).

11. L. P. Mc Master, Macromolecules, $\underline{6}$, 760 (1973).

12. L. P. Mc Master "Aspects of liquid-liquid phase transition phenomena in multicomponent polymeric systems" Adv. Chem. Sci. N. 142, Amer. Chem. Soc. Washington D.C., 1975.

13. P. J. Flory, Discuss. Farady Soc., $\underline{49}$, 7 (1970).

14. R. H. Lacombe and I. C. Sanchez, J. Phys. Chem., $\underline{80}$, 2568 (1976).

15. R. L. Imken, D. R. Paul and J. W. Barlow, Polym. Eng. and Sci. $\underline{16}$, 593 (1976).

16. R. L. Scott, J. Chem. Phys. $\underline{17}$, 279 (1949).

17. F. E. Bailey and J. V. Koleske from "Polyethylene oxide", pp.142 Acedemic Press, N.Y., 1976
J. Brandrup and E. H. Immergut "Polymer Handbook", Interscience Publiscience Publication, N. Y., 1975.

18. T. K. Kwei, G. D. Patterson and T. T. Wang, Macromolecules, 9, 780 (1976).

19. T. G. Fox, Bull. Am. Phys. Soc., $\underline{2}$, 123 (1956).

20. P. Tormala, Europ. Polym. J., $\underline{10}$, 519 (1979).

21. J. Boon and J. M. Azcue, J. Polym. Sci.A-2, $\underline{6}$, 885 (1968).

22. M. L. Williams, R. F. Landel and J. D. Ferry, J. Am. Chem. Soc. $\underline{77}$, 3701 (1955).

23. G. Vidotto, D. L. Levy and A. J. Kovacs, Kolloid Z.Z. Polym. $\underline{230}$, 289 (1969).

24. J. I. Lauritzen and J. D. Hoffman, J. Appl. Phys., $\underline{44}$, 4340 (1973).

LE ROLE DES COPOLYMERES SEQUENCES ET GREFFES DANS

LES ALLIAGES DE POLYMERES

G. Riess

Laboratoire de Chimie Macromoléculaire, Ecole Nationale
Supérieure de Chimie de Mulhouse, Centre de Recherches
sur la Physicochimie des Surfaces Solides, Mulhouse
(France)

Some approaches to new polymeric materials are
shown in the case of polymer blends prepared by adequate
combination of 2 or more polymers of different types.

A summary of our research work concerning resin-ela-
stomer blends is given, by considering especially the
blends of HIPS or ABS type, which are of special interest
as impact resistant polymeric materials. In the case of
blends formed by 2 incompatible polymers poly A and
poly B, we have shown that the addition of block or graft
copolymer Cop A-B is necessary in order to regulate the
dispersion degree of one phase in the other, the interfacial adhesion and the morphology.

After having recalled the preparation techniques of
polymer blends and the principles of emulsification by
block and graft copolymers, we have shown the emulsifying
effect of these copolymers for the system:
polystyrene (PS) - polyisoprene (PI) -
poly(styrene-b-isoprene) (Cop SI).
The optimum emulsification conditions have thus been
defined and we shown that the most efficient block
copolymers as emulsifiers are those having a 50/50
composition and a higher molecular weight than the corresponding homopolymers. The different techniques we

developped for the localization of a copolymer in a 2
phase system have also been reviewed. The interfacial
adhesion in such systems has been characterized by the
"occupation density" which corresponds to the number of
anchoring points promoted by the block copolymer per unit
surface of the dispersed phase. Furthermore, we have
defined the conditions leading to optimum mechanical properties in the case of impact resistant polymer blends.

Finally, we approached the problem of morphology for
polymer blends by considering the case of triblock copolymers Cop SIM:
poly(styrene-b-isoprene-b-methylmethacrylate).
The principle of "emulsification in the solid state",
which is the starting point for the study of polymer
blends, has been extended to systems of 2 different block
copolymers and a homopolymer. For these type of blends

$$\text{poly A} \qquad (\text{Cop A-B})_1 \qquad (\text{Cop A-B})_2$$

it has been possible to generalize the emulsifying effect
of block copolymers by showing that a given copolymer
$(\text{Cop A-B})_1$ can be emulsified under proper conditions in
a matrix of poly A by using a second one $(\text{Cop A-B})_2$.
This type of approach leads to an interesting range of
morphologies in the case of polymer blends.

INTRODUCTION

De nouveaux matériaux polymères, intéressants du point de vue
de leurs propriétés, peuvent être obtenus par la réalisation d'alliages, c'est-à-dire par combinaison adéquate de deux ou de plusieurs
polymères de nature différente. En considérant le cas le plus simple
de 2 constituants polymères, on peut avoir suivant les caractéristiques thermodynamiques du système:
- soit un mélange à une seule phase si les polymères sont compatibles
- soit un mélange à deux phases s'ils sont incompatibles.

Dans le premier cas, celui des polymères compatibles, on forme
un système thermodynamiquement stable dont les propriétés sont
généralement intermédiaires entre celles des constituants initiaux.

De tels systèmes, compatibles dans un large domaine de compositions
et de températures, sont cependant peu fréquents, voire des exceptions. Le plus souvent donc, en combinant 2 polymères, on obtient
en raison de leur incompatibilité des systèmes biphasiques, en considérant la nature chimique des phases.

Des performances mécaniques intéressantes peuvent être atteintes
pour de tels alliages en réglant de façon adéquate:
- la morphologie dy système
- le degré de dispersion d'une phase dans l'autre
- l'adhésion interfaciale.

En nous limitant à un système formé par 2 polymères amorphes
incompatibles

poly A   et   poly B

nous nous proposons de résumer les travaux que nous avons effectués
dans le domaine des alliages de polymères et qui ont été publiés par
ailleurs[1-14]. Ces recherches on notamment permis de montrer qu'un
copolymère séquencé ou greffé Cop A-B est un constituant essentiel
des alliages pour régler la morphologie, le degré de dispersion et
l'adhésion interfaciale.

A titre d'exemple, nous considèrerons plus spécialement les
alliages de polymères résistants au choc, c'est-à-dire ceux formés
par une phase continue résine et comportant une dispersion d'élastomère. Il s'agit par conséquent de matériaux du type "polystyrène
choc" ou "résines ABS" qui sont formés:
- d'un polymère A du type résine (polystyrène, copolymère statistique
  styrène-acrylonitrile)
- d'un polymère B du type élastomère, notamment de polybutadiène
  ou de polyisoprène
- d'un copolymère séquencé ou greffé correspondant Cop A-B.

Pour de tels alliages, la phase continue résine peut conférer
au matériau la rigidité et un point de ramollissement élevé, tandis
que la phase élastomère peut augmenter de façon importante la résistance au choc. On peut ainsi passer d'un polymère cassant tel que
le polystyrène à un matériau ductile en respectant les conditions
requises de morphologie, de degré de dispersion et d'adhésion interfaciale qui sont caractéristiques des systèmes biphasiques.

1. MÉTHODES DE PRÉPARATION DES ALLIAGES DE POLYMERES

En considérant le système

$$\text{poly A} \quad \text{Cop A-B} \quad \text{poly B}$$

les alliages peuvent être préparés:
- par dissolution dans un solvant commun suivie soit d'une précipitation, soit de l'évaporation du solvant
- par mélange mécanique à une température supérieure à la température de transition vitreuse (Tg) des différents constituants
- par coprécipitation de mélanges de latex, c'est-à-dire de dispersions de polymères soit en milieu aqueux, soit en milieu organique
- par coprécipitation d'un latex et d'un polymère soluble dans la phase dispersante du latex constituée généralment par un solvant organique[15].

Pour une approche du problème des alliages au stade laboratoire, nous avons essentiellement retenu la première mèthode qui consiste à dissoudre les divers constituants dans un solvant commun, c'est-à-dire un solvant dont le paramètre de solubilité δ_S est compris entre ceux des polymères poly A et poly B, soit

$$\delta_A < \delta_S < \delta_B$$

Par évaporation contrôlée du solvant, on peut notamment obtenir des films d'alliages pour lesquels on peut déterminer par microscopie optique ou par microscopie électronique la morphologie et le degré de dispersion d'une phase dans l'autre.

Un test relativement simple dans ce cas consiste également à évaluer la transparence des films obtenus à partir des mélanges de polymères. C'est ainsi qu'un film constitué d'un système a deux phases, apparaîtra trouble si les dimensions de la phase dispersée sont supérieures à 800-1.000 $\overset{\circ}{A}$ environ, soit approximativement 1/5 de la longueur d'onde de la lumière visible ($\lambda/5$), et si par ailleurs les indices de réfraction des deux polymères son différents. Par contre, si pour le mélange des mêmes polymères les film apparaît transparent, on peut dire que les dimensions de la phase dispersée sont inférieures à 800-1.000 $\overset{\circ}{A}$ et, dans ce cas, la microscopie électronique par exemple permet de distinguer un système à deux phases d'un mélange compatible.

Dans la pratique industrielle, notamment lors de la préparation de "polystyrène choc" et de résines ABS, on polymérise par voie radicalaire, soit du styrène, soit un mélange de styrène et d'acrylonitrile, en présence d'un élastomère, généralement du polybutadiène (PBut), dissous dans ces monomères vinyliques. Au cours de la polymérisation il se forme du polystyrène (PS), respectivement du copolymère statistique poly(styrène-co-acrylonitrile) désigné par SAN. Du fait des réactions de transfert de chaîne sur PBut on assiste par ailleurs à la formation in situ de copolymère greffé.

poly(butadiène-g-styrène)   Cop PBut-PS

poly [butadiène-g (styrène-co-acrylonitrile)]   Cop PBut-SAN

La cinétique de polymérisation et les caractéristiques des différents polymères formés au cours de ces réactions de greffage ont été étudiés en détail dans notre laboratoire au cours de ces dernières années[16-26]. Nous avons entre autres pu montrer que le copolymère greffé ainsi formé in situ est un des constituants essentiels pour conférer des propriétés mécaniques intéressantes aux alliages.

2. EFFET EMULSIFIANT DES COPOLYMERES - INFLUENCE DU COPOLYMERE
   SUR LE DEGRE DE DISPERSION

L'étude détaillée des copolymères séquencés ou greffés Cop A-B en présence de deux liquides non miscibles a et b, sélectifs de chacune des séquences du copolymère, a permis de mettre en évidence leur caractère émulsifiant très général[27-31]. Nous avons notamment pu montrer, que ces copolymères séquencés ou greffés, contrairement aux copolymères statistiques, permettent d'obtenir des émulsions stables eau/huile, huile/eau, huile/huile, ce qui prouve leur analogie avec les tensioactifs.

Un cas particulièrement typique de l'effet émulsifiant des copolymères est donné pour le système de deux polymères incompatibles

poly A   poly B

en solution dans un solvant commun, soit un solvant organique qui peut être un monomère, soit de l'eau.

Le système, qui se sépare en deux phases du fait de l'incompatibilité des polymères en présence, peut être émulsifié à l'aide de copolymères séquencés ou greffés. On obtient ainsi des "émulsions

polymères" stables du type huile/huile ou eau/eau[32]. Dans le cas
d'une "émulsion polymère huile/huile" la phase continue de l'emulsion
est formée de la solution (poly A + solvant), tandis que la phase
dispersée comporte la solution (poly B + solvant).

Comme suite logique de cette étude, nous avons également examiné
le rôle des copolymères séquencés ou greffés Cop A-B en présence
des homopolymères correspondants poly A et poly B lors de la préparation des alliages de polymères, notamment par la technique des
films décrite précédemment. Dans le but de vérifier le caractère
émulsifiant des copolymères, c'est-à-dire leur aptitude à régler le
degré de dispersion d'une phase dans l'autre, nous avons étudié de
façon détaillée les systèmes modèles:

polystyrène (PS) - polyisoprène (PI) - poly(styrène-isoprène)
Cop SI

polystyrène - polyméthacrylate de méthyle (PM) - poly(styrène-mé-
thacrylate de méthyle) Cop SM

Pour cette étude, nous avons mis en oeuvre des homopolymères
et des copolymères préparés par voie anionique, c'est-à-dire des
polymères bien définis du point de vue masse moléculaire, composition
et structure. La Figure 1 donne à titre d'exemple les diagrammes
ternaires correspondants à ces alliages

PS - PI - Cop SI

Dans ces diagrammes nous avons indiqué l'aspect des films en
fonction de la composition des mélanges, ce qui permet de délimiter
des zones où les films sont opaques et d'autres où les films sont
transparents. Il apparaît ainsi que pour des alliages comportant
du copolymère séquencé Cop SI on peut obtenir des films transparents,
montrant que ce type de copolymères joue le rôle d'émulsifiant et
permet de régler le degré de dispersion d'une phase dans l'autre.
Nous avons en effet pu vérifier par microscopie électronique que les
films transparents correspondent encore à des systèmes biphasiques,
pour lesquels les dimensions des particules dispersèes sont de
l'ordre de quelques centaines d'Angstroems.

Au cours d'une étude systématique, nous avons pu préciser le
pouvoir émulsifiant des copolymères séquencés en fonction de leur
composition et de leur masse moléculaire. Ces résultants sont
résumés dans la Figure 1.

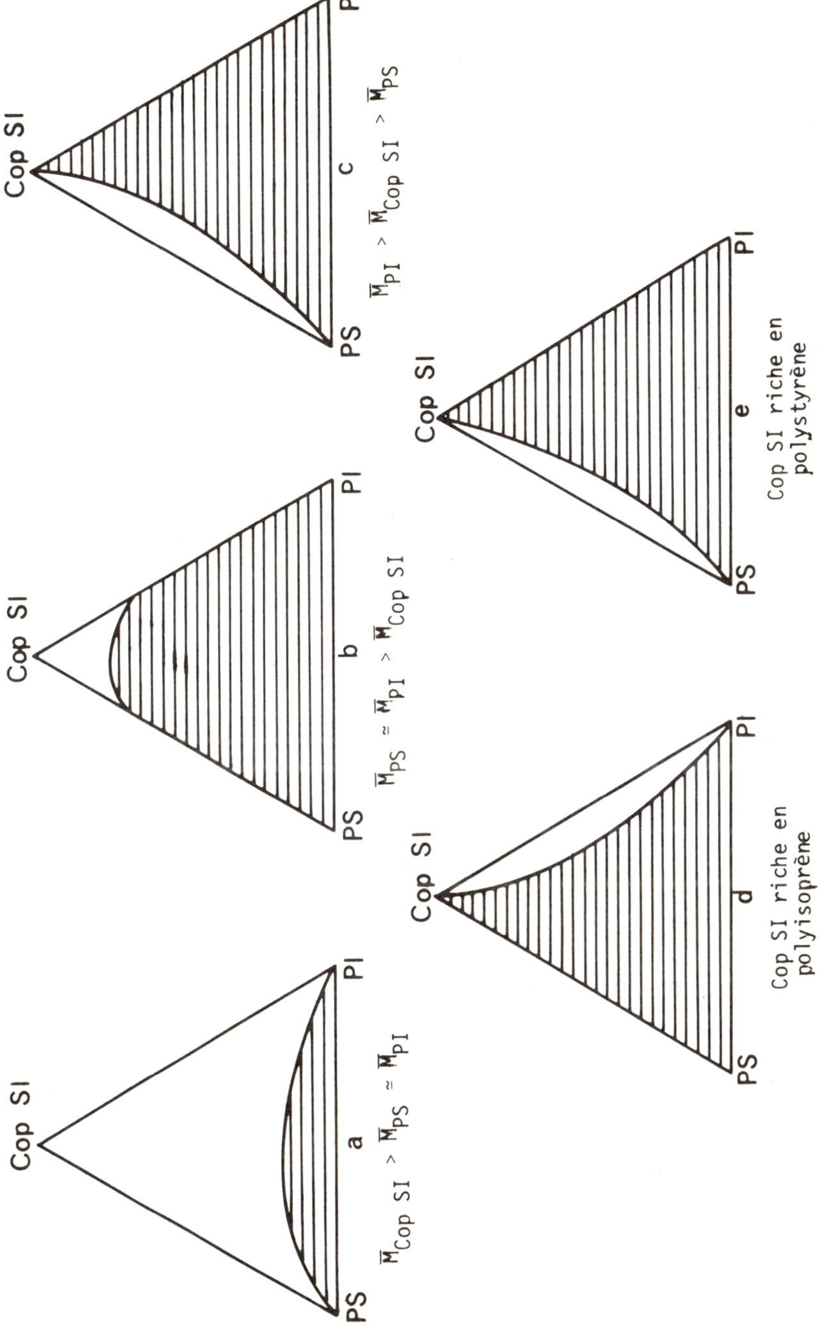

Figure 1. Effet émulsifiant de copolymères sequences. Systèmes ternaires PS-PI-Cop SI. a - b - c: Cop SI composition 50/50.

En considérant tout d'abord les systèmes comportant un copolymère biséquencé Cop SI de composition 50/50, on peut noter, en comparant les Figures 1a et 1b, que l'effet émulsifiant est meilleur lorsque les homopolymères correspondants ont des masses moléculaires à celle du Cop SI, soit

$$\overline{M}_{Cop\ SI} > \overline{M}_{PS} = \overline{M}_{PI}$$

On remarque également que si les masses moléculaires des homopolymères sont similaires, on obtient des diagrams pratiquement symétriques. Dans ces conditions, on peut en déduire que le Cop SI émulsifie aussi bien le PS que le PI. Cette symétrie disparaît par contre pour

$$\overline{M}_{PS} < \overline{M}_{Cop\ SI} < \overline{M}_{PI}$$

comme l'indique la Figure 1c. Dans ce cas, seulement de faibles quantités de PI peuvent être émulsifiées et le degré de dispersion est optimum si la phase continue du système est formée de polystyrène provenant du PS homopolymère et de la séquence correspondante du Cop SI.

Un même type de comportement apparaît pour les Figures 1d et 1e, qui représentent les diagrammes obtenus pour des mélanges comportant du Cop SI dont la composition s'écarte nettement d'une valeur de 50/50. Ainsi, un Cop SI à forte teneur en PI, donnera préférentiellement des émulsions finement divisées si le polyisoprène forme la phase continue du système. Inversement, un Cop SI à forte teneur en PS aura un meilleur pouvoir émulsifiant si le polystyrène constitue la phase continue de l'alliage. Ce comportement est donc à rapprocher de celui trouvé précédemment pour les émulsions huile/huile.

Pour comparer l'efficacité du Cop SI dans les différents systèmes, il faut par conséquent tenir compte:
- de la composition du Cop SI
- des masses moléculaires du Cop SI ainsi que des homopolymères PS et PI.

Il est donc commode de définir par

$$\alpha = \frac{\overline{M}\ du\ PS\ homopolymère}{\overline{M}\ de\ la\ séquence\ PS\ du\ Cop\ SI}$$

$$\beta = \frac{\overline{M} \text{ du PI homopolymère}}{\overline{M} \text{ de la séquence PI du Cop SI}}$$

Dans ces conditions, l'optimum d'efficacité du Cop SI en tant qu'émulsifiant sera donné par

$$\alpha = \beta \leqslant 1$$

ce qui correspond à des Cop SI de composition molaire 50/50 et dont les masses moléculaires des séquences sont supérieures ou égales à celles des homopolymères PS et PI.

Ces résultats, qui sont en accord avec ceux obtenus pour le système

$$\text{PS} - \text{Cop SM} - \text{PM}$$

ont été confirmés par la suite par Skoulios et coll.[33]. Signalons également qu'une explication de ces phénomènes a pu être donnés par Meier[34] en se basant sur les vitesses de précipitation des différents polymères en présence lors de l'évaporation du solvant et notamment pendant les premiers stades de la formation du film. L'"emulsification à l'état solide" et l'organisation des phases ne peut en effet se produire dans les conditions optimales que si le système se trouve à un moment donné dans un état de viscosité relativement faible, comme c'est par exemple le cas des solutions de polymère. Il en est de même lors de la préparation du "polystyrène choc", c'est-à-dire en polymérisant du styrène en présence de polybutadiène. Dans ce cas, le polystyrène et le copolymère greffé sont formés in situ dans un milieu, dont la viscosité, relativement faible au départ, augmente avec le taux de conversion. L'émulsification et l'organisation des phases ont donc essentiellement lieu tant que le taux de conversion en polystyrène, qui conditionne la viscosité du système, reste en-dessous d'un certain seuil. C'est ainsi que la morphologie du produit final est pratiquement celle qu'on observe déja à un taux de conversion de 30-40% c'est-à-dire lorsque le système devient figé du fait de sa viscosité.

Cet ensemble de travaux que nous avons effectués a donc permis de définir les conditions requises pour une bonne émulsification de la phase dispersée élastomère pour les alliages

$$\text{PS} \quad \text{Cop SI} \quad \text{PI}$$

qui conduisent généralement pour les conditions favorables à une
morphologie du type sphérique (type a) Figure 2.

3. LOCALISATION D'UN COPOLYMERE SEQUENCE DANS UN SYSTEME A DEUX PHASES

Ayant ainsi défini les conditions optimales d'émulsification,
il nous a paru intéressant de vérifier que les copolymères séquencés
les plus efficaces en tant qu'émulsifiant, sont effectivement
localisés à l'interface des deux phases. Ce problème est également
d'importance pour la réalisation des alliages de polymères, pour
lesquels un copolymère localisé à l'interface peut assurer un ancrage
entre les deux phases.

La microscopie interférentielle est une première méthode que
nou⌣ avons mise en oeuvre pour localiser un Cop SI dans un système

$$\text{polystyrène (PS) - polyisoprène (PI)}$$

dont les indices de réfraction sont respectivement 1,59 et 1,51.
Par détermination de l'indice de réfraction de chaque phase, il
devient possible d'en déduire leur composition, c'est-à-dire les
teneurs respectives en PS et en PI. Si le Cop SI se solubilise
ainsi dans la phase PS, on observera une diminution de l'indice de
réfraction par rapport à celui de l'homopolymère PS pur. Inversement
si le Cop SI se solubilise dans la phase PI, on notera une augmen-
tation de l'indice de réfraction. Il s'en suit que le copolymère
est à l'interface, lorsque les deux phases du système PS-PI-Cop SI
présentent exactement les indices des homopolymères correspondants.
Signalons que cette méthode est cependant limitée aux systèmes dont
les tailles de particules sont de l'ordre d'une dizaine de microns.

Une technique similaire, mise au point par Y. Jolivet[9], consiste
à réticuler sélectivement la phase PI par irradiation γ et à déterminer le taux de polystyrène et le polyisoprène dans la phase soluble
et dans la phase réticulée. Les résultats les plus intéressants
ont cependant été obtenus en utilisant un copolymère fluorescent de
structure:

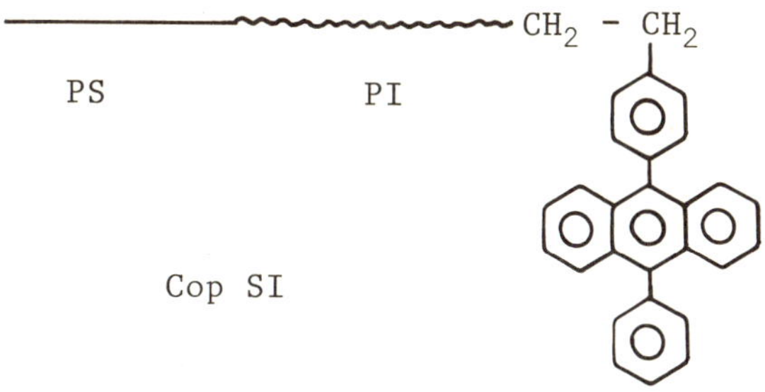

PS   PI $CH_2 - CH_2$

Cop SI

obtenu par voie anionique en faisant réagir un Cop SI "vivant" avec
le 9 phényl-10-anthracène styrène. Un tel copolymère incorporé
dans un mélange peut être localisé par microscopie à fluorescence
UV.

Dans une autre approche de ce problème, nous avons utilisé la
microsonde de Castaing qui permet par example de détecter sur des
coupes de polymères de très faibles teneurs en halogènes. Des Cop SI
ont ainsi pu être localisés dans des alliages après chloration
sélective de la séquence PI des copolymères. Des taux de chloration
de 1 à 2% du PI se sont avérés suffisants pour permettre la détection
du Cop SI. La mise en oeuvre de ces différentes techniques a permis
de confirmer que les Cop SI se trouvent préférentiellement à l'interface des deux phases, si leur composition est de l'ordre de 50/50
et si les masses moléculaires de leurs séquences sont égales ou
supérieures à celles des homopolymères correspondants.

En accord avec les résultats trouvés précédemment, on peut donc
en conclure que les copolymères les plus efficaces en tant qu'émulsifiant sont également ceux qui ne présentent pas de solubilité
préférentielle dans l'une ou l'autre phase. Compte tenu de ces
données, nous avons pu caractériser l'adhesion interfaciale dans
un alliage de polymères à l'aide de la densité d'occupation D qui
est le nombre de points d'ancrage par unité de surface. La densité
d'occupation est ainsi définie par

$$D = \frac{\text{nombre de liaisons A-B du Cop}}{\text{surface des particules dispersées}}$$

en considérant un volume unitaire d'alliage.

Cette grandeur est directement accessible par l'expérience,

étant donné qu'on peut connaître la concentration des Cop A-B dans
l'alliage, ainsi que le nombre et le diamètre des particules de phase
dispersée. Ces dernières caractéristiques sont généralement déter-
minées par microscopie optique ou électronique sur des coupes minces.
Par l'étude détaillée de différents alliages, il a été possible de
montrer qu'il existe une corrélation directe entre les propriétés
mécaniques du matériau et la densité d'occupation D, qui caractérise
l'adhésion interfaciale dans un système biphasique[35,36].

En plus de ce paramètre, il est apparu au cours de nos recherches sur les alliages de polymères résistants au choc que des caractéristiques mécaniques optimales sont généralement atteintes dans
les conditions suivantes:
- la masse moléculaire du polymère formant la phase continue de
  l'alliage doit être la plus élevée possible, de préférence supé-
  rieur à 100.000 pour augmenter la résistance au choc, mais d'un
  autre côté il est nécessaire qu'elle reste inférieure à 500.000
  pour faciliter la mise en oeuvre, notamment par injection.
- la phase dispersée formée par l'élastomère doit être réticulée
  ou avoir une masse moléculaire élevée favorisant l'enchevêtrement
  des chaînes ($\overline{M}_n$ > 100.000).
- la taille des particules doit être de préférence de l'ordre de
  quelques micromètres dans le cas d'une matrice polystyrène.
- la teneur en phase dispersée doit être de l'ordre de 20 à 30% en
  volume.

4. MORPHOLOGIE DES ALLIAGES DE POLYMERES

Craig[37] a proposé 3 structures types en ce qui concerne la
morphologie de systèmes biphasiques résine-élastomère. Ces 3
structures a, b et c sont représentées schématiquement par la Figure
2.

La structure a, correspondant à des particules sphériques
d'élastomère dispersé dans une matrice formée par la résine, est
généralement obtenue pour des polystyrènes choc ou des ABS préparés
par émulsion, c'est-à-dire à partir de latex de polybutadiène.
Cette structure est également celle que nous avons envisagée lors
de l'étude des systèmes modèles décrits dans les paragraphes pré-
cédents.

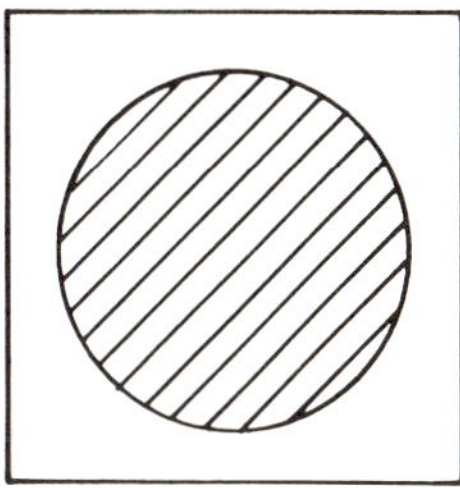

type a
structure sphérique

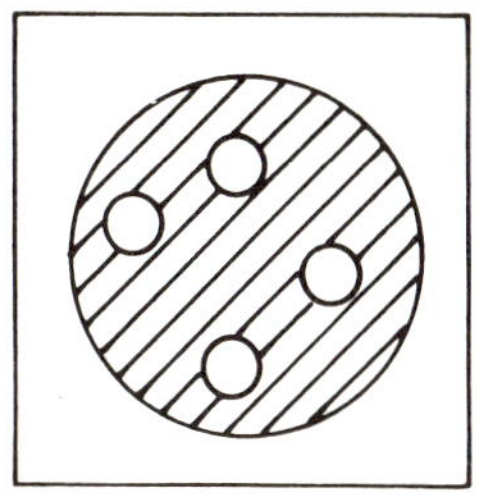

type b
structure d'émulsions
multiples

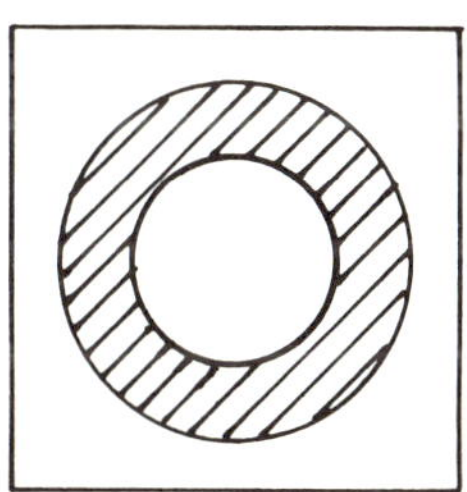

type c
''structure en couronne''

Figure 2. Morphologies d'alliages de polymères.
          Structures schematiques de systèmes a 2 phases.

La structure b, comportant des inclusions de résine dans la
phase élastomère dispersée, est celle qu'on rencontre pour les polystyrènes choc et les ABS préparés par le procédé masse, c'est-à-dire
par polymérisation de monomères vinyliques (styrène, acrylonitrile..)
en présence d'un élastomère tel que le polybutadiène. L'inversion
de phase qui se produit lors de ce processus de polymérisation
conduit à ces émulsion multiples. Cette morphologie présente l'avan-
tage de pouvoir accroître le volume de la phase disperée pour une
teneur relativement réduit en élastomère.

La structure c, que est du type capsule, correspond à une
enveloppe d'élastomère, d'épaisseur définie, entourant une particule
de résine de diamètre donné.

Une telle morphologie a pu être obtenue en mettant en oeuvre
des copolymères triséquencés

poly(styrène-b-isoprène-b-méthacrylate de méthyle)

designés par Cop SIM[12-14].

Suivant les masses moléculaires relatives des 3 séquences, il
a été possible d'obtenir, aussi bien pour le copolymère pur qu'en
présence de proportions définies d'homopolymères PS, PI ou PM, des
structures du type capsule où l'enveloppe est constituée de polyisoprène, la particule dispersée étant soit du polystyrène (PS),
soit du polyméthacrylate de méthyle (PM).

Une approche particulièrement intéressante à de nouveaux types
de morphologies a été montrée récemment au cours de la thèse de
S. Marti[15,38]. Par extension des systemès

poly A   Cop A-B   poly B

précédemment étudés, nous avons ainsi considéré les alliages
comportant 2 copolymères séquencés différents en présence d'un
homopolymère, qui sont par conséquent du type:

poly A $(Cop\ A-B)_1$ $(Cop\ A-B)_2$

Une étude systématique a de cette façon été entreprise pour
les alliages

PS $(Cop\ SI)_1$ $(Cop\ SI)_2$

à base de polystyrène (PS) et de deux copolymères séquencés

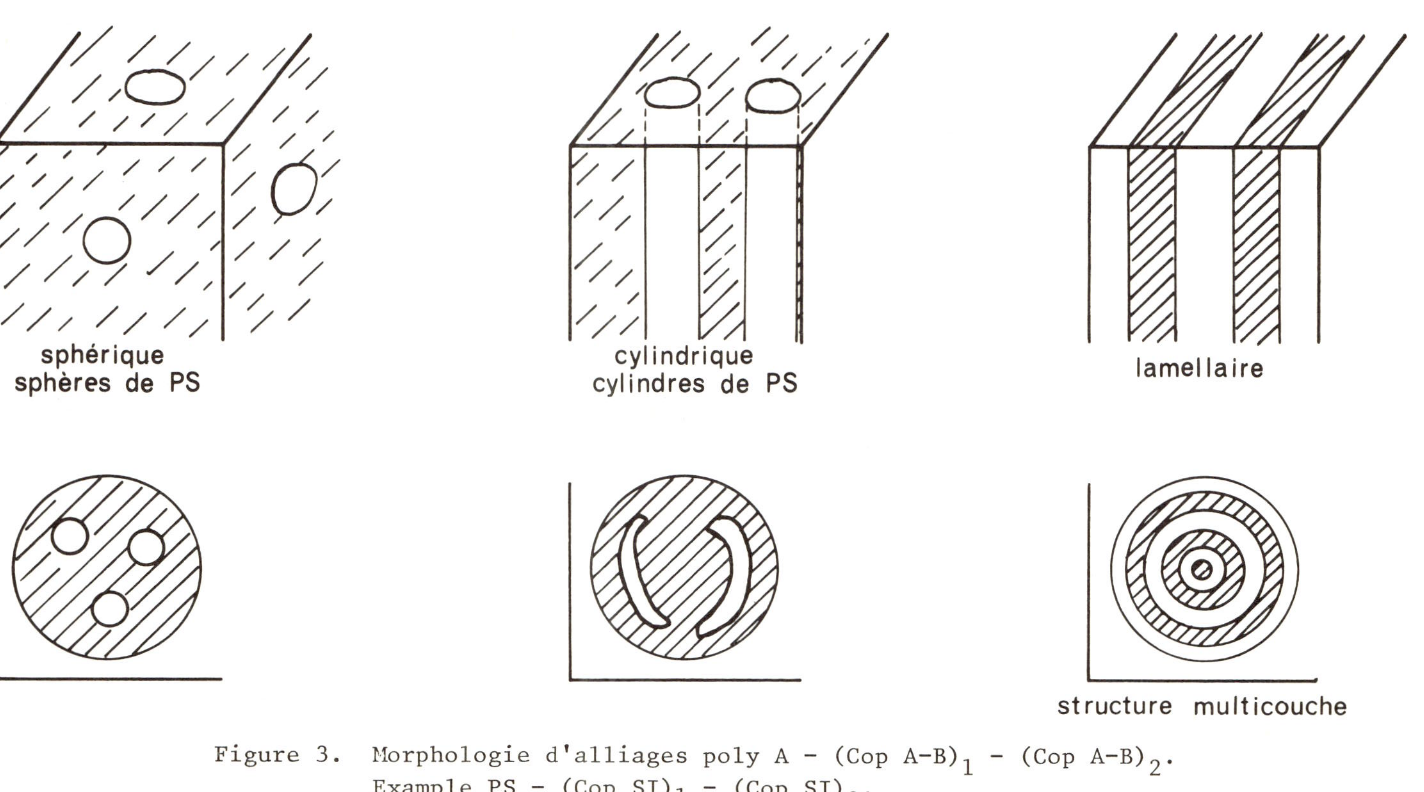

Figure 3. Morphologie d'alliages poly A − (Cop A−B)$_1$ − (Cop A−B)$_2$. Example PS − (Cop SI)$_1$ − (Cop SI)$_2$.

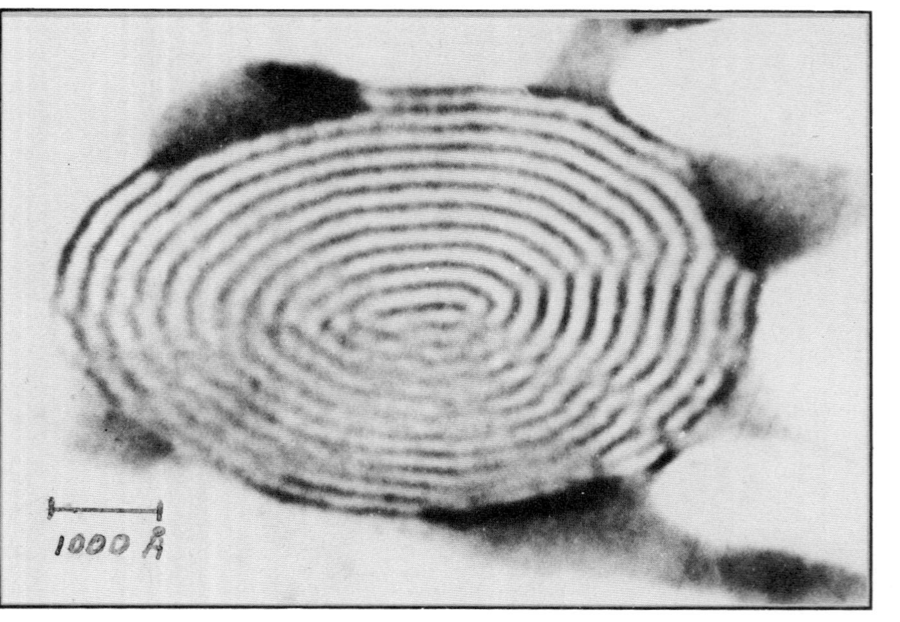

Figure 4. Alliages de polymères – Structure multicouche PS – $(\text{Cop SI})_1$ – $(\text{Cop SI})_2$.

poly(styrène-b-isoprène) Cop SI. Au cours de ces recherches, nous
avons notamment pu généraliser la notion d'effet émulsifiant des
copolymères en montrant qu'un copolymère séquencé donné peut être
émulsifié par un autre dans des conditions bien définies. En considérant plus spécialment les alliages à phase continue PS, il est
apparu que la phase dispersée d'un tel alliage peut être formée par
un des copolymères séquencés, soit le $(Cop\ SI)_1$, émulsifié sous
forme de fines particules dans la matrice PS à l'aide du $(Cop\ SI)_2$.
Etant donné que le $(Cop\ SI)_1$, suivant sa composition, peut donner
lieu aux structures mésomorphes du type sphérique, cylindrique ou
lamellaire, la phase disperée peut présenter une telle morphologie
caractéristique.

Dans la Figure 3 nous avons représenté de façon schématique
la morphologie des alliages à phase continue PS et comportant une
dispersion sous forme de particules sphériques de $(Cop\ SI)_1$ de
différentes structures mésomorphes.

Ainsi si le $(Cop\ SI)_1$ présente à l'état pur une structure
mésomorphe du type lamellaire, il peut être dispersé dans la matrice
PS à l'aide du $(Cop\ SI)_2$, pour conduire à des particules généralement
sphériques ayant une morphologie caractéristique du type multicouche
(Figure 4c).

Les structures les plus régulières sont obtenues dans ces conditions si
- la masse moléculaire du PS est supérieure à celles des deux copolymères mis en oeuvre
- la masse moléculaire du $(Cop\ SI)_2$, jouant le rôle d'émulsifiant,
  est supérieure à celle du $(Cop\ SI)_1$ formant la phase dispersée
- les compositions des $(Cop\ SI)_1$ et $(Cop\ SI)_2$ sont similaires.

En respectant ces critères, c'est-à-dire en absence de "solu-
bilisation" de PS dans le Cop $(SI)_1$, on peut vérifier que l'épaisseur des différentes couches polystyrène et polyisoprène dans la
particule dispersée correspond aux dimensions des lamelles telles
qu'elles apparaissent pour la structure mésomorphe du $(Cop\ SI)_1$
pur. La Figure 4 illustre cette morphologie du type multicouche
"structure oignon" pour desparticules dispersées dans une matrice
PS. Des travaux que nous avons entrepris récemment[39], en parallèle
à ceux de Echte[40], ont par ailleurs permis de montrer que ces nou-
velles morphologies pour le polystyrène choc peuvent également

être obtenues en polymérisant, dans des conditions bien définies, du styrène en présence de copolymères séquencés. Ces approches ouvrent par conséquent des perspectives intéressantes pour l'obtention d'une nouvelle gamme de morphologies dans le cas des alliages de polymères.

BIBLIOGRAPHIE

1. G. Riess, J. Hohler, C. Tournut et A. Banderet, Die Makromol. Chem., 101, 58 (1967).
2. J. Kohler, G. Riess et A. Banderet, European Polym. J., 4, 173 (1968).
3. J. Kohler, G. Riess et A. Banderet, European Polym. J., 4, 187 (1968).
4. G. Riess, J. Kohler, C. Tornut et A. Banderet, Rubber Chemistry and Technology, 42 (2), 447 (1969).
5. A. Banderet, J. Periard et G. Riess, Int. Symp. Macromol. Chem. Prepr., 4, 313 (1969).
6. G. Riess, J. Periard, J. Kohler, Y. Jolivet et A. Banderet Revue Gén. du Caoutchouc et des Plastiques, 48 (4), 431 (1971).
7. G. Riess, Ind. Chim. Belg., 37, 1097 (1972).
8. G. Riess et Y. Jolivet, American Chemical Society Los Angeles, Preprint, 15 (1), 266 (1974).
9. G. Riess et Y. Jolivet, Adv. in Chem. Series, 142 (22), 243 (1975).
10. G. Riess, Progr. Colloid & Polym. Sci., 57, 262 (1975).
11. G. Riess, S. Marti, J.L. Refregier et M. Schlienger, "Polymer Alloys" Ed. D. Klempner, K.C. Frisch, Polymer Sci. and Technol. vol. 10, Plenum Press, 1977.
12. G. Riess, Die Angew. Makromol. Chem., 60/61, 21 (1977).
13. G. Riess, Toughening of Plastics, 3.1 (1978).
14. G. Riess, M. Schlienger et S. Marti, Macromolecules (sous press).
15. S. Marti, Thèse Mulhouse, 26/11/1977.
16. J.L. Locatelli et G. Riess, Die Angew. Makromol. Chem., 27, 201 (1972).
17. J.L. Locatelli et G. Riess, Die Angew. Makromol. Chem., 26, 117 (1972).
18. J.L. Locatelli et G. Riess, J. Polym. Sci., 11, 3309 (1973).
19. J.L. Locatelli et G. Riess, Die Angew. Makromol. Chem., 28, 161 (1973).
20. J.L. Locatelli et G. Riess, Die Angew. Makromol. Chem., 32, 101 (1973).

21. J.L. Locatelli et G. Riess, Die Angew. Makromol. Chem., $\underline{32}$, 117 (1973).

22. J.L. Locatelli et G. Riess, Die Makromol. Chem., $\underline{175}$, 3523 (1974).

23. J.L. Refregier, J.L. Locatelli et G. Riess, Eur. Polym. J., $\underline{10}$, 139 (1974).

24. J.L. Locatelli et G. Riess, Die Angew. Makromol. Chem., $\underline{35}$, 57 (1974).

25. G. Riess et J.L. Locatelli, Adv. in Chem. Series, $\underline{142}$ (17), 186 (1975).

26. G. Riess, C. Beslin, J.L. Locatelli et J.L. Refregier "Polymer Alloys" Ed. D. Klempner, K.C. Frisch, Polymer Sci. and Technol. vol. 10, Plenum Press, 1977.

27. J. Periard, A. Banderet et G. Riess, Polymer Letters, $\underline{8}$, 109 (1970).

28. J. Periard et G. Riess, Kolloid-Z. u. Z. Polymere, $\underline{248}$, 877 (1971).

29. G. Riess, J. Periard et A. Banderet, "Colloidal and Morphological Behaviour of Block and Graft Copolymers", Ed. G.E. Molau, Plenum Press, 1971.

30. D. Marti, J. Nervo et G. Riess, Progr. Colloid & Polym. Sci., $\underline{58}$, 114 (1975).

31. J. Periard et G. Riess, Colloid & Polym. Sci., $\underline{253}$, 362 (1975).

32. M. Ossenbach-Sauter et G. Riess, C. R. Acad. Sci. Paris, $\underline{283}$, 269 (1976).

33. A. Skoulios, P. Helfer, Y. Gallot et J. Selb, Makromol. Chem., $\underline{148}$, 305 (1971).

34. D. Meier, Communication privée.

35. J. Periard, A. Banderet et G. Riess, Die Angew. Makromol. Chem., $\underline{15}$, 37 (1971).

36. J. Periard, A. Banderet et G. Riess, Die Angew. Makromol. Chem., $\underline{15}$, 55 (1971).

37. T.O. Craig, J. Polym. Sci. Chem. Ed., $\underline{12}$, 2105 (1974).

38. S. Marti et G. Riess, Makromol. Chem., $\underline{179}$, 2569 (1978).

39. P. Gaillard, Thèse en cours de préparation.

40. A. Echte, Angew. Makromol. Chem., $\underline{58\text{-}59}$, 175 (1977).

SYNTHESIS AND PROPERTIES OF NEW TOUGHENED RESINS BASED

ON ETHYLENE-PROPYLENE COPOLYMERS

S. Cesca, S. Arrighetti, A. De Chirico, A. Brancaccio

ASSORENI, Polymer Research Laboratories

San Donato Milanese, Italy

Toughened resins (ACS) displaying interesting physical properties can be synthesized by grafting styrene and acrylonitrile (SAN) onto EPM provided that suitable reaction conditions are chosen. ACS resins exhibit better ageing behaviour than classical toughened plastics which contain unsaturations in the elastomeric back-bone. The mechanical properties are satisfactory owing to the network-like morphology of the rubbery phase and the level of graft achieved in the ACS resins studied.

INTRODUCTION

The poor weathering resistance of high impact resins like ABS, which are obtained by copolymerizing styrene and acrylonitrile (SAN) in the presence of polybutadiene, has been so far the most limiting factor to their wide use in outdoor application. The decrease of their mechanical properties with both time and temperature is mainly caused by the oxygen attack on the unsaturation of the grafted rubber[1,2]. One way for improving the scarce stability of the aforesaid resins has been the replacement of the polybutadiene backbone with more stable rubbery substrates containing a low amount of unsaturation. The most employed materials were ethylene-propylene based terpolymers (EPDM)[3,4] although wholly saturated materials like ethylene copolymers (EVA) and homopolymer (LDPE) have been also proposed[5,6].

In previous articles[7] we have described the synthesis of high impact-resistant resins by grafting SAN onto triene based ethylene terpolymers (ATS resins). It was shown that the mechanical properties of ATS resins were better than those of ABS in spite of the very small amount of unsaturated termonomer contained in the starting elastomer. On the other hand, the very low content of unsaturation was responsible for their excellent ageing behaviour.

In this paper we report on the synthesis of toughened resins (named ACS resins) obtained by grafting SAN onto wholly saturated ethylene-propylene copolymers (EPM) as a continuation of our previous investigations carried out with EPTM[7].

EXPERIMENTAL

Materials

Styrene, acrylonitrile, toluene, n-heptane and iso-octane were purified according to procedures described in previous works[7]. Ter-butyl-5-ethyl-perhexanoate (TBPH) was a commercial product and was used as received. Benzoyl peroxide (BPO) was crystallized from toluene. Methyl-ethyl-ketone (MEK) and acetone were distilled before use. EPMs were synthesized with the procedure described elsewhere[8].

Procedure

Grafting reactions were carried out under inert atmosphere in a pressure resistant reactor. The reactor consisted of a 800 cm^3 cylindrical glass vessel topped with a stainless steel plate. The plate was provided with a stainless steel stirrer, a thermometer well, a drop-funnel and a raised threaded port. A curved stainless steel tubing, reaching the bottom of the reaction vessel could be fitted to the threaded port in order to remove samples of suspension formed during the grafting-polymerization runs. The glass cylinder was immersed in a heated oil bath. A hydrocarbon solution of EPM was introduced into the reaction vessel with the appropriate amount of styrene (STY) and acrylonitrile (ACN). The reaction star-

ted by adding the initiator and was stopped by adding some cm^3 of
isopropanol containing 5% of a phenolic antioxidant.

Both solvent and monomers could be added to the reaction medium
during the polymerization by dropping them from the funnel with the
aid of a nitrogen pressure.

ACS resins were recovered by pouring the resulting suspension
into isopropanol (5 times the volume of the suspension). A fine
white powder was collected after filtering and drying overnight
under vacuum at 50°C.

Analyses

The different products of the grafting reaction, namely ungrafted SAN, ungrafted EPM and EPM-g-SAN, were separated by extraction of ACS resins with solvent. The procedure of extraction of
free SAN with MEK or acetone was described in a previous paper[7a].

Ungrafted EPM was extracted with n-heptane in a Soxhelet apparatus for 48 h. In order to facilitate the solubilization of EPM
by the solvent, ACS resins were previously diluted with commercial
SAN, mixed on a 150 x 300 two rolls mill (3' at 160°C) and finely
powered in a grinding mill. These operations facilitate the separation of ungrafted EPM from the grafted one. In fact, the rollmill
treatment induces formation of rubbery as well as resinous domains
in ACS resins (see Results and Discussion). Grafted EPM is anchored
to the resinous phase whereas the ungrafted one can be easily extracted by the selective solvent.

The degree of graft (G) as well as the grafting efficiency (E)
were claculated as described elsewhere[7].

The intrinsic viscosity measurements of SAN were carried out
in MEK at 30°C.

GPC measurements were carried out by two Waters Instruments,
Models 200 and 440, using THF as elution liquid at a flow of 1
ml/min and as column packing materials Styragel and μ-Styragel
respectively. The GPC Model 440 was also equipped with a detector
working at a wave length of 254 nm and can give two elution chroma-

tograms at the same time for EPM-g-SAN. The first chromatogram
corresponds to refractive index variation for whole graft polymer
and the second one to ultraviolet (uv) absorption of SAN grafted
onto EPM; both chromatograms are a function of elution volume and
then of molecular weight.

The mechanical properties of ACS resins were determined following ASTM methods. All materials were homogeneized and 0,2 phr of
an antioxidant agent (Irganox 1076, a Ciba Geigy product) was added
by mixing on a 150 x 300 two roll mill at 180°. Test specimens were
cut from sheet compression moulded at 200°C and 4 MPa.

The morphology of ACS resins was determined according to the
procedure described elsewhere[7b].

RESULTS AND DISCUSSION

Synthesis of ACS resins

The grafting reaction onto EPM was carried out by means of the
so called "solution procedure" since the reactive monomers and the
elastomer were dissolved in a suitable solvent. However, only the
grafting reactions carried out in toluene maintain the character of
nearly perfect solutions throughout the reaction time. Aliphatic
hydrocarbons such as iso-octane and n-heptane as well as their mixtures with toluene are poor solvents of SAN which precipitates progressively from the reaction mixture together with the most part
of grafted EPM. The presence of grafted EPM since the early stage
of the reaction, prevents SAN clogging because EPM-g-SAN acts as
an emulsifying agent[9]. Actually, the grafting process takes place
in solution-suspension with the aforesaid solvents.

The choice of a proper solvent may be of crucial importance
for such a type of reactions as it is shown in Figure 1 where the
values of the grafting efficiency (E) and the degree of graft (G)
as well as the amount of grafted elastomer are plotted as a function
of the monomer/EPM ratio. Two aliphatic solvents, i.e. iso-octane
and n-heptane, were used. The higher values of G and E observed
when grafting reactions were carried out in iso-octane could be
due to the ability of the iso-octane molecule to yield free radicals
which contribute, through transfer reactions, to the formation of

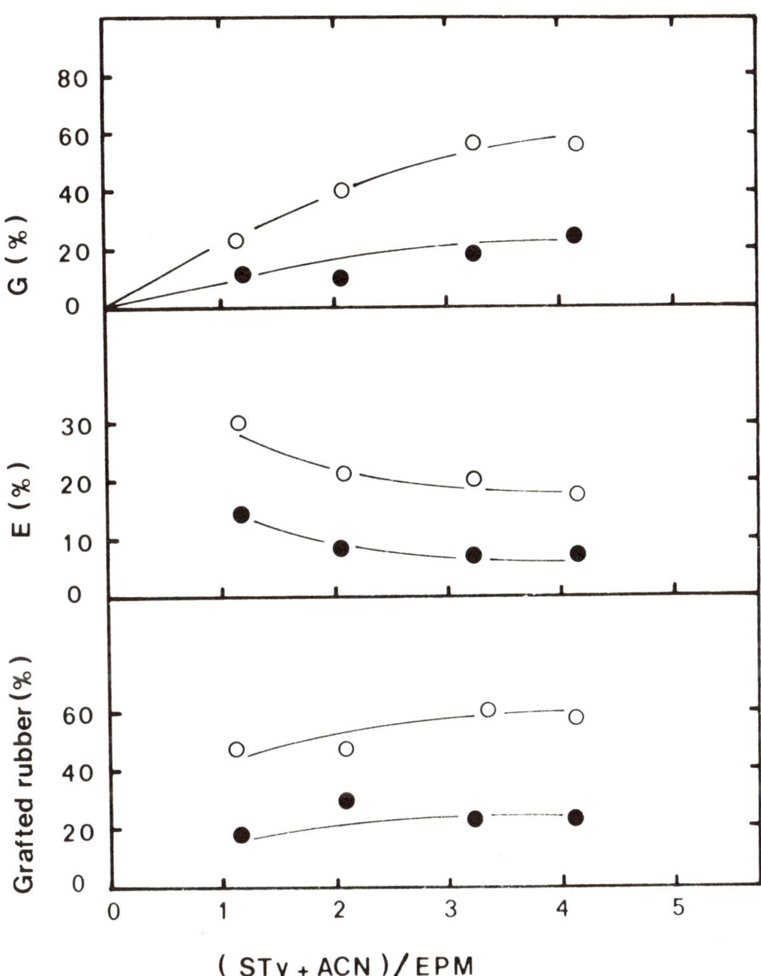

Figure 1.   Influence of the monomers ratio on EPM grafting.
Conditions: T = 90°C; EPM = 60 g/l; TBPH = 3 g/l;
time = 1h; solvent = n-heptane (●); iso-octane (o)

radical sites on EPM chains. Similar results have been obtained by
Alberts et al[5] when some olefins, e.g. propylene or isobutene, were
used in grafting LDPE and EVA.

The typical course of the overall process (including the formation of grafted SAN as well as ungrafted one) carried out in iso-octane is shown in Figure 2 for three different values of the ACN:STY
ratio in the feed.

The copolymerization is a very fast process, the rate of which
is strongly dependent on the relative amount of ACN in the feed.
Levels of ACN near to 50 wt % in the feed makes the reaction to
reach 80% conversion within 10 minutes.

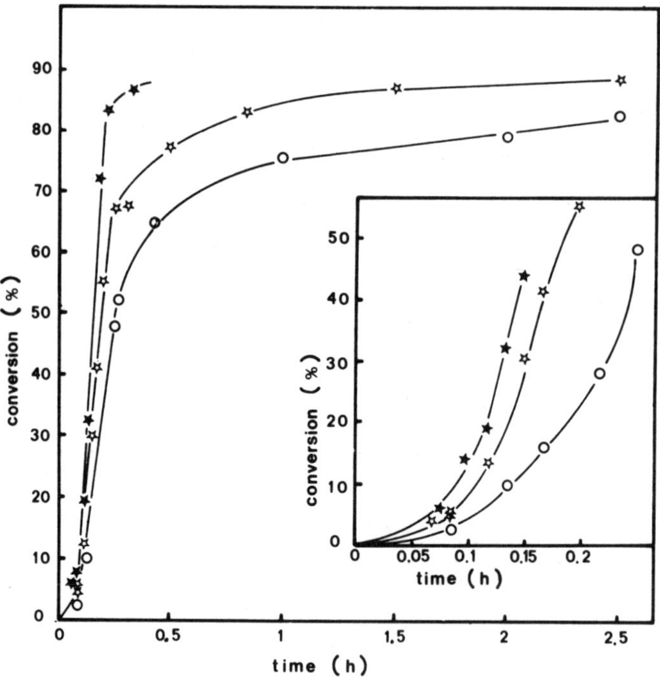

Figure 2. Dependence of monomers conversion on the reaction time.
           Conditions: monomers = 135 g. 1; EPM = 42 g. 1; T = 90°C;
           ACN/STY = 1 (★), 0,67 (✪), 0,47 (O).
           Solvent = iso-octane.

This high copolymerization rate is favoured by the presence of
aliphatic solvents, the role of which, based on the unability of
aliphatic solvents to dissolve SAN, has been already described in
previous papers[3,7]. However, after having reached a certain degree
of conversion the value of which depends upon the monomer ratio,
the copolymerization rate decreases as a consequence of the reduced
amount of monomers present in the reaction medium. The ACN content
in the feed influences the overall process by favouring the precipitation of the elastomeric backbone[3] and raising the overall polymerization rate[10]. The induction time which indicates the intermediate stage between the solution and the suspension state of the
reaction medium is, as a consequence, shortened by increasing the
ACN content of the feed.

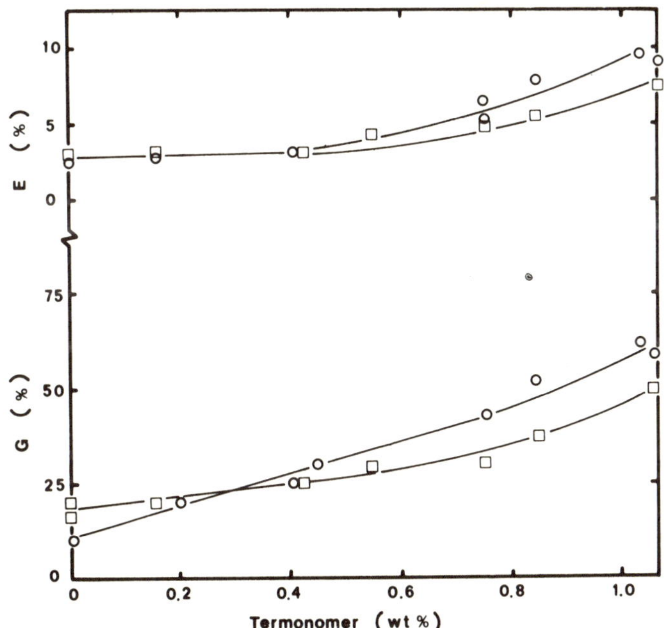

Figure 3.   Dependence of the grafting efficiency (E) and degree of
            graft on termonomer content in EPTM grafting.
            Conditions: STY = 215 g/1; EPTM = 27,5 g/1; T = 70°C;
            time = 6 h; solvent = n–heptane–toluene (1:1 wt);
            (O) = BPO; (□) = TBPH.

Wholly saturated elastomeric backbones such as EPMs are generally scarcely reactive toward radical grafting reactions unless suitable reaction conditions are chosen for enhancing their reactivity. Figure 3 shows the sharp decrease of the grafting efficiency as well as the degree of graft by reducing the content of unsaturation in an ethylene-propylene based terpolymer. The trend of the diagrams of Figure 3 (for decreasing amounts of termonomer) yields an intercept which corresponds to the degree of graft of a completely saturated copolymer (EPM).

The values of G and E, under the adopted conditions, are quite low and not sufficient to impart to the resulting resins satisfactory technological properties[7].

Figure 3 shows that the two different initiators display different reactivity. T-butyl-5-ethyl-perhexanoate (TBPH) seems to be a little more efficient than benzoylperoxide (BPO) in promoting the grafting process when the termonomer content is very low. This difference of reactivity has been verified (Figure 4) for different concentrations of the two aforesaid initiators. The value of E increases as TBPH concentration decreases while the reactivity of TBO is rather poor and no noticeable variation of the value of E is observed. This result reflects the different selectivity in the grafting process displayed by TBPH and BPO.

The opposite trend is conversely shown by the amount of grafted elastomer which increases, as expected, by increasing the amount of initiator.

The important and different role played by some initiators is shown in Table 1 which collects the values of G and E obtained in the synthesis of ACS resins. The grafting temperature and initiator concentration were chosen in such a way that the same concentration of free radicals arising from the different initiators was formed within a fixed reaction time.

The different reaction temperatures corresponding to the best decomposition temperature of each initiator, make the comparison of the runs of Table 1 rather difficult. Peroxides having high decomposition temperatures show a high efficiency in the grafting process, mostly as a consequence of the increased value of the constant of the hydrogen abstraction reaction from EPM chains, according to

TABLE 1

REACTIVITY OF DIFFERENT INITIATORS IN THE SYNTHESIS OF ACS RESINS

Conditions: STY + ACN = 215 g/1;   EPM = 20,6 g/1;   time = 6 h;   solvent = n-heptane-toluene (0,55 : 0, 45 wt)

Run	Initiator (mmol/1)		T°C	Grafted rubber (wt %)	G (%)	E (%)
1	AIBN	5,7	85	$\simeq 0$	$\simeq 7$	$\simeq 1,1$
2	BPO	6,83	"	28,6	11,4	1,86
3	TBPH	7,78	"	37,3	23,7	3,7
4	P1335	7,6	114	51,3	73,3	11,4
5	BPB	6,85	120	66	33,3	5,5
6	DTBP	12,05	130	63	82,4	14,1

BPO = Benzoylperoxide,   TBPH = t-butyl-5-ethylperhexanoate;   AIBN = Azo-bis-iso-butyro-nitrile; P1335 = t-butyl-2,2,2- trimethylperhexanoate;   BPB = t-butyl-perbenzoate,   DTBP = Di-t-butylperoxide.

Arrhenius equation. Nevertheless, the more enhanced reactivity
among those initiators is shown by di-t-but-peroxide (DTBP) which
release two t-butoxy radicals per molecule of DTBP, whereas the highest grafting efficiency among the initiators employed at 85°C is
shown by TBPH which can achieve a grafting efficiency twice higher
than that of BPO. This difference has to be ascribed to the noticeable tendency of t-butoxy radicals to abstract hydrogen atoms
from a hydrocarbon substrate[11]. Benzoyloxy radicals are otherwise
much more reactive towards unsaturation (see Figure 3) by either
adding to double bonds or abstracting allylic hydrogen atoms[12] .

The dependence of G, E and the amount of grafted EPM on the
temperature of reaction is reported in Figure 5 for three values of
the monomers/elastomer ratio (R). The amounts of grafted EPM and
the degree of graft are located on the upper curves when high concentrations of monomers are used. The values of E are always higher by working at lower concentration of monomers and steadily decrease by increasing the reaction temperature on account of the con-
current contribution of two factors: 1) the larger amount of the
istantaneous radical concentration due to the increased rate of the
initiator decomposition; 2) the larger amount of $CH_3^{\cdot}$ radicals coming from the β-scission of primary radicals.

The depressing action asserted on the values of E by increasing
amounts of radicals has been already shown by Figure 4 (see also
ref. 7b). As far as the behaviour of $CH_3^{\cdot}$ radicals is concerned, it
is worthnoting that they possess a high tendency to double bond addition and scarce ability to hydrogen abstraction. Furthermore,
steric effects are not so important for methyl radicals as for t-butoxy ones. Therefore, the increased concentration of $CH_3^{\cdot}$ radicals,
caused by working at higher temperatures, promotes the formation
of free SAN instead of grafted SAN and the values of E are consequently depressed.

The values of G don't follow the same trend of E because in
the computation of this parameter the conversion of monomers is
used. In fact:

$$G = \frac{[M] \ conv.}{[EPM] \cdot 100} \cdot E$$

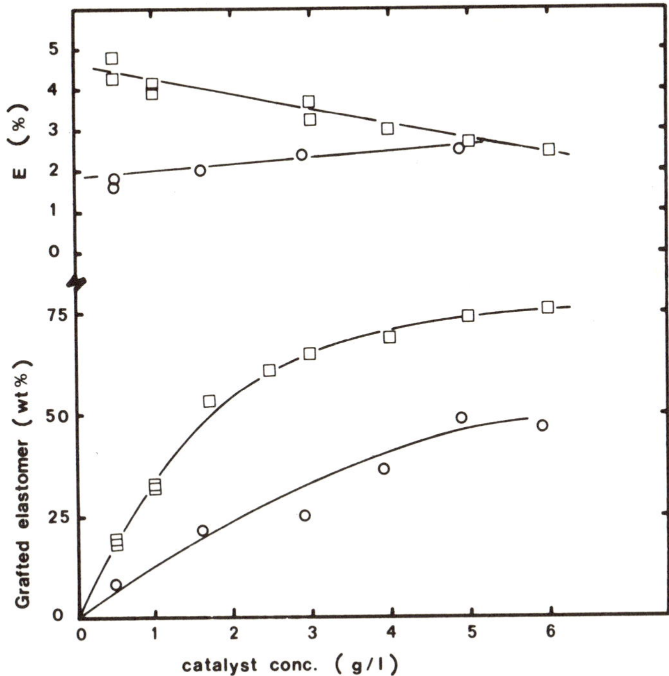

Figure 4.   Dependence of the grafting efficiency (E) and the amount
of grafted EPM on the concentration of BPO and TBPH.
Conditions and simbols as in Figure 3.

where $[M]$ = monomers conc.(g/l)
       $[EPM]$ = copolymer conc. (g/l)
At low reaction times, as those of Figure 5 are, the increasing conversion of the monomers by increasing the reaction temperature,
overcomes the effect of the aforesaid decrease of E.

Reactivity of the elastomeric backbone
____

     The previous figures show that an amount of elastomer, ranging
between 20 and 80% and depending on the adopted experimental condi-
·tions, doesn't undergo the grafting process. The dependence of
both the amount of grafted EPM and the molecular weight of the ungrafted elastomer on the reaction time is illustrated in Figure 6.
The grafting reaction on EPM proceeds rapidly during the first period of reaction and then an almost constant amount of ungrafted
EPM is attained. This result is due to the decrease of the mono-

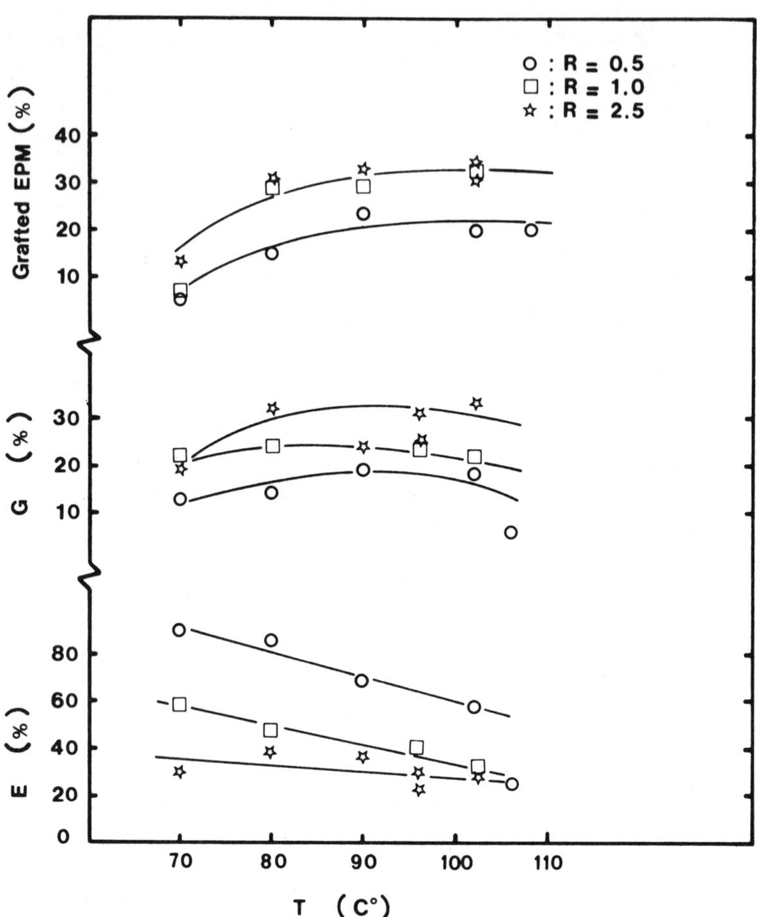

Figure 5. Grafting of EPM at different temperatures. Conditions as in Figure 1 except TBPH = 1 g/1; solvent = iso-octane.

mers concentration and also to the increasing influence of the heterogeneity of the system at high conversion when EPM chains having
progressively lower MW were observed to remain ungrafted. Actually
the intrinsic viscosity of ungrafted EPM steadily decreases with
the reaction time (Figure 6).

Two hypotheses may be put forward for explaining the second
diagram of Figure 6: 1) beside the grafting reaction there are secondary degradation reactions involving EPM chains whose extent depends on the reaction conditions; 2) the elastomeric chains having
higher molecular weight undergo the grafting reaction preferentially. In order to verify the two alternative hypotheses, four different grafting reactions have been carried out with the same sample
of EPM having $\overline{M}_v$ = 300.000 and $\overline{M}_v/\overline{M}_n$ = 5.3. Four different ratios
of grafted SAN/grafted EPM have been obtained in the produced ACS
resins.

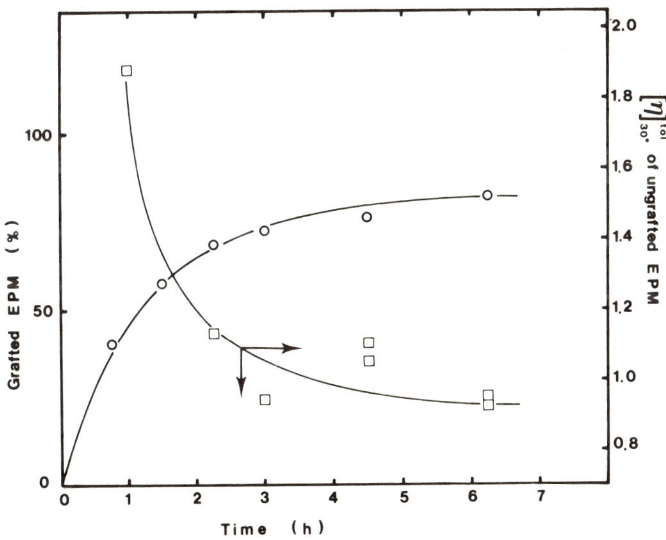

Figure 6. Dependence of the amount of grafted EPM and the intrinsic viscosity of ungrafted EPM on the reaction time.

After separation of the components by selective extractions of
the raw resins, the molecular weight of both EPM-g-SAN and the elastomeric backbone of the graft copolymer were calculated according
to the method described in reference 13.

The results, collected in Table II, show that the molecular
weight of the elastomeric backbone of the graft copolymer is always
higher than that one of the starting EPM and the molecular weight
of the elastomeric backbone decreases by increasing the amount of
grafted SAN.

The results of Table II have been confirmed by comparing the
GPC elution chromatograms of both the ungrafted and grafted EPM contained in the second sample of Table II with that of the starting
elastomer (Figures 7,8). Figure 7 shows that the GPC elution curve
of the ungrafted EPM is contained in that of the starting EPM. The
dashed area, which corresponds to the fraction of the starting EPM
which has been grafted, indicates that the grafting reaction occurs
prevailingly on the elastomer chains having higher MW.

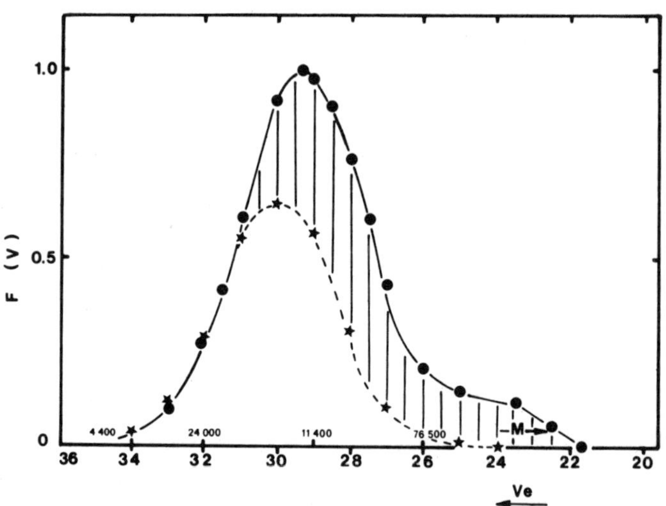

Figure 7. Comparison between the GPC elution curve of the ungraf-
           ted EPM contained in an ACS resin and that of the star-
           ting EPM. (●) starting EPM; (✳) ungrafted EPM

TABLE II

INTRINSIC VISCOSITY AND MOLECULAR WEIGHTS OF THE PARTS CONSTITUTING ACS RESINS

Sample	$\dfrac{\text{Grafted SAN}}{\text{Grafted EPM}}$	$[\eta]_{30°}^{\text{MEK}}$ extracted EPM	$[\eta]_{30°}^{\text{tol}}$ extracted EPM	$[\eta]_{30°}^{\text{THF}}$ EPM-g-SAN	$\overline{M}_v$ elastomeric back-bone
1	1,18	0,52	0,96	2,50	340.000
2	0,83	0,80	1,20	3,00	520.000
3	0,63	0,62	1,30	3,50	440.000
4	0,14	0,40	1,44	1,28	930.000

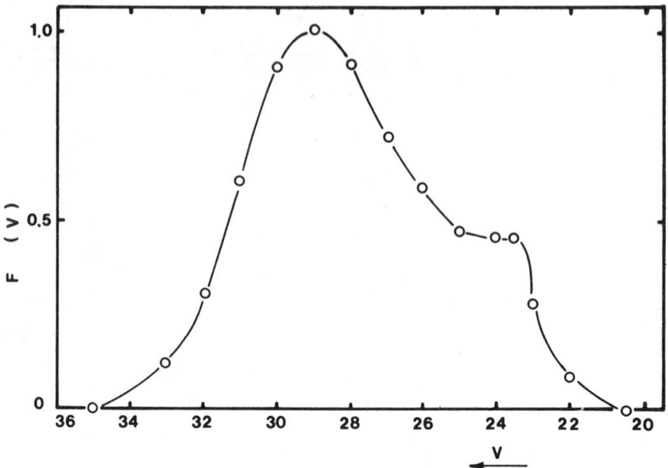

Figure 8. GPC elution curve of the grafted EPM contained in an
           ACS resin.

The GPC elution curve of the grafted EPM (Figure 8) follows
the same trend and shows a broad shoulder in the region of higher
MWs. The GPC elution curves of ungrafted EPM obtained at 5 different reaction times during the synthesis of a typical ACS resin,
are plotted in Figure 9. They shift toward lower molecular weights
as a function of the grafting time but, each of them does not exceed,
at lower MW, the elution curve of starting EPM.

We may deduce from these results that, under our conditions,
there is not any degradation of starting EPM and the elastomer
chains displaying higher MW are preferentially grafted during the
first time of reaction.

Morphology of ACS resins

An ACS resin containing 14 wt % of elastomer has been examined
with scanning electron microscope. The micrograph (Figure 10) show
a structure consisting of a continuous and thick elastomeric net-

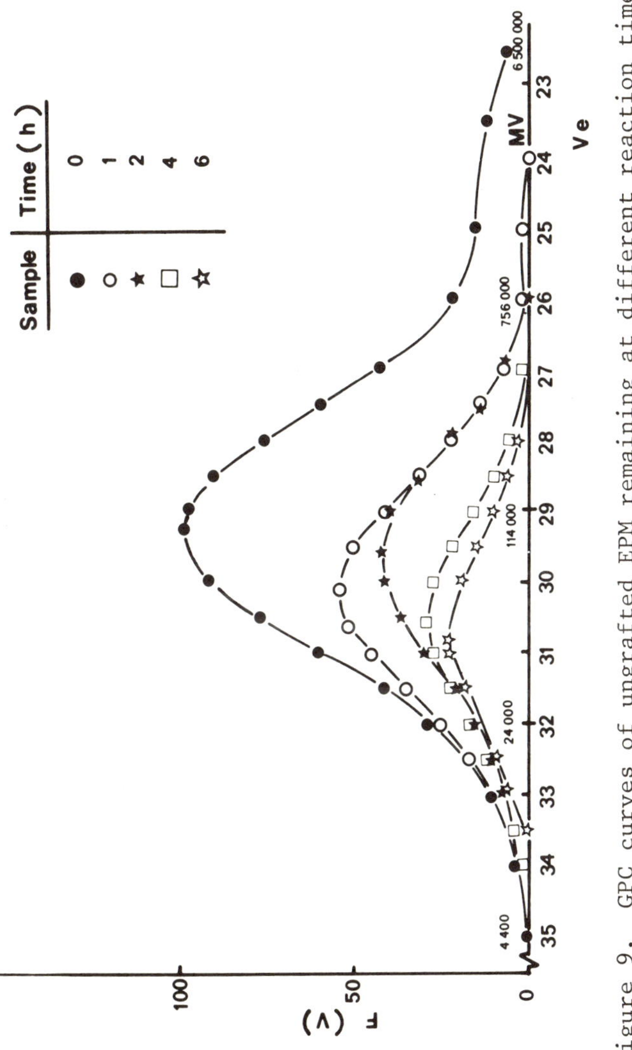

Figure 9. GPC curves of ungrafted EPM remaining at different reaction times.

work similar to that exhibited by ATS resins[7]. The elastomeric network is connected with a continuous SAN matrix in such a way that the two phases seem to penetrate one into the other. This is a very particular morphology as the most known rubber modified polymers are usually made by a rigid continuous matrix with a rubbery phase dispersed as isolated particles.

A thermomechanical treatment on a laboratory mill induces (Figure 11) a small change of structure with apparent separation of the rubbery phase. This behaviour, unknown in ATS resins which shows a thicker network of the rubbery phase after thermomechanical treatments, may be due to the total absence of gel content in ACS resins. In fact, measurements of gel content at 25°, indicates for ACS resins very small amounts of insoluble material.

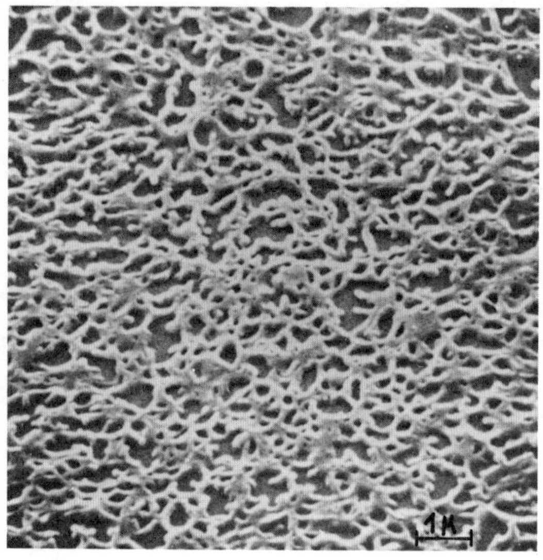

Figure 10.   Morphology of a typical ACS resin (SEM micrograph after treatment with acetone).

The total absence of gel does not influence adversely the technological properties of ACS resins, since their performance is mostly due to the morphological situation, where the grafted rubber forms a continuous network. In fact, a decay of the technological performance in such a type of resins has been observed[7a] whenever

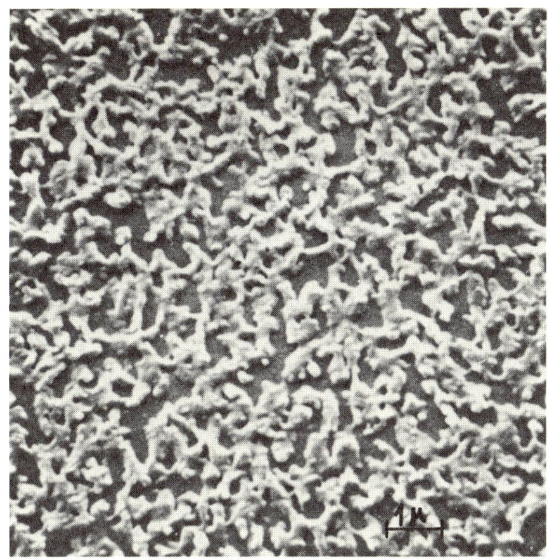

Figure 11.   Morphology of the ACS resin of Figure 10 after thermo-
             mechanical treatment in a laboratory mill.

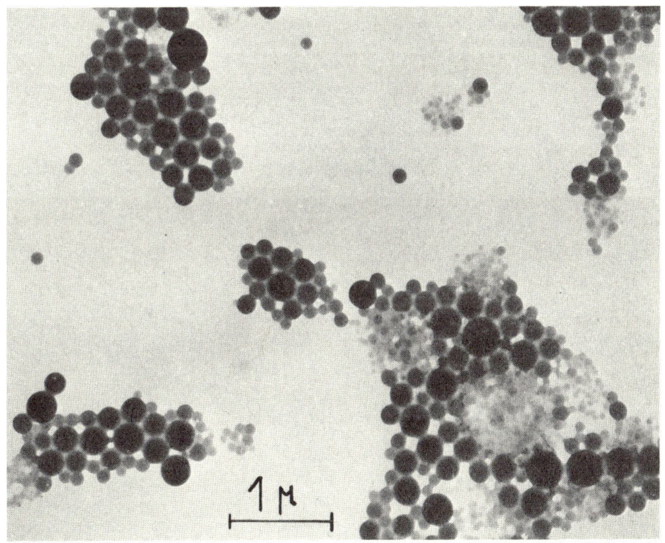

Figure 12.   T.E.M. micrograph of an ACS polymerization emulsion.
             Solvent: toluene-n-heptane 1:1 by weight.

attempts had been made to increase their gel content. On the contrary, the technological properties of ABS resins are improved by
partial crosslinking of the rubbery phase which is dispersed as isolated particles[14].

A typical sample of ACS resin, drawn from the reaction vessel,
has been examined by trasmission electron microscopy (Figure 12).
The micrograph shows microspheres with average diameter of 0,25 μ.
The same sample, treated with acetone for dissolving free SAN (Figure 13), shows that the SAN phase is surrounded by grafted EPM.

This is another peculiar characteristic of ACS resins morphology as a result of the different solubility of the two components
of the graft copolymer. In fact, SAN is less soluble than EPM in
the reaction solvents[7] and precipitates first from the reaction medium, whereas the more soluble EPM, which is bound up with SAN remains on the surface of the precipitated particles and in contact
with the solvent.

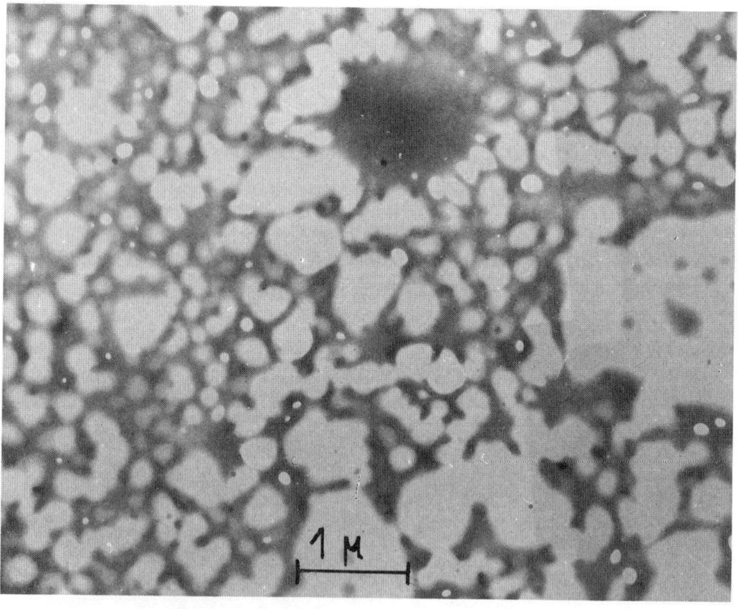

Figure 13.   T.E.M. micrograph of the ACS of Figure 12 after washing
             with acetone.

TABLE III

IR CARBONYL ABSORBANCE AT 1710 cm^{-1} OF SOME TOUGHENED RESINS AGED FOR DIFFERENT TIME IN W.O.M.

Ageing Time (h)	$\log (\frac{1}{1_o} / mm)$		
	ACS	ATS	ABS
0	1,02	1,14	1,85
24	1,18	1,56	4,10
48	1,23	1,77	6,81
117	1,65	2,64	--
200	1,79	3,83	--

TABLE IV

	Rubber content (w/w %)	Izod Impact (J/m)	Modulus (E) (MPa)	Rockwell Hardness (R)	HDT[1] (°C)	Torque[2] (Nm)
ATS	10	150	2800	110	95	18
"	15	350	2100	103	91	20
ACS	12,5	235	2370	108	92	19
"	11	180	2760	111	93	18
ABS	15	130	2400	106	84	17
"	8	9	2970	111	86	19

1) Heat distortion temperature

2) Measured after 20 min of mixing in Brabender Plastograf (190°C – 50 rpm).

Ageing behaviour and mechanical properties

The main advantage of a completely saturated elastomeric backbone such as that employed in the synthesis of ACS resins, is the excellent outdoor ageing performance with respect to both commercial ABS, as well as ATS resins (above mentioned). In fact, the results displayed in Table III by comparing ACS, ABS and ATS resins show that the resistance in weatherometer experiments, (I.R. evaluation as an index of the oxidation degree of the elastomer) increases with the exposition time when the content of unsaturation in the starting elastomer decreases. Some technological properties of ACS resins are collected in Table IV in comparison with the corresponding ones of ABS resins. The network morphology of the rubbery phase as well as the degree of graft achieved in ACS resins is the main reason of the satisfactory values of the impact-resistance, while the high values of hardness and flexural modulus are mainly due to the overall content of SAN in the resin.

CONCLUSIONS

Thermoplastic resins with good physical properties can be synthesized by grafting SAN onto wholly saturated materials such as ethylene-propylene copolymers provided that suitable reaction conditions are chosen for enhancing the reactivity of the elastomer.

Peculiar parameters in this type of synthesis are the choice of a proper solvent together with a radical initiator which release t-butoxyl radicals on account of its enhanced tendency to abstract hydrogen atom from saturated hydrocarbon backbone.

Other important parameters of synthesis have been shown to be the temperature of reaction and the concentration of monomers.

ACS resins exhibit better ageing behaviour than either ABS or other similar resins which contain unsaturations in the starting elastomer, even though at very low level[7]. The mechanical properties of ACS resins are satisfactory owing to the network-like morphology of the rubbery backbone and the relatively high degree of grafting achieved during the synthesis.

REFERENCES

1. B.D. Gesner, J. Appl. Polym. Sci., $\underline{9}$, 3701 (1965).

2. J. Stabenow, F. Haaf, Angew. Makromol. Chem. 29/30, 1 (1973).

3. C.L. Meredith, Rubber Chem. and Techn., $\underline{44}$, 1130 (1971).

4. F. Severini, E. Mariani, E. Cerri, A. Pagliari, Chimica & Industria, $\underline{55}$, 270 (1973).

5. M. Alberts, M. Bartl, R. Kuhn, Adv. Chem. Ser., $\underline{142}$, 214 (1975).

6. M. Alberts, M. Bartl, V. Steffen, Kautshuk und Gummi Kunstoffe, $\underline{10}$, 725 (1977).

7. a) S. Arrighetti, A. Brancaccio, S. Cesca, G. Giuliani, Chimica & Industria, $\underline{59}$, 483, 605, 685 (1977).

    b) E. Montani, R. Vitali, Chimica & Industria, $\underline{60}$, (1), 4 (1978).

8. S. Cesca, S. Arrighetti, W. Marconi, Chimica & Industria, $\underline{50}$, 171 (1968).

9. G.E. Molau, J. Polym. Sci., $\underline{3}$, 1267 (1965).

10. R.F. Blanks, B.N. Shah, J. Polym. Sci., A1, $\underline{14}$, 2589 (1976).

11. H. Atarot, A. Faucitano, European Polymer J., $\underline{12}$, 169 (1976).

12. H.A.J. Battaerd, G.W. Fregeor, "Graft Copolymers", Interscience, New York, 1967.

13. R. Collins, M.B. Huglin, R.W. Richards, Europ. Polymer J., $\underline{11}$, 197 (1975).

14. P. Zitck, S. Mysik, J. Zelinger, Angew. Makromol. Chem. 6, 116 (1968).

MACROMOLECULAR ALLOY SYSTEMS:
A CONTRIBUTION TO THE DETERMINATION OF THE STRUCTURE
OF IMPACT RESISTANT POLYAMIDE – POLYOLEFIN – ALLOYS

G. Illing

Dr. Illing KG. Makromolekulare Chemie
D-6114 Groß-Umstadt, West Germany

Synthetic polyamides such as polycaprolactam and
poly-hexamethylene-adipamide are hard, brittle moulding
masses in the dry state. The impact and shock resistance
in the dry state can be improved considerably by alloying
with polyolefins. The work done on the determination of
the structure of such macromolecular multicomponent systems
is described. The studies show that optimum results are
obtained when polyolefin particles ranging from 0.5 to 5
microns are embedded in the continuous polyamide matrix
and if an adhesion of the non-miscible components is
present at the interface due to the block-graft polymers.

COMPOSITION AND DEFINITION OF MACROMOLECULAR MULTICOMPONENT SYSTEMS

Macromolecular substances, i.e. high polymer substances are
generally incompatible. Even homopolymers with the same chemical
composition – but with different molecular weights – do not produce
homogeneous materials. Thus dotted film or so-called "strings of
pearls" are obtained from a chemically uniform substance such as
polyethylene or polycaprolactam with very different molecular
weights, i.e. the very high molecular weight molecules are embedded
as so-called gel particles in the low molecular weight matrix.
The desired homogeneity is only obtained by very intensive mixing
with high shearing forces. When subjected to large shearing forces,

however, changes in the molecular structure occur, i.e. decomposition and block-graft reactions take place. This applies all the more, if an attempt is made to mix such incompatible macromolecular substances as polyamides and polyolefins. On the other hand, it is to be expected that the relatively hard and brittle polyamides in the dry state, such as polycaprolactam or polyhexamethyleneadipamide, would produce materials which had a good stiffness, hardness, heat resistance, improved impact resistance and shock resistance, when combined with the soft but firmly elastic polyolefins, such as polyethylene, polyethylene-acrylicacid- or polyethylene-vinyl-acetatecopolymers and polyisobutylene.

Due to the incompatibility of the extremely different polymer components, viz. polyamides and polyolefins, the presence of graft copolymers made from these components is necessary to produce a minimum of interphase adhesion and thus to obtain a material which appears homogeneous and which also has an improved impact resistance in the cold and when dry. This type of organic materials is known as polymer alloys[1-4,18].

As a rule, polymer mixtures are incompatible i.e. they are not a polymer combination with a molecular mixing and thus one phase, but instead they are in actual fact usually heterogeneous mixtures of various high molecular weight polymers. In a continuous phase of the main component, polymer particles of the other polymer component are embedded as a discontinuous phase. The result of this is, that the material properties of the polymer blends are not only determined by the chemical composition and the molecular structure of the individual components but also, to a very large extent, by the morphology of the heterogeneous phase and the bonds on the phase interface.

In the case of polyamide - polyolefin alloys, the polyamide component forms the homogeneous phase in which one or more other olefin polymers are incorporated in the form of a finely distributed discontinuous phase, with primary or secondary valency bonds at the phase interface. Different products are obtained depending on the method of preparation. They differ in the shape, size and distribution of the polyolefin particles. First the opinion was held that a finely dispersed distribution was utmost significance for the production of homogeneous products. In actual fact, materials with a relatively homogeneous appearance were produced if the particles of the finely dispersed polyolefin component were less than one

micron in size. If the impact resistance of these physical mixtures
is compared to polymer alloys which contain graft copolymers as contact points at the interface, it is observed that the resistance to
impact or shock is only improved if block-graft polymers are present
at the interface as primary valencies (refer to Table 1).

TABLE 1

Properties of PA-6/PE mixtures and alloys

Properties	Unit	PA-6/PE Mixture	PA-6/PE Alloy
Density	g/cm^3	1,10 - 1,12	1,10 - 1,12
Tensile strength	N/mm^2	46 - 53	51 - 56
E-module	N/mm^2	2800 - 2900	2800 - 2900
Impact strength	KJ/m^2	2 - 4	8 - 12
Shock resistance	Joule	1 - 4	16 - 20

     Similar results are obtained with the analogous system polysty-
rene-rubber[5]. Merz[6] and Davenport[6b] already suspected that graft
polymers act as emulsifiers or contact points at the phase interface.
This was clearly proved by Vollmert[5] and Willersinn[7] on the macromolecular multi-phase system polystyrene-rubber. The macromolecular
composite system polystyrene-rubber is very similar to the system
polyamide - polyolefin. Both consist of several phases due to the
incompatibility of the hard and soft components. In the macromolecular composite system polystyrene-rubber, the soft rubber component
is dispersed and embedded in the continuous hard phase of polystyrene. In the macromolecular composite system polyamide - polyolefin,
the soft polyolefin component is incorporated in the continuous,
hard polyamide phase. In both cases, the continuity of the hard
phase guarantees the desired stiffness, tensile strength and dimensional stability under heat, of the composite system. The finely
distributed soft components which are bonded to the matrix at the
interface by primary and secondary valencies, act as energy absorbers
which absorb the energy of impact or shock to which the composite

system may be subjected and convert it to heat and thus render it
harmless. The disperse distribution of the soft component does not
lead to any deterioration or change in the other physical properties
of the continuous, hard phase, worth mentioning. Measurements of
the impact strength and shock resistance of composite systems made
of polyamides and polyolefins show a dependence on the size and
nature of the embedded soft component and on the strength of the bond
between the discontinuous and continuous phases at the phase interface.

SIZE DISTRIBUTION OF THE POLYOLEFIN PARTICLES IN THE POLYAMIDE PHASE

As already mentioned, a polyamide - polyolefin mixture with a
relatively good appearance and homogeneity is obtained when the
dispersed polyolefin phase consists of very fine particles with a
small size range. Gloss and surface quality improve with decreasing
particle size. Coarse polyolefin particles give a dull surface and
even form small silver skins on the material surface which can be
pulled off. This appears particularly when there is a large range
of particle sizes. If no chemical affinity is present at the phase
interface between the continuous PA phase and the embedded, soft
polyolefin particles, then the particle size of the soft component
depends on the mixing process and is thus not fixed but is changed
during processing by the shearing forces present. This unfortunate
relation of the particle size and distribution is typical for physical mixtures. Only when partial grafting or very polar components,
e.g. olefin copolymers containing carboxyl groups, are used the
dispersed phase can be fixed as a result of the affinity (ionic
bonds) which is then present. The resistance to shock and impact
strength is also influenced by the particle size. It goes through
a maximum. In the case of a coarse dispersion relatively soft and
for a very fine dispersion hard products with poor shock resistance
and impact strength are obtained. The coefficient of friction is
considerably lower for a coarse distribution of the polyethylene
or polyolefin particles. The more the dispersed phase is fixed by
the corresponding high shearing forces or by the addition of cross
linking substances, e.g. radical formers such as peroxides[8], the
more remarkable will be the increase in impact strength and shock
resistance.

INFLUENCE OF BLOCK-GRAFT POLYMERS ON THE SHOCK RESISTANCE AND THE
IMPACT STRENGTH

During the preparation of mixtures of polyamides and polyolefins,
particularly in the case of polyethylene, a higher impact strength
is found with increasing mixing, i.e. with increasing shearing forces,
used to disperse the soft component in the hard phase. As a rule
5 - 12% of the soft polyolefin component are added to the polyamide.
Only special products contain significantly smaller (up to 0.5%) or
larger quantities (up to 40% or more). It is not possible, however,
to obtain homogeneous products when the polyethylene content is more
than 15% without using a special mixer or without the addition of
graft copolymers to improve the homogeneity. This shows that no
homogeneous mass, which can be used as a material, is obtained without
a bond between the embedded soft phase and the surrounding, continuous hard phase. Since the increasing homogeneity and the simultaneous improvement in the impact strength resulting from increased
mixing can only be explained by the formation of block-graft polymers, the question arises as to how these products are formed.

When the polyamides are mixed with the polyolefins in the melt,
larger or smaller shearing forces arise, depending on the intensity
of the mixing. I has been known for a long time[9] that these forces
split a larger or smaller number of the macromolecular chains. This
is defined as a mechanochemical decomposition. When the macromolecules are split mechanically, chain radicals are produced which try
to bond with other radicals or to attack complete polymer chains,
whereby new radicals are produced at the points of attack, which in
their turn attempt to form bonds with other radicals.

As can be seen from the reaction diagram in Figure 1, in this
way branched macromolecules are formed. When polymer mixtures are
produced from the polymer component A (polyamide), e.g. polycaprolactam and B (polyolefin), e.g. polyethylene or ethylene copolymers,
"A" and "B" radicals are formed. This has already been proved by
several experimental studies[10,11]. It was observed that the radical
density depends on the shearing strain. Polycaprolactam tends to
form radicals, in particular, more than polyethylene. The "A" radical chains can react with themselves, i.e. with other "A" radical
chains or with "B" radical chains, whereby bulk polymers are formed
(see Figure 2). In addition, radical chains can react with intact
polymer chains of both types, to form radicals inside the chain,

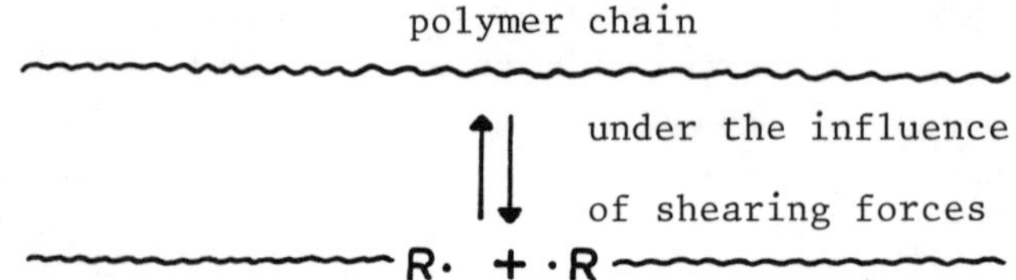

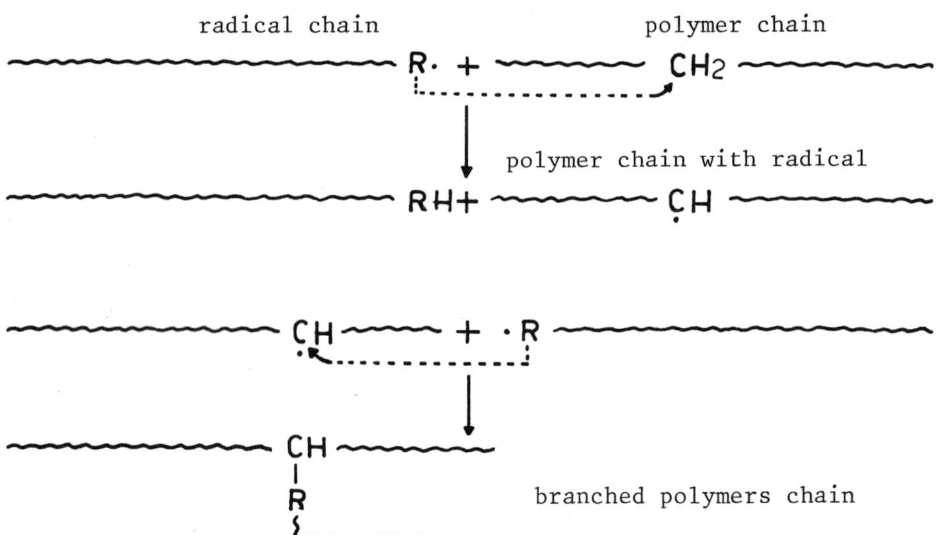

Figure 1. Formation and reactions of radical chains.

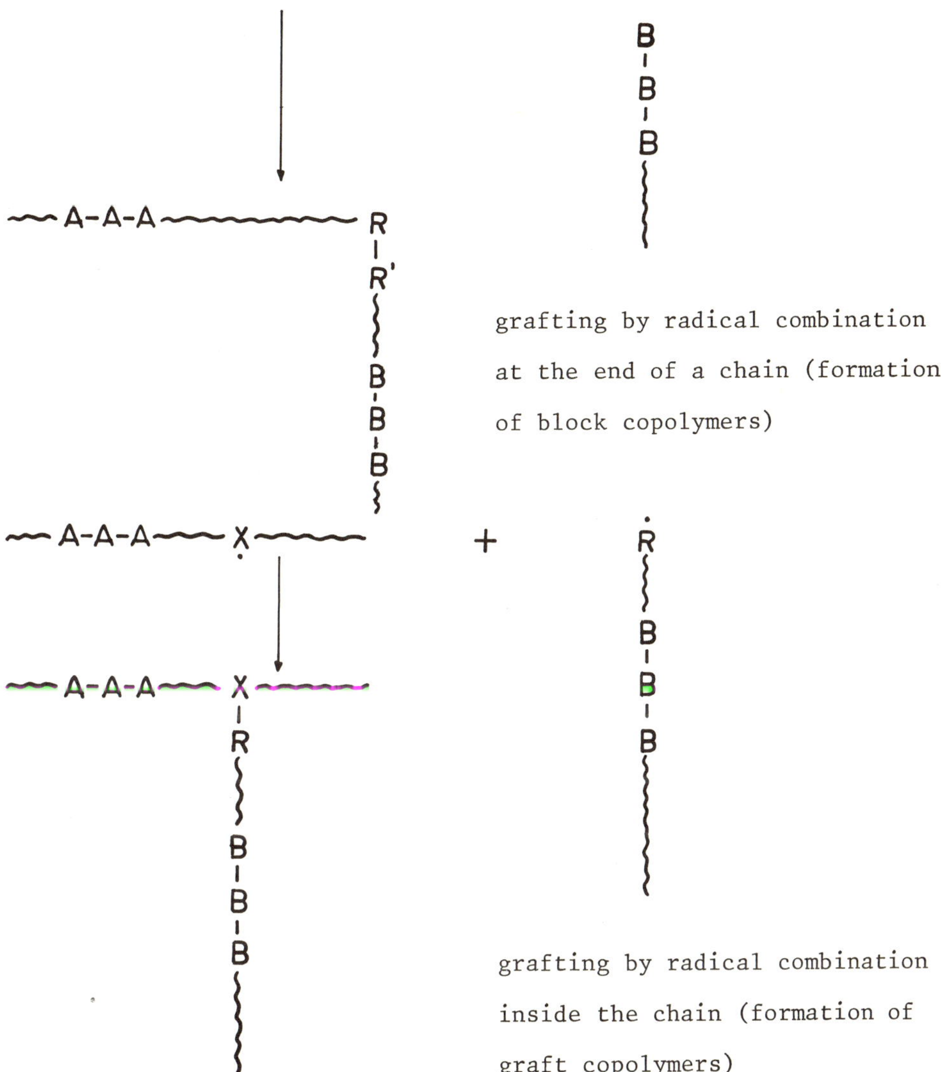

Figure 2. Formation of bulk and graft copolymers due to radical combination in melts of binary polymer mixtures, under the influence of shearing forces.

whereby branched or interlaced A and B polymers as well as various
types of graft copolymers are produced (cf diagrams in Figure 3).
Since radicals are amongst the most reactive groups in organic chemistry, these reactions cannot be inhibited in any way, in particular
the reaction of the radicals amongst themselves takes place very
rapidly. Thus if the presence of chain radicals in the PA-PE melts
has been verified, graft copolymers from PA and PE are certainly
also present.

The formation of graft copolymers from PA and polyolefin can
be promoted by adding peroxides to the molten mixture[8]. Since
peroxide molecules readily decompose they form radicals easily which
attack the polymer chains and lead to the formation of radicals.
Another method of promoting graft copolymer formation is to modify
the polyamide, e.g. by the introduction of nitroso groups[12], or the
polyethylene.

In the melt, provided that shearing forces are sufficiently
large, polyolefins containing carboxyl groups, e.g. ethylene-acryli-
cacid are converted to block-graft-polycondensates by reaction with
the amino groups or carbonamide groups of the polyamide and represent
a special case of block-graft polymers ("re-amidation" reaction).
In order to obtain products from the polyamides and polyolefins for
technical use, i.e. homogeneous mixtures with an even and smooth
surface, a high impact strength and shock resistance in a dry state
and in the cold,an improved flow characteristics for injection
moulding, blockgraft copolymers must not only be present in traces
but in sufficient number to ensure the stabilisation of the dispersed, discontinuous phase at the phase interface. If this is the
case, the mixture is known as a polymer alloy, consisting of a continuous phase, the main component A, an embedded discontinuous phase
B and graft copolymers made of segments of the components A + B and
which are present mainly at the phase interface. The correctness
of this definition of a polymer alloy can be tested experimentally
in that the characteristic properties of alloys are measured for
mixtures of the polymer components A + B with and without the addition of graft copolymers. This was verified for the composite system
styrene - rubber by Vollmert[5], Willersim[7], Riess[13] and Cherdron[14].
Analogous results were obtained for the system polyamide - polyolefin,
as will be shown in the experimental section.

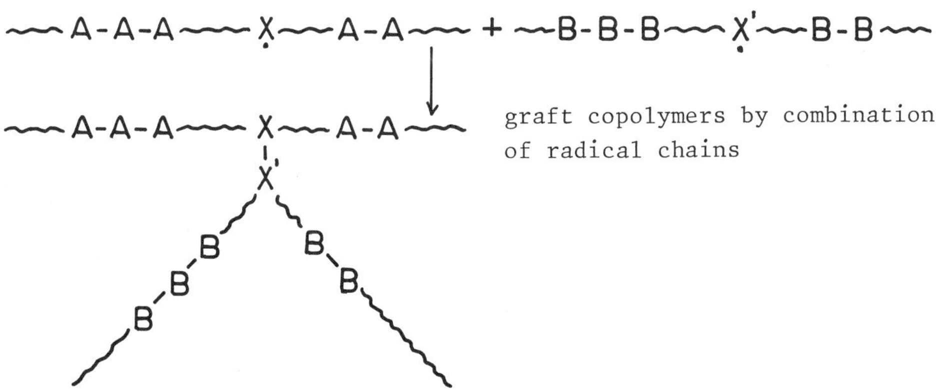

Figure 3. Possibilities of graft copolymer formation by radical combination in binary polymer mixtures.

INVESTIGATION OF THE COMPOSITE SYSTEM POLYAMIDE – POLYOLEFIN

1. The chemical analysis

It is very difficult to separate polyamide – polyolefin alloys,
i.e. polymer mixtures containing graft copolymers, into the components and to characterize the fractions obtained analytically. This
is because polymer composite systems are never uniform with regard
to at least two or more factors, viz. the chemical chain structure
and the molecular size. In case of the composite system polyamide –
polyolefin, there is a particularly high degree of incopatibility.
The incompatibility of the two polymer components is so large that
not even a common solvent for two components can be found. PA-6 or
PA-66 is soluble in sulphuric acid, formic acid, adipic acid or
cresol, while polyethylene is insoluble in all of them. PE on the
other hand, dissolves in xylene, toluene, cyclohexane, tetraline or
other hydrocarbons in which PA-6 or PA-66 are completely insoluble.

If one attempts to dissolve a mixture of PA-6 and PE which is
practically free of graft polymers in pure formic acid, the polyamide
dissolves completely within 2 to 4 hours, while the PE precipitates
in the form of coarse, white flakes (Figure 4). If the same experiment is carried out with a PA-6/PE alloy, i.e. with a polyamide –
polyethylene mixture which was subject to strong shearing forces
during mixing so that it contains a certain proportion of graft copolymers, or with a mixture of polyamide and polyethylene to which
graft copolymers have been added, a colloidal solution is obtained
(Figure 5). Its stability and degree of turbidity is a measure of
the quantity of graft copolymer in the mixture. This is known as
the Molau-test.

The fact that stable emulsions can be formed when two different,
incompatible polymers are added to the same solvent is due to the
dispersion action of graft copolymers according to the generally
acknowledged studies by Molau[15,13,14]. These graft copolymers act
as protective colloids. Since the graft copolymers which stabilize
the polymer-polymer-dispersion are only found on the surface of the
colloidally dispersed phase, relatively small quantities of graft
copolymers (approx 1 - 2%) already suffice to ensure a stabilizing
effect.

According to Molau, the emulsifying action of the graft copoly-

Figure 4. "Solutions" of alloys (left) and mixtures of PA-6 and
           PE (right) in formic acid (cf text).

mers depend on the length of the graft sidechains. The emulsion
stabilizing effect only comes into force clearly from a certain chain
length onwards. The Molau model of the polymer-polymer-emulsion
(Figure 5) agrees with our ideas on the structure of polymer alloys,
whereby the polymer phase B which is embedded dispersly in the continuous polyamide phase A, is rendered stable and compatible by the
graft copolymers. According to Molau's model, the graft copolymers
are in the interface of the emulsion droplets, so that each chain
section A is joined to phase A and each chain section B with phase
B, or embedded in this phase.

The phase copolymers always have the tendency to go to the
interface since their various chains get the most complete solvation
here and this arrangement corresponds to the most stable state thermodynamically. The graft copolymers can only be completely accomo-
dated in the phase interface, however, when the phase boundary is

large enough. The smaller the size of the dispersed particles
(droplets) in the discontinuous phase, the larger the phase interface
so that this results in a stabilizing effect for the polymer-polymer
emulsion.

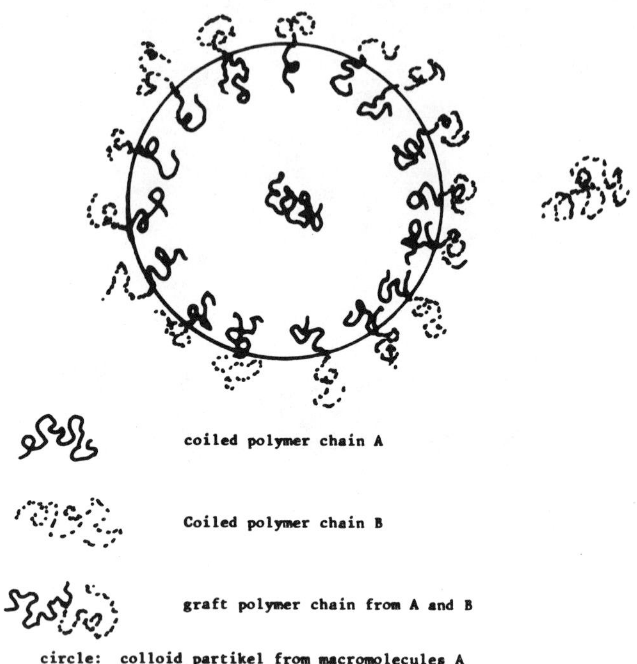

Figure 5. Diagrammatic representation of the stabilizing effect
           of bulk and graft copolymers on emulsions.

1.2 Fractionated extracts

In the case of physical mixtures of PA-6 and PE, the PA component can be dissolved quantitatively with formic acid and the PE
component with xylene or toluene[1].

Physical mixtures are obtained when PA is mixed with PE powder
or if PE powder is stirred into molten PA briefly. When mixing is
longer and more intensive, a small amount of graft polymers is
formed already, visible in the form of a slight turbidity of the
solution of 10% in formic acid (slight positive Molau test reaction).
In the case of PA-6/PE alloy, i.e. of a composite mixture consisting

of PA-6, PE and blockgraft polymers, which show a strong positive
Molau test reaction, the complete separation by extraction with
formic acid or xylene is no longer possible.

When a soxhlet apparatus is used, the PA-components are in the
formic acid fractions and this can be verified by the CHN values of
the individual fractions. The PE-component is found in the xylene
or toluene fractions and can be detected with the aid of IR-spectra
(Figure 6). The number and intensity of the bands characteristic
for the PA is greater, the higher the proportion of block-graft polymers, i.e. the more positive the Molau test reaction.

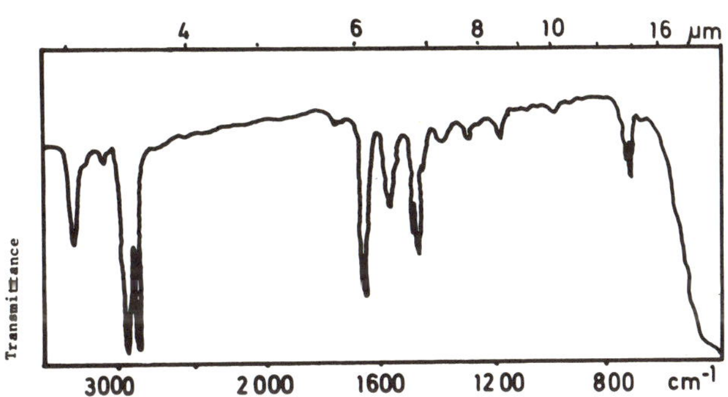

Figure 6.   IR-Spectrum of the toluene extract of a PA-6/PE composite
(alloy) after removing the solvent (film on NaCl).

1.3 IR-Spectroscopy

IR-Spectroscopy does not permit to differentiate between PA/PE
mixtures and PA/PE alloys. There is also no difference between pure
PA-6 and PA/PE alloys or mixtures, since the PE component does not
have any new, characteristic bands with regard to the PA-6 or PA-66
(Figure 7).

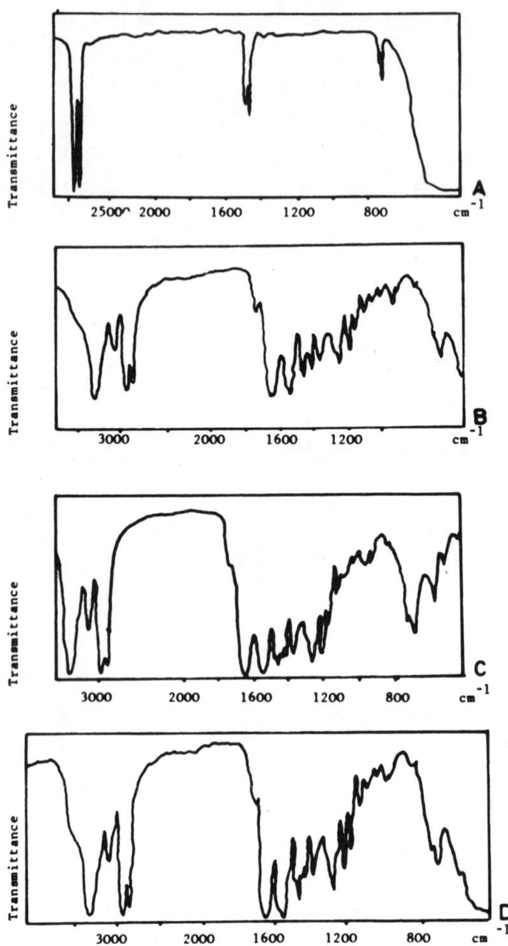

Figure 7.   IR spectrum of PE (A), PA-6 (B), mixture of PA-6/PE (92:8)
            (C) and a composite made of PA-6 and PE (92:8) (D).
            Films (A,B) or sheets (C,D) on and between NaCl respecti-
            vely.

Figure 8. Moulded part made of PA-6/PE alloys (right) and mixtures (left).

1.4 <u>Microscopic structure</u>

From microtome sections under the microscope it can clearly be seen that the PE or polyolefin particles are distributed in a more or less finely dispersed manner in the continuous phase. By staining the PA component with Neocarim, the microscopic structure is easily visible and it is possible to determine the size and size distribution of the polyolefin particles in the polyamide phase. It is remarkable that the microscopic structures of the physical mixtures hardly differ from those of alloys. Only in the case of thermoplastic processing, it can be seen that the structure is fixed at the phase interface for alloys due to the influence of the graft copolymers. For a physical mixture made of the same components and having the same composition but practically no graft copolymers (negative Molau test) the structure will be different after it has passed through the molten state. The soft components are now dispersed more coarsely and a clear tendency to agglomeration and demixing can be observed. The lack of homogeneity in molded parts, the appearance of so-called silver skins, the flaking off near the sprue, are typical sings of a physical mixture, in contrast to alloys (Figure 8), which do not have these properties and always produce uniform moulded parts with a homogeneous appearance.

1.5 <u>Torsion oscillation measurements according to DIN 53 445</u>

In torsion oscillation measurements the mechanical damping in relation to temperature is measured. A test oject having the dimensions 1 x 10 x 40 mm is set into motion at various temperatures. As a result of the internal friction, part of the energy used to change the shape of the object is converted to heat which results in the free oscillations with time.

As a measure of the mechanical damping of the material, the natural logarithm of the ratio of two amplitudes one oscillation period apart, i.e. the so-called logarithmic decrement of the mechanical damping, is used. The greater this value, the more rapidly the oscillations decreases. The damping maximums thus indicate a change in the internal structure and softening ranges. The amorphous, crystalline plastics, such as the polyamides, have characteristic curves on a damping - temperature diagram, whereby the maximums indicate secondary conversion ranges of the amorphous components (Figure

9). The softening range of the crystalline components is indicated
by an infinite increase of the log of the damping decrement at
higher temperatures.

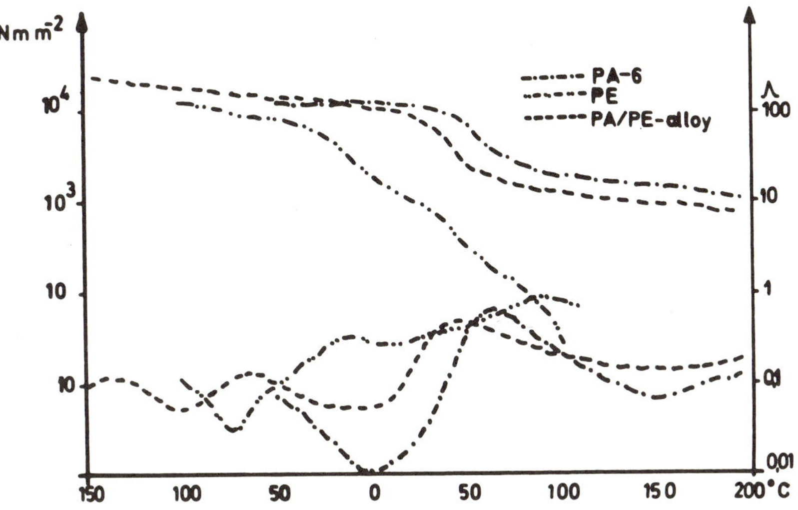

Figure 9. Modulus of rigidity curves (top) and log decrement of
           the mechanical damping of PA-6, PE and a PA-6/PE
           alloy.

     Polyamide – polyolefin alloys have characteristic curves in
the damping temperature diagram, due to the energy absorbing action
of the embedded soft component in the surrounding, hard polyamide
phase. The height of the maximum or the area below the curve is
at the same time a measure of the proportion of mechanical energy
which is converted to heat when the mechanical-thermal behaviour
is determined in the oscillation experiment. The heat generated
under the experimental conditions is due to the molecular relaxation
phenomenon.

     The increase in area under the two maximums for the polyole-
fin-alloy 6-PA B 70 L as compared to pure PA can clearly be seen
and these also indicates the glass temperatures at +45°C and -64°C.
In the composite system discussed here, the loss of mechanical

energy takes place in the polyolefin phase and thus the loss of
mechanical energy and the impact strength and shock resistance will
be all the larger, the higher the polyolefin content of the entire
system. The conditions are similar and even clearer for the system
styrene - rubber[16]. From the torsion oscillation measurements, the
G-module in relation to the temperature can be determined at the
same time as the log decrement of the mechanical damping. This is
a measure for the stiffness.

The softening range can be recognized by a large decrease in
the G-module. If the curves for the modulus of rigidity of PA/PE
mixtures as compared to PA/PE alloys and pure polyamides are inspec-
ted, the influence of the soft polyolefin component can be seen in
the low G-module values, in particular before the softening range.
Reproducible measurements are only obtained when homogeneous ojects
are measured.

1.6 Measurements of the impact strength and shock resistance

The degree of toughness can be determined by various methods.
According to DIN 53 453 specifications, the loss of energy of a
pendulum is measured. On falling the pendulum passes a slit where
a standard size rod of the product being measured has been laid
crosswise so that the rod is broken by the falling pendulum. Since
polyamides are very tough by nature, a wedge shaped or round notch
is cut into the test specimen. Since the two most important polyamides for injection moulding, viz. PA-6 and PA-66 are only brittle
in a dry state, the impact strength measurements as compared to the
PA-polyolefin alloys are only of significance in the dry state,
i.e. when the moisture content is less than 0.1%.

The notched impact strength values obtained for the PA-6 and
PA-66 used for injection moulding, vary between 3 and 4 kJ/m^2 for
the products on the market. The values for polyamides alloyed with
polyethylenes or polyolefins are in the range of 8 - 24 kJ/m^2
(Table 2).

The value for the Ultramid were taken from the BASF specification sheets, those for Durethan from data supplied by Bayer, and

for PA-6 B 70 SK and PA-6 BC 80 SE from Dr. Illing KG. specification sheets.

TABLE 2

Notch impact strength according to DIN 53 453

non-alloy	PA-6 injection moulding materials	alloyed PA-6 injection moulding materials	
Ultramid B 3	$3 - 4$ kJ/m^2	Ultramid B 3 L	12 kJ/m^2
Durethan BK 30S	$3 - 4$ kJ/m^2	Durethan VP KL	20 kJ/m^2
PA-6 B 70 SK	$3 - 4$ kJ/m^2	PA-6 BC 80 SE	38 kJ/m^2

In order to be able to determine the influence of the graft copolymer content on the thoughness, the pendulum experiments were carried out on perforated test objects according to DIN 53 448. The values with standard notched rods deviated too much.

From the data reported in Table 3 it can be seen that the impact resistance when dry, increases with an increase in the alloyed polyolefin component. If the water content is increased, these differences become smaller since the mobility of the polymer chains in the amorphous zone of the polyamide matrix increases. This leads to the formation of further soft, tough zones which increase the impact strength, while the hard, brittle zones are limited to the crystalline zones of the polyamide only. It can clearly also be seen that a mixture which only has a weak Molau test reaction, as proof of the fact that there are hardly any graft copolymers present at the phase interface, hardly changes the impact strength.

The tests carried out with a falling pin gave analogous results. According to this method the shock resistance is determined by dropping a striking pin weighing 2 kg from various heights on to a 2 mm thick disc to be tested. Table 4 shows the results found on discs with a 0,09 % moisture content.

TABLE 3

Impact strength of PA-6/PE alloys for different moisture contents

Products	Impact strength value according to DIN 53 448 measured with flexible rods, ϕ 6 mm	Moisture content of the samples in %			
		0,15	3	6	8
PA-6 + % alloyed polyolefin component					
	0 %	64	90	145	190
	2 %	75	164	190	190
	4 %	118	163	196	205
	8 %	142	205	222	231
	12 %	165	222	238	242
	20 %	216	241	250	252
PE/PE mixture with 8% PE, weak Molau test result		74	98	150	192

TABLE 4

Shock resistance of PA-6/Polyolefin composites

Product	Shock resistance
PA-6 + alloy components 0 %	2 Joule
2 %	3 Joule
8 %	16 Joule
12 %	26 Joule
20 %	26 Joule

STRUCTURE AND IMPACT RESISTANCE

From these results it can be seen that the presence of a soft,
tough polyolefin phase in the polyamide phase, which is hard and
brittle in the dry state, does not suffice to increase the impact
strength, i.e. to give it the character of a hard, resistant material. This applies even if the polyolefin phase is very finely
dispersed. What really is necessary is a certain minimum compatibility of the two phases, which is provided by the PA/PE or PA/po-
lyolefin graft copolymers. Experiments with ionomeric ethylene-
acryliacid salts[17] and with active fillers and high shearing forces[18]
have shown that the compatibility can also be increased considerably
via secondary valency bonds, ionic or H-bonds. A certain degree
of compatibility at the phase interface is necessary for all macromolecular composite systems which are incompatible, in order to
transfer the energy of an impact on the surrounding matrix of the
hard phase rapidly enough to the embedded soft phase, where it is
rendered harmless by being converted into heat.

If this transfer to the soft phase is not possible immediately,
the energy absorbed remains in the continuous hard phase and can
lead to the formation of tension peaks at unfavourable points
which cause the rupture of primary or secondary valency bonds and
thus to fractures and cracks in the moulded part. With an increasing water content of the PA phase, the mobility of the PE chains
in the amorphous zones increases. The impact strength increases,
i.e. more and more impact energy can already be converted to wave
energy and thus to heat in the continuous phase, without breaking
the hard phase. PA-6 and PA-66 are partly crystalline polyamides
with a hard, brittle, crystalline proportion of approx. 35 - 45 %.
In polyamides of this nature, a homogeneous impact resistant mixture
is present when the amorphous parts form the soft, tough component
due to the absorption of water and the crystalline parts form the
hard, brittle component.

EXPERIMENTAL SECTION

1. Molau test

300 mg of the sample to be examined are decomposed with 3 ml pure
formic acid in a test tube. The polyamide component dissolves

within 1 - 4 hours depending on the fineness and molecular weight
of the sample used. Fillers settle at the bottom, polyethylene or
ethylene copolymers swim at the top. If an alloy is present, that
means that the sample contains graft copolymers in addition to the
PA and PE or PE copolymers, a stable, milky turbid, colloidal solution is obtained.

2. <u>Preparation of block-graft polycondensates from caprolactam and an ethylene acrylic acid copolymer</u>

<u>A</u> parts of caprolactam and <u>B</u> parts of ethylene-acrylic acid-
copolymer with an acrylic acid content of 8% are placed in an autoclave which can be heated and stirred. The autoclave is sealed and
rinsed three times with 5 bar nitrogen. Subsequently it is heated
to 250°C within 2 hours and the pressure than increases to 12 - 15
bar. Decompression is then carried out within one hour. The reaction
is completed after leaving for 5 hours at 250°C on rising with nitrogen with no pressure. A very tough, highly viscous melt is obtained
which is pressed out of the autoclave with 18 - 20 atm. of nitrogen.
The properties of prepared copolymers are shown in Table 5.

TABLE 5

Properties of a block-graft polycondensates made of caprolactam and
ethylene-acrylic acid copolymer

	2%	+4%	+6%	+8%
Melting point	210–212	208–212	206–210	204–208
K-value	68,4	67,2	66,7	68
Relative viscosity	2,45	2,36	2,34	2,4
Shock resistance in a dry state	6 Joule	8 Joule	16 Joule	20 Joule
Molau test	partly positive[+]	positive	positive	positive

[+] The Molau test result is strongly positive as soon as the raw
polymerizate is moulded in a molten state after passing through
an extruder or injection moulding machine.

In the raw polymerizate, the grafted parts are distributed heterogeneously since all the reactions are carried out without intensive stirring. There is a considerable increase in the shock resistance of the individual samples in the dry state, with an increase in the ethylene-acrylic-acid copolymer content.

This study was made with the collaboration of Prof. Dr. Bruno Vollmert (Polymer Institute of the University, (T.C.) in Karlsruhe).

Part of the analytical work, in particular the IR-spectroscopic examinations were carried out by Dr. Udo Knecht at Giessen University.

Prof. Dipl. Ing. Walter Hellerich (Heilbronn Technical College) measured the mechanical properties.

Dr. Siegfried Schaaf (Emser Werke, Domat/Switzerland) and Dr. Johannes Arndt (Dr. Illing KG, Gross-Umstadt) corrected the manuscript and gave me numerous valuable hints for which I would like to express my thanks.

REFERENCES

1. G. Illing, Kunststoff und Gummi,$\underline{8}$, 275 (1968).
2. G. Illing, Kunststofftechnik, $\underline{10}$, 92 (1978).
3. G. Illing, Kunststofftechnik, $\underline{8}$, 413 (1969).
4. G. Illing, Kunststofftechnik, $\underline{11}$, 125 (1972).
5. B. Vollmert, Angew. Makromol. Chem. Chem., $\underline{3}$, 1 (1968).
6. E. H. Merz, G. C. Claver, M. Baer, J. Polym. Sci., $\underline{22}$, 325 (1956).
    E. Davenport, L. W. Hubart, M. R. Petitt, Brit. Plastics, 230, 549 (1959).
7. H. Willersinn, Makromol. Chem., 101, 296 (1967).
8. H. Craubner, G. illing, DBP, $\underline{1}$, 131, 883 (1962).
9. Encyclopedia of Polymer Science and Technology vol. 2, 503, vol. 4, 676 ff., Interscience Publishers, New York, 1965.
10. G. Illing, D. Zettler, Belg. Patent, 633, 017 (1963).
11. Mitteilung aus dem Deutschen Kunststoff-Institut Darmstadt N. 4, 1 (1969).
12. G. Illing, A. Hrubesch, Osterreich, Patent 232, 724 (1962).

13. G. Riess, "Influence des copolymers séquencés et graffés dans les alliayes de polymeres", Vortrag auf dem 6, Internat. Makromolekularen Symposium, Interlaken, Schweiz, Juni 1978.
14. H. Cherdron: Aufbau u. Eigenschaften von Polymer-Mischungen Vortrag auf dem 6. Internat. Makromolekularen Symposium, Interlaken, Schweiz, Juni 1978.
15. G. E. Molau, J. Polym. Sci. A3, 1267 (1965).
        Kolloid Z. und Z. fur Polymere, $\underline{238}$, 493 (1970).
16. K. H. Illers, Ber. Bunsenges. Physik. Chem., $\underline{70}$, 353 (1966).
17. G. Illing, DE-OS, $\underline{1}$, 645, 198 (1970).
18. G. Illing, DE-AS, $\underline{2}$, 140, 041 (1976).
19. A. Ullmann, Encyclopedia of Polymer Science and Technology, vol. 4, Band 15, S 183 ff. Verlag Chemie, Weinheim (1978).

SYNTHESIS OF INTERFACIAL AGENTS AND THEIR USE

IN NYLON 6/RUBBER ALLOYS

M. Avella, R. Greco, N. Lanzetta, G. Maglio,
M. Malinconico, E. Martuscelli, R. Palumbo and G. Ragosta

Istituto di Ricerche su Tecnologia dei Polimeri e
Reologia, C.N.R., Arco Felice (Napoli)

INTRODUCTION

One of the major deficiency of rigid thermoplastic materials
as polystyrene, polyvinylchloride, isotactic polypropylene and polyamides is the resistance to impact, especially at lower temperatures.
This limitation is commonly remedied by blending these polymers with
rubbers.

The formulation of rigid plastic/rubber alloys may lead to materials with substantial different overall morphologies and properties
according to, whether the two components are compatible or not.

Compatibility should produce solution of the rubber in the
rigid component, with the result of a gradual and proportional modification of properties with composition.

When the two components are incompatible, generally the system
separates into two phases i.e. a rigid thermoplastic phase that may
incorporate a certain amount of rubber and a rubber phase with or
without dispersed rigid polymer particles in it. Generally the
major component tends to form a continuous matrix which controls
most properties such as modulus and thermal mechanical behaviour.
The dispersed phase, in which prevails the rubbery component, tends
to form small domains improving impact strength, stress cracking
resistance and ductility of the matrix.

According to recent studies the rubbery domains presumably stop the craze growth which can yield the formation and development of failure cracks during impact[1]. Energy is in fact absorbed by the formation of crazes, without a macroscopic fracture: the rubbery domains acting as stress concentrators, keep a barrier to the extension of crazing[2].

The following factors play a decisive role assuring an optimum toughness in an heterogeneous blend system:

i)   size, shape and internal constitution of rubbery domains
ii)   volume fraction of rubber
iii) interfacial tension and adhesion between the disperse phase and matrix[3].

High interfacial tension contributes, together with high viscosities, to the inherent difficulty of imparting a sufficient degree of dispersion and to facilitate gross separation or stratification process. Poor adhesion, between dispersed particles and matrix, leads to preferential failure pathways and hence to very weak and brittle materials with respect to their mechanical properties[4]. The presence of block or graft copolymers of appropriate chemical structure in heterogeneous blends provides a lowering of the interfacial energy of the phases and an improvement of the interfacial adhesion[4]. The final morphological features is a reduction of a particle sizes of the dispersed phase in the blend. It is belevied that the graft or block copolymer tends to dispose itself at the interface, contributing to stability against segregation during processing, too[5]. Graft copolymers, able to act as interfacial agents, can be generated directly during the mixing, adding to a binary alloy a third component bearing functional groups reactive towards one of the components and soluble in the other. It is well established, now, that in order to improve the impact resistance of plastic materials, grafting reactions are a necessary factor.

Polyamides, as many other semicrystalline polymers, have poor impact resistance especially at temperatures below T_g and in the dry state. Their impact strength may be improved developing satisfactory methods to incorporate a rubber-toughening agent. Rubber modified nylon 6 and nylon 66 containing, as a disperse phase, a cross-linked elastomer grafted to the matrix, were prepared by Chompff[6]. These materials have improved tensile elongation and impact strength both for dry and as-molded specimens. Kray and Bellet[7] described in their patents the synthesis of a graft poly-

amide-hydrocarbon polymer. This copolymer (I), obtained reacting
an ethylene/ethylacrylate copolymer with ε-caprolactam at 255°C,
shows elongation at break and impact strength values higher than
those of nylon and nylon/(I) blends. Nylon/polyolefins alloys have
also studied in order to improve impact and shock resistance in
nylon items[8,9,10]. Ide and Hasegawa[11] have shown that dispersion
degree and the mechanical properties of nylon-6/polypropylene and
nylon-6/polystyrene blends may be markedly improved adding a certain
amount of a maleic anhydride-grafted polyolefin. The formation of
graft polymers between maleic anhydride in polyolefins and terminal
amino groups of nylon 6 generated during the mixing seems to be
responsible of this behaviour.

In the present paper we describe the modification of random
ethylene-propylene copolymers (EPM) by grafting functional groups
reactive towards $-NH_2$ end groups. The preparation and the properties
of nylon 6/EPM blends are also described. The properties of these
blends are compared both with those containing a third component
(i.e. a modified EPM) and with those constituted only by nylon 6 and
the modified EPM.

SYNTHESIS OF INTERFACIAL AGENTS

We have modified Dutral Co/54 (Du), an amorphous ethylene-
propylene copolymer, 53% of C_2 by weight (Montedison Spa) by
grafting onto the backbone groups such as $-POCl_2$, $-PO(OCH_3)_2$ or
succinic anhydride able to react, during the melt mixing, with the
$-NH_2$ end groups of the polyamides according to the following scheme
analogous to that reported by Ide and Hasegawa in their paper on
polypropylene/nylon 6 blends[11]:

$$
\begin{array}{c}
\left\{ \begin{array}{l} CH_2-C\diagup\!\!\!\!\diagdown O \\ CH-C \end{array} \right\}\!\!O \;+\; H_2N-(Nylon) \;\longrightarrow\; \left\{ \begin{array}{l} CH_2-C-NH-(Nylon) \\ CH-C-OH \end{array} \right\}
\end{array}
$$

Alternatively, the Dutral modified as above, was also used as
substrate for the preparation of a Dutral-g-nylon 6 copolymer
taking advantage of the effect of ester and anhydride groups on
initiating the anionic polymerization of ε-caprolactame.

a) Chlorophosphonation of Dutral

The grafting of the $-POCl_2$ group on Dutral was performed in
PCl_3 by a modification of the procedure reported for EPM[12] which
involves a free radical chain reaction of the polymer with PCl_3
and oxygen according to the stoichiometry:

$$RH + 2PCl_3 + O_2 \rightarrow RPOCl_2 + POCl_3 + HCl$$

$$(I)$$

The phosphonylation degree was followed by titrating the developed
HCl. When the desired amount of grafting was reached, PCl_3 was
distilled off while anhydrous o-dichlorobenzene was slowly added.
Chlorophosphonyl groups were subsequently converted to more stable
dimethylphosphonate groups $-PO(OCH_3)_2$ by reaction of (I) with
anhydrous methanol during 24 hours. The polymer was recovered by
precipitation in methanol.

The $Du-g-PO(OCH_3)_2$ was characterized by elemental analysis and
IR spectroscopy. In a typical run the content of phosphorous found
by elemental analysis was 1,35% (calcd 1,67%) thus indicating an
average distribution of one phosphonyl group every 130 chain carbon
atoms. Infrared absorption peaks characteristic of methylphosphonate
groups, were observed at 820 (w) 980 (m), 1040-1060 (m), 1190-1210
(m) and 1255 (sh) cm^{-1}.

b) Preparation of Dutral-g-succinic anhydride (Du-g-SA).

The addition of maleic anhydride to hydrocarbon polymer chains
is generally accomplished both by thermal and free radicals
reactions[13]. In the presence of an organic peroxide the reaction
proceeds by extraction of an hydrogen radical from the hydrocarbon
chain which subsequently generates a σ bond with a carbon atom of
the double bond of the anhydride. In the case of EPM copolymers
are present methyl (1), methylene (2) and methine (3) hydrogen in
different relative amounts. Although the case of hydrogen abstraction is 3>2>1, the insertion of the hanydride does not occur preferentially at the $-CH$ carbon atom owing to the different population
of 1,2 and 3 in Dutral[14]. We have carried out the grafting reaction
by refluxing a xylene solution of Dutral with the appropriate amount
of benzoyl peroxide and an excess of maleic anhydride.

In a typical procedure, 7.98 g of Dutral and 9.20 g (92 mmol) of freshly sublimed maleic snhydride were dissolved in 400 ml of dry xylene by refluxing 2 hours; successively 0.79 g (3.26 mmol) of benzoyl peroxide in 23 ml of xylene were added within 15 minutes and the solution was refluxed for additional 3 hours. The Dutral-g-succinic anhydride (Du-g-SA) was precipitated in dry acetone, repeatedly washed with 100 ml portions of acetone and dried in a vacuum oven. Modified polymers with different amounts of succinic anhydride (up to 6% by weight) were prepared, depending on the weight ratio BP/Dutral. The anhydride contents were determined by titration of a o-dichlorobenzene solution of the polymer with a dilute (0.05M) solution of NaOH in o-dichlorobenzene/ethanol mixture (9/1 v/v). The IR spectra performed on films obtained by slow evaporation of a toluene solution showed absorptions at 1865, 1785 and 1225 cm^{-1} characteristic of the anhydride group. The modified polymers were stored in a dry atmosphere in order to avoid the hydrolysis of the anhydride group into carboxylic groups. The hydrolysis is practically complete within 24 hours at 100% RH as shown by the disappearance of the bands at 1785 and 1865 cm^{-1} and by the presence of a new strong peak at 1710 cm^{-1} characteristic of a carboxylic group.

c) <u>Synthesis of Dutral-g-Nylon</u>

Grafting of Nylon 6 chains by anionic polymerization onto modified polystyrene and polyethylene has been described by Matzner et al.[15] and by D. Braun and Eisenlhor[16] respectively. The grafting of Nylon 6 onto Dutral was performed by us, using the Du-g-SA, ε-caprolactame (CL) and sodium caprolactame (NaCL) as initiator. The reaction was carried out in the bulk under nitrogen atmosphere and N-acetylcaprolactame (A-CL) was added as activator in order to improve the polymerization rate. The reaction of ε-caprolactame with the anhydride leads to the formation of the N-carbamoylcaprolactame group, able to initiate the anionic ring opening polymerization of the lactame.

Using samples of Du-g-SA of medium grafting degree (3% SA) we obtained Du-g-nylon 6 copolymers having nylon contents 10÷20% by weight depending on the molar ratio NaCL/ACL. The highest grafting degree 20% was obtained by reacting 2.00 g of Du-g-SA with 12.0 g of CL, 0.044 g of NaCL and 0.140 g of ACL at 180°C to yield a mixture of Du-g-nylon (2.15 g) and high molecular weight nylon 6 which was separated from the graft copolymer by repeated formic acid extractions at room temperature. Only trace amounts of Du or Du-g-SA were found in the mixture by xylene extraction. The Du-g-nylon 6 copolymers were characterized by elemental analysis (N content), IR spectroscopy and by DSC analysis. The IR spectra showed characteristic amide bands at 3300,1640 and 1545 cm^{-1}. The DSC thermogram showed, in the case of 20% graft copolymer, an endotherm at 493 K, very close to the T_m of nylon 6, suggesting the formation of polyamide side chains of relatively high molecular weight. This finding is in good agreement with the high viscosities of the nylon 6 homopolymer extracted (η_{inh}=0.6÷1.00 dl/g in m-cresol at 25°C). The obtained Du-g-nylon 6 copolymers are insoluble and do not flow in the melt and this indicates some degree of crosslinking.

PREPARATION AND MECHANICAL PROPERTIES OF ALLOYS

The different alloys, whose compositions are indicated in Table 1, have been prepared at 260°C in a Brabenderlike apparatus (Rheocord manufactured by Haake inc.) by melt mixing nylon 6 with Dutral and/or Du-g-SA. Successively, from such alloys sheets of 0.7 and 3.2 mm. of thickness were obtained by compression molding at 230°C and 180 atm. From the 0.7 mm. thick sheets, dumbell shaped specimens have been cut to perform mechanical tensile tests by an Instron Machine. From the 3.2 mm. thick sheets, parallelepidal specimens (3.2x13x50 mm.) were obtained by using a milling machine to perform Izod impact tests. A notch of 0.75 mm. has been made on such specimens. The materials have been conditioned to obtain in all the samples the same amount of absorbed water (about 3% by weight). The procedure is: the specimens were immersed in water at 90°C for a given period of time depending on their thickness (1hr/mm. of thickness) and then kept for 5 days in a polyethylene sealed envelope. In this way the materials absorbed the same amount of water as if they had been exposed to a 50% RH until equilibrium conditions were reached. Stress-strain curves, for all the blends reported in Table 1, have been obtained by an Instron machine, at room temperature and at a cross-head speed of 10mm/min. Also Izod impact tests have been performed on these blends.

The results in terms of the modulus E, the ultimate tensile strength σ_u, the elongation at break ε_u and of the Izod impact resistance are reported in Table 1 as a function of the blend composition.

TABLE 1

Mechanical properties of the investigated blends as function of composition.

Code	Ny	A1	A2	B1	B2	C1	C2
Compo-sition	Ny 100%	Ny 90% Du 10%	Ny 75% Du 25%	Ny 87% Du 10% DugSA 3%	Ny 90% Du 5% DugSA 5%	Ny 90% DugSA 10%	Ny 80% DugSA 20%
E (Kg/cm^2)	5.2×10^3	2.8×10^3	4.0×10^2	3.5×10^3	3.5×10^3	3.3×10^3	2.4×10^3
σ_u (Kg/cm^2)	5.8×10^2	3.1×10^2	8.5×10^1	3.1×10^2	3.4×10^2	2.7×10^2	1.6×10^2
ε_u	3.0	1.5	0.4	1.1	1.3	0.7	0.2
R (J/m^2)	1.2×10^4	1.9×10^4	1.4×10^4	2.3×10^4	4.5×10^4	1.0×10^5	$1.2 \times 10^5 (*)$

(*) This value is probably still lower than the actual one.

Such statement is supported by the fact that the specimen failure did not occur in corri-spondence of the notch section but elsewhere, because of same inhomogeneity present in the blend specimen.

For all the blends the necking phenomena exhibited by the Nylon homopolymer at the operating conditions, disappears showing that the cold-drawing of the specimens is strongly hindered by the presence of the other components in the blend.

The type A alloys consist of mechanically melt-mixed blends of Nylon (Ny) and of a rubbery ethylene-propylene copolymer (Du). Their Young modulus decreases appreciably as the Nylon content decreases. This can be due to the softening effect of the rubbery copolymer, which lowers on the other hand also the overall crystallinity fraction of the whole blend, which depends, of course, only on the Ny content. Furthermore since Du and Ny are probably incompatible, type A alloys should be characterized by a Ny matrix containing Du dispersed particles with a poor adhesion between the two phases. Such morphology could hinder the Ny homopolymer cold-drawing, making disappear the neck formation as previously mentioned. Also the ultimate properties σ_u and ε_u strongly decrease as the Du content increases. The Izod impact strength R values of binary blends are on the contrary slightly higher than those of pure Ny. The fact that Du imparts only a slight toughening to Ny is to be probably ascribed to the poor adhesion between the disperse phase and the matrix.

The type B blends consist of Nylon, Dutral and of the copolymer Du-g-SA. During the mechanical mixing at 260°C the $-NH_2$ Nylon chain ends may easily react with the anhydride groups giving a graft copolymer which tends to act as an interfacial agent. The Young modulus E, the strength σ_u and the elongation at break ε_u have values only slightly lower than those corresponding to pure Ny. Instead the Izod impact strenght R increases noticeably passing from pure Ny to blend B1 and then to blend B2. The results of Table 1 show only slight differences in the tensile mechanical properties of type A1 and B1 blends which have almost the same Ny content. A possible interpretation of the R improvement obtained passing from blend B1 to B2 could be given by the fact that the ratio of Du-g-SA/Du in blend B1 (30/100) is too low with respect to the same ratio in blend B2 (50/50). Therefore the graft copolymer in blend B1 may cover a small extent of the interface between the polyamide and the rubber phase, whereas in blend B2 its content may be sufficient for an effective interfacial action. All the other mechanical properties do not vary appreciably with respect to the type A1 blend.

The behaviour of type C blends seems more interesting; in fact

E, σ_u and ε_u decreases slightly, whereas R increases strongly as
the Ny content decreases. The lowering of the Young modulus can be
attributed to the softening effect of the rubbery copolymer domains
as for type A and B blends. The σ_u and ε_u decrease can be due to
the quasi-network formed by the graft copolymer interconnections
throughout the Ny matrix which makes the materials less resistant
to high deformation values. The increase in R values is presumably
due to the reduction of the rubbery domains and to the strong adhesion existing between the dispersed phase and the Ny matrix.

CONCLUSIONS

    The results presented in this paper can only be considered as
preliminary data and starting point to develop this research further.
The simple inclusion of an incompatible rubber in Nylon 6 (type A
blends) doesn't seem to improve significantly the impact strength.
The best results have been obtained for the binary blends Ny/Du-g-SA
which however require a large amount (10-20%) of the expensive
Du-g-SA additive. As far as the ternary type B blends are concerned
it is to be remarked that the impact behaviour could be improved by
chosing a suitable amount of Du-g-SA with respect to the Dutral
content and the best processing conditions to optimize the grafting
reaction. The work is in progress in our Institute to find the
right conditions of processing and the best compositions in order
to obtain overall morphologies able to impart the best properties
to the material.

ACKNOWLEDGEMENTS

    The authors wish to acknowledge SNIA VISCOSA and MONTEDISON
S.p.A. for supplying polymers and Dr. Casale and Prof. Dall'Asta
(SNIA VISCOSA) for stimulating discussions. One of us, M. Avella,
acknowledges SNIA VISCOSA for financial support.

REFERENCES

1. R. D. Deanin, ACS Polymer Preprints, <u>14</u>, (2) 728 (1973).
2. C. B. Bucknall, "Toughened Plastics", Applied Science Publ. LTD
    London, 1977.

3. C. B. Bucknall, A. S. Argon and M. M. Salama, Phil. Mag., 36, 1217 (1977).

4. D. R. Paul, "Interfacial agents for polymers" from "Polymer Blends", by. D. R. Paul and S. Newman Eds, Academic Press, New York, 1978.

5. a) W. M. Barentsen, D. Heikens and P. Piet, Polymer, 15, 122 (1974).
    b) D. Heikens, N. Hoen, W. M. Barentsen, P. Piet and H. Landau J. Polym. Sci., Polymer Symposium, 62, 309 (1978).

6. A. J. Chompff "Rubber modified polyamides", U.S. Patent 3, 380, 948 (April 29, 1975).

7. R. J. Kray and R. J. Bellet "Grafted polyamides hydrocarbon polymers and processes to prepare them", French patent 1, 470, 255 (Febr. 17, 1967).

8. D. Braun and U. Eisenlohr, Kunststoffe, 65, 139 (1975).

9. G. Illing, Kunststoffe, 7, 275 (1968).

10. U.S.A. Pat. 248.229 (1962).

11. F. Ide and A. Hasegawa, J. Appl. Pol. Sci., 18, 963 (1974).

12. C. Leonard Jr., W. E. Loeb, J. H. Mason and J. A. Stenstron, J. Polym. Sci., 55, 799 (1961); J. Appl. Polym. Sci., 5, 157 (1961).

13. J. Le Bras, R. Pautrat and C. P. Pinazzi, in "Chemical Reactions of Polymers" (High Polymers vol. XIX), E.M. Fettes Ed., Chapter II-G, p. 202-219, Wiley Interscience, New York, 1964.

14. F. P. Baldwin and G. Ver Strate, Rubber Chem. Technol., 45, 710 (1972).

15. a) M. Matzner, D. L. Schober, R. N. Johnson, L. M. Roberson and J. E. Mc Grath, "Permeability of Plastic Films and Coatings", H. B. Hopfenberg ed., Plenum Publ. Corp., New York, 1974.
    b) M. Matzner, D. L. Schober and J. E. Mc Grath, Europ. Polym. J., 9, 469 (1973)

16. D. Braun and U. Eisenlohr, Angew. Makromol. Chem., 58/59, 227 (1977).

TENSILE PROPERTIES AND MORPHOLOGY OF COPOLYMER MODIFIED
BLENDS OF POLYSTYRENE AND POLYETHYLENE

S.D. Sjoerdsma, A.C.A.M. Bleijenberg and D. Heikens
Eindhoven University of Technology, Netherlands

1. INTRODUCTION

The mechanical behaviour of heterogeneous polymer-polymer blends
depends on the mechanical properties of the polymers, on the morphology of the system and on the degree of the mechanical contact between the phases. The three aspects will be explained and illustrated on the basis of results of investigations on the system poly-
styrene/low density polyethylene (PS/PE) to which graft or blockcopolymers of PS and low density polyethylene (PS-g-PE or PS-b-PE) were
added. Although different kinds of mechanical experiments at different speeds and temperature have been carried out, only results
of tensile experiments at room temperature will be treated here.

In section 2 results of tensile experiments of graft copolymer
modified PS/PE blends will be discussed in terms of the morphology
of the blends which in turn depends on the concentration of the dispersed phase and on the amount of graft copolymer added.

In section 3 will be shown that curves of modulus versus concentration gives information about the morphology of the blend and
about the number of phases present. Deformation mechanism as revealed by measurement of volume changes during tensile deformation
is discussed in section 4, together with the anomaly of the Poisson
ratios in case of weak mechanical interactions between the phases.

2. TENSILE EXPERIMENTS AND MORPHOLOGY OF GRAFT COPOLYMER MODIFIED
   PS/PE BLENDS[1,2,3,4]

Materials

Blends were made from polystyrene, Styrene 666E (Dow), low
density polyethylene, Stamylan 1510 (Dutch State Mines) and especially prepared graft copolymer.

This graft copolymer of PS and PE was prepared by a Friedel-
Craft reaction between the homopolymers dissolved in cyclohexane
under the influence of $AlCl_3$. The reaction was stopped by adding
a T_x isopropanol. Small amounts of substance consisting out of un-
reacted PS and PE were removed from the precipitated reaction product
by sequential extraction with ethylacetate and n-heptane. The pure
graft copolymer consisted of 47 percent VPS as determined by infrared
spectroscopy and pyrolysis experiments and was soluble in hot cyclohexane.

Preparation of blends

Blends of PS, PE and of graftcopolymer were made on a laboratory
mill at 190°C. The circumferential speed for the two rolls was 6
and 18 cm sec^{-1} and the distance between them 0.08 cm. The blending
time was 7 minutes. Melt-index measurements of the pure components
subjected to the same procedure showed that practically no degradation took place during the blending operation. Blends of PS and PE
were made containing 0, 2.5, 5, 10, 25, 40, 50, 60, 80, 90, 95, 100
percent by weight of PE. In two PS-rich series, i.e blends containing 2.5, 5, 25 and 40 percent of PE, 5 and 30 percent of the dispersed phase PE was replaced by the graft copolymer. In two PE-rich
series, i.e. blends containing 60, 75, 90 and 95 percent of PE, 5
and 30 percent of the dispersed PS was replaced by graft copolymer.

Tensile experiments

Tensile deformation experiments were carried out on an Instron
tensile tester on samples prepared according to ASTM D-1708. The
rate of deformation was 0.40 min^{-1}.

Stress strain curves of PS-PE blends depend on the morphology, and thus somewhat on the conditions of the melt-blending. By standardizing the melt-blending the results were well reproducible.

Results and discussion

Morphology

The dispersed phases consisting of PE in the PS-rich blends and of PS in the PE-rich blends formed spheres in blends containing less than 30 percent or more than 60 percent of PE. Between these percentages semi-continuous to continuous phases are present. Under blend preparation conditions PE seems to have a stronger tendency to form these continuous phases than PS. Replacing 5 to 30 percent of the dispersed materials by graft copolymer, appears to broaden the inversion region ranging from 30–60 percent of PE to one ranging from 20–75 percent of PE.

As shown in Figure 1 the dimensions of PE particles in PS-rich blends and of PS particles in PE-rich blends diminish greatly upon addition of graft copolymer to the blends. The fact that already small amounts of graft copolymer (5 percent of the dispersed phase concentration) have these strong effects in an indication for the presence of these copolymers on the interface. The olefin segments will "dissolve" in the PE particle phase whereas the PS-chains will protrude in the PS matrix and _vice versa_ (Figure 2). It could be expected also that in this way the strength of the interface would be enhanced. Macroscale adhesion experiments carried out with hot sealed PS and copolymer films and with hot sealed PE and copolymer films and on laminates of three films of PE, of copolymer and of PS confirmed the adhesive properties of the copolymer[1].

Deformation behaviour in tensile tests

For blends with PE contents between 0 and 40 percent the yield elongation as funtion of PE concentration goes through a very slight minimum, the elongation at break however passes through a profound maximum (Figure 3). The increase of the elongation at break with increasing PE content can easy be explained by the increase of the number of PE particles which will induce more crazing and void formation. Whereas PS breaks in a brittle way, stress-strain curves

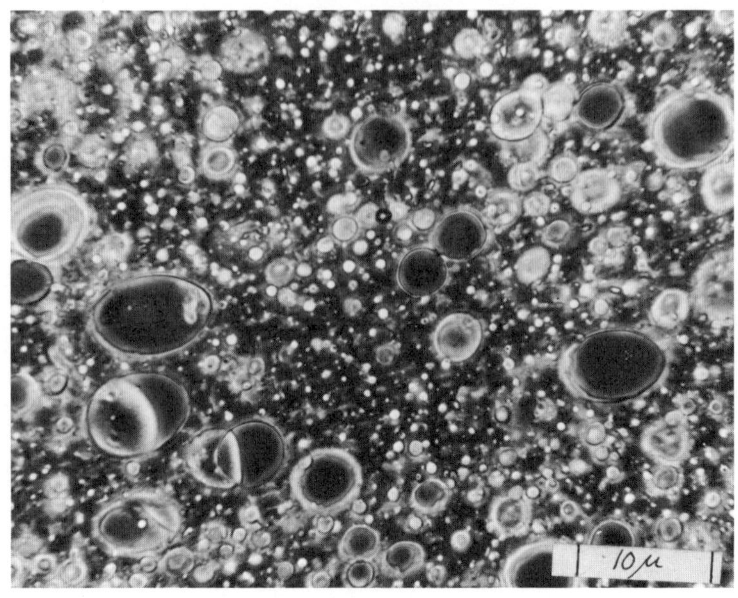

(a)

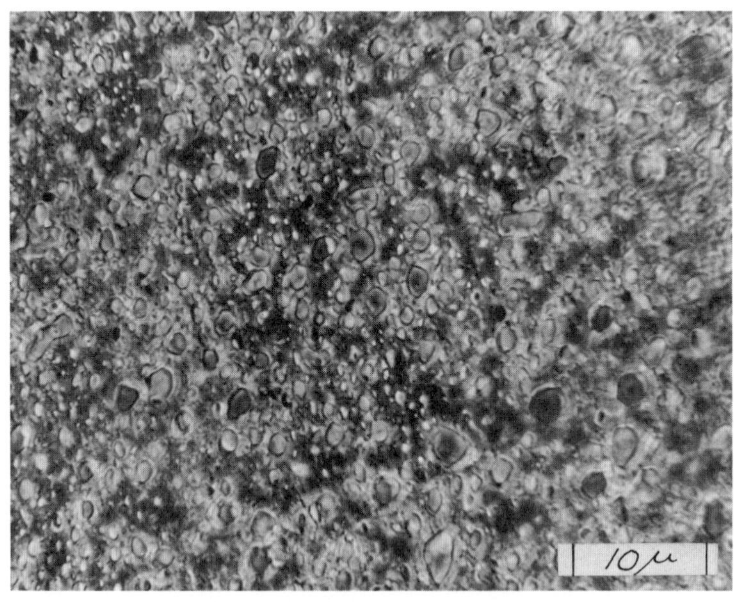

(b)

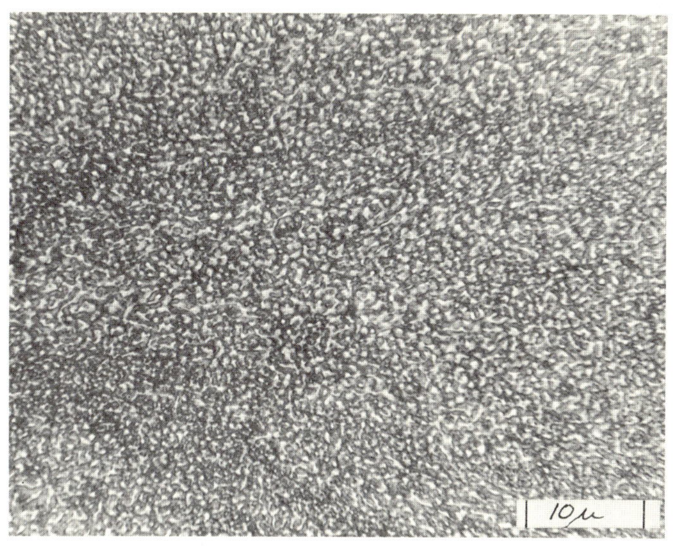

(c)

Figure 1. Phase contrast micrographs of blends of PS/PS-g-PE/PE
of compositions:

1a. 75/0/25   1b. 75/1.25/23.75   1c. 75/7.5/17.5

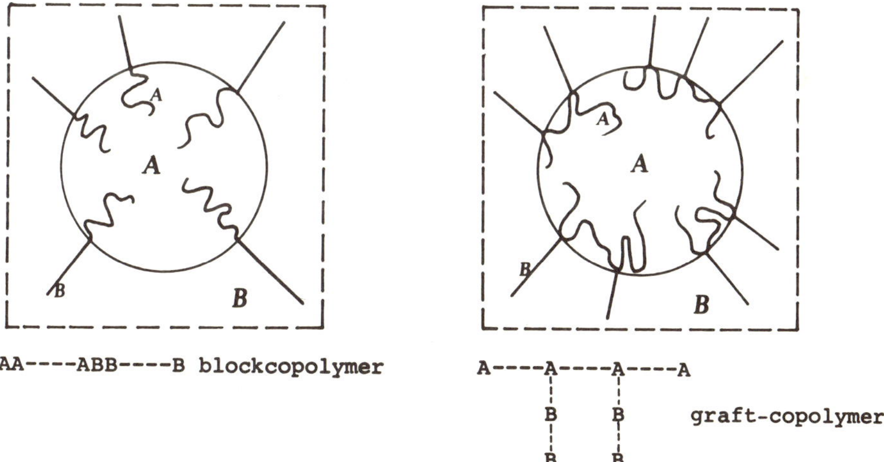

Figure 2. Block and graft copolymer at the interface on a particle
in polymer blends.

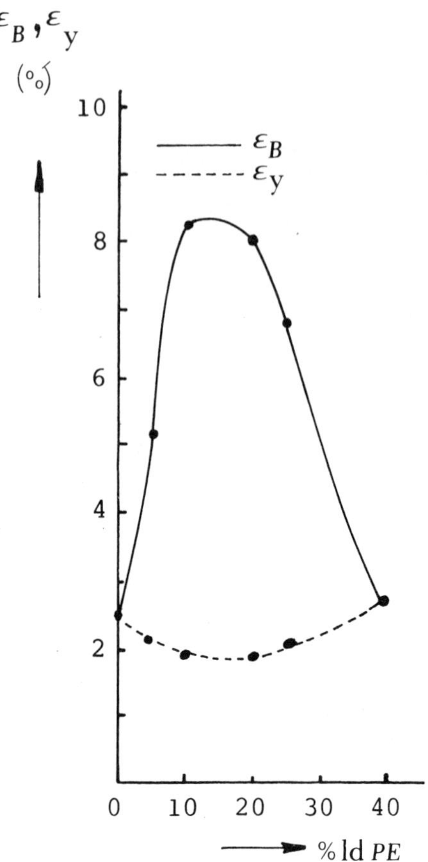

Figure 3. Elongation at yield ε_y and at break ε_B of PS/PE blends.

of blends containing 5-25 percent of PE show an upper and a lower
yield point. In a blend containing 40 percent of PE, however, so
many crazes will be formed that they run into each other causing
the failure of the material before yielding. Replacing 5 or 30
percent of the PE in blends containing between 0-40 percent of PE
by copolymer leads to an increase in tensile and yield strength
(Figure 4) of 20-40 percent. The increase in the tensile and yield
strength can be explained by better adhesion between the dispersed
phase PE and the PS-matrix and thus by an increase of effective

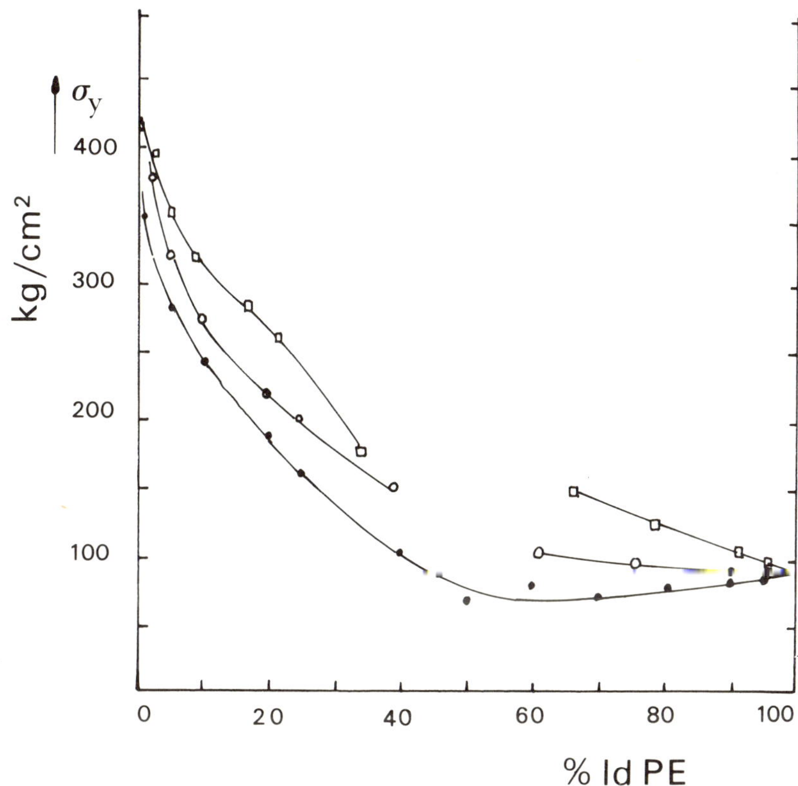

Figure 4.   Yield stress of PS/PE blends.   Series (·) contains
no graft copolymer.   Series indicated (o) and (□)
contained 5 and 30 percent of graft copolymer PS-g-PE.
The latter percentages expressed relative to the
dispersed phase material PS (or PE) + graft copolymer.

load bearing cross section. These large effects indicate again that
the graft copolymer must be present at the interfaces between the
PE-particles and the PS-matrix. Fracture surfaces of PS-rich PS/PE
blends show surface irregularities on PE particles when graft copolymer was present in the blends while smooth surfaces are observed
in blends containing no copolymer (Figures 5 and 6). These irregularities on the surfaces are interpreted as indications of connections between PS and PE that were broken during the fracture process.
In case of copolymer containing PE-rich blends fibrous PE connections
often survive the fracture process and the PS particles remain connected to the PE matrix (Figure 7 and 8).

The yield elongation of PE-rich blends containing 0-30 percent
of PE decreases with increasing PS content apparently by the formation of the many oblong voids in the direction of stress around the
non deformed PS particles. The tensile and yield strength of PE-
rich blends increases with addition of graft copolymer as can be
expected of the increased adhesion between the phases which leads
to larger bearing cross section.

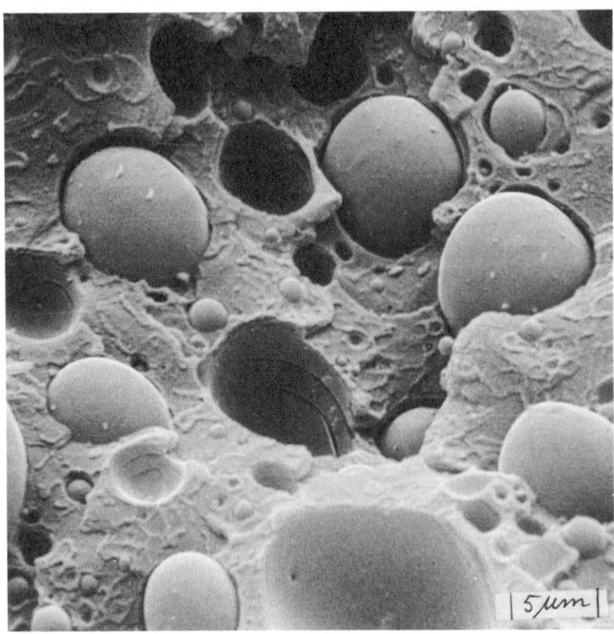

Figure 5.   Smooth surfaces of PE particles on a fracture surface
            of a PS/PE blend containing no graft copolymer (PS/PE,
            75/25).

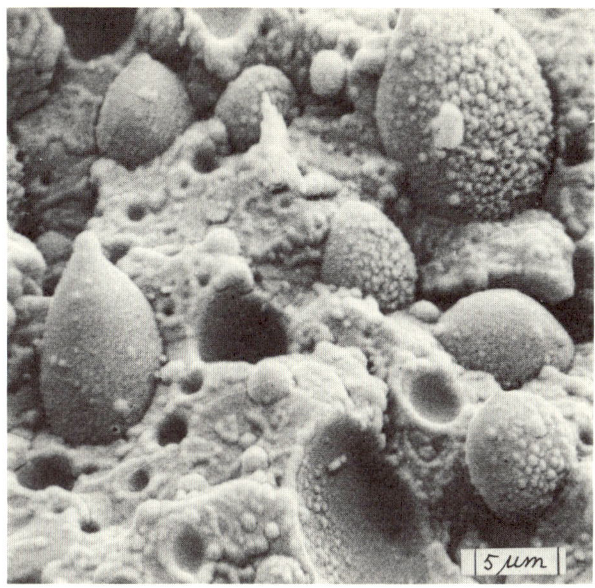

Figure 6. Irregularities on surfaces of PE particles on a fracture surface of a PS/PS-g-PE/PE blend (75/1.25/23.75).

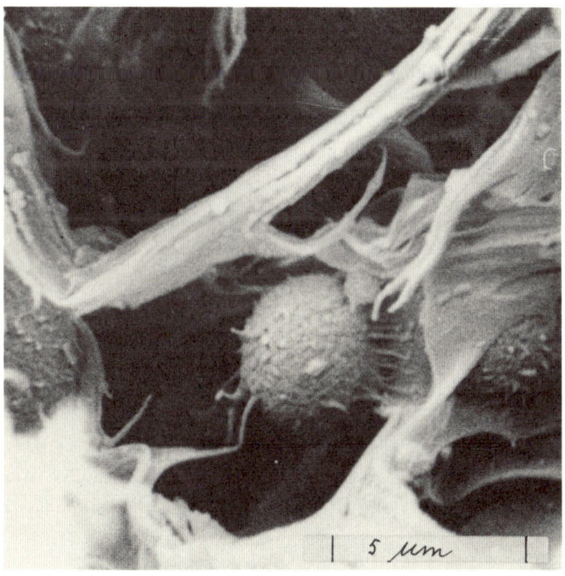

Figure 7. Smooth surfaces of PS particles on a fracture surface of PS/PE blend containing no graft copolymer (PS/PE, 30/70).

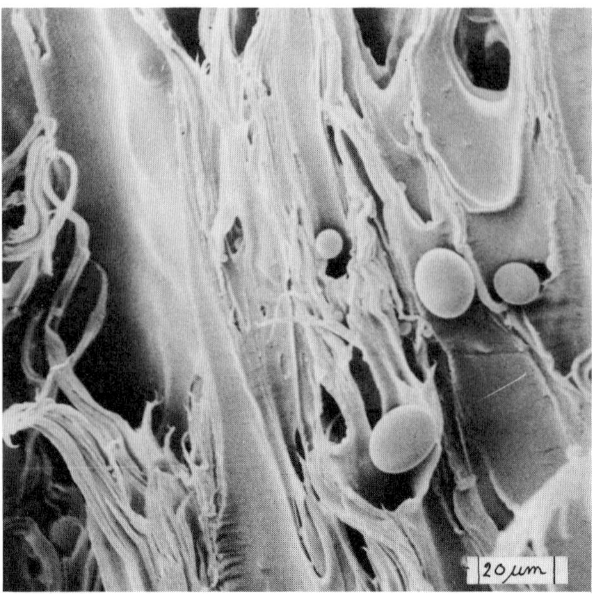

Figure 8.   Surfaces of PS particles with fibrous PE protrusions on a fracture surface of a PS/PS-g-PE/PE blend (17.5/7.5 /75).

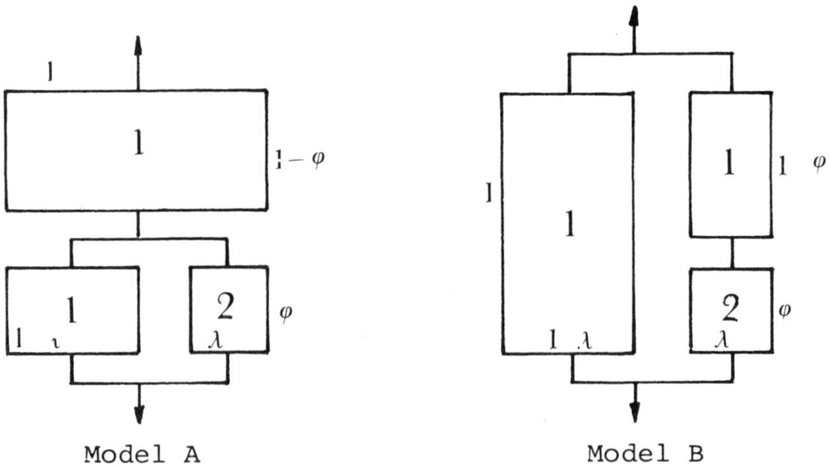

Figure 9.   Two dimensional Takayanagi models a and b (partly parallel and partly series).

Tensile modulus

For the modulus-concentration behaviour as found for PS-PE
blends the two dimensional Takayanagi models could be tried[5]. The
Takayanagi models can be looked upon as combinations of the well
known series and parallel models of Paul[6] (Figure 9). However, Bohn[7]
discussed the Takayanagi models as two dimensional unit cell descrip-
tions of a material consisting of a regular array of dispersed phase
in a matrix.

Barentsen modified the two dimensional models by proposing three
dimensional versions (Figure 10) cubes with sides λ dispersed regular in a matrix and supposed these to be a better description of a
three dimensional real blend (Figure 10)[1,8]. As intuitively the
model will hold only for dispersed systems, the two combined series
and parallel models will produce four curves, two for PS dispersed
in a PE matrix and two for PE dispersed in a PS matrix. The equations are, for a system with a dispersed phase 2 in a matrix 1,

$$\frac{E_A}{E_1} = \frac{1 + (R - 1)(1 - \lambda_2^2)}{1 + (R - 1)\left| 1 - \lambda_2^2(1 - \lambda_2)\right|} \tag{1}$$

$$\frac{E_B}{E_1} = \frac{1 + (R - 1)\lambda_2(1 - \lambda_2^2)}{1 + (R - 1)\lambda_2} \tag{2}$$

and for a system with a dispersed phase 1 in a matrix 2,

$$\frac{E_A}{E_2} = \frac{1 + (R - 1)\lambda_1^2}{1 + (R - 1)\lambda_1^2(1 - \lambda_1)} \tag{3}$$

$$\frac{E_B}{E_2} = \frac{1 + (R - 1)\{1 - \lambda_1(1 - \lambda_1^2)\}}{1 + (R - 1)(1 - \lambda_1)} \tag{4}$$

$R = E_1/E_2$, and λ_1^3 and λ_2^3 are the volume fractions of the dispersed
phases 1 and 2, respectively. Curves have been calculated for PS/PE

blends (E_{PS}/E_{PE} = R $\tilde{=}$ 12) for the relative moduli with E_{PS} = 1 and
are presented in Figure 11 together with the experimental results.
As has been shown in a later paper the good fit of the model for
PS-rich blends is partially caused by the very low modulus value of
PE[9]. The values for blends containing PS, PE and small amounts of
graft copolymer, not presented in Figure 11, are also represented
by the same curve if, as the concentration of PE is taken the sum
of the concentrations of free PE and of bound PE present in the
graft copolymer. This result is thus in accordance with the model
of Figure 2 where all copolymer is present at the PS/PE interface.
Another interesting feature is, at present, the form of the curve.
The curve at high PS content is concave. If the concentration of
PE gets higher than 25 percent, however, the curve bends downwards,
and this coincides with the beginning of phase inversion as observed
in micrographs of cross sections and in scanning electron microscope
pictures (SEM) of fracture surfaces (Figure 12). This formation of
semi-continuous phases of PE for concentrations higher than 25 percent leads to progressively low values of the modulus. So here the
sudden decrease of the modulus (as can be seen the modulus concentration curve of Figure 11) will be a clear indication for phase
inversion. This observation will be quite general. The use of the
modulus concentration curves as a means for studying the morphology
and also for determining the number of phases present in a system,
will be treated in the next section.

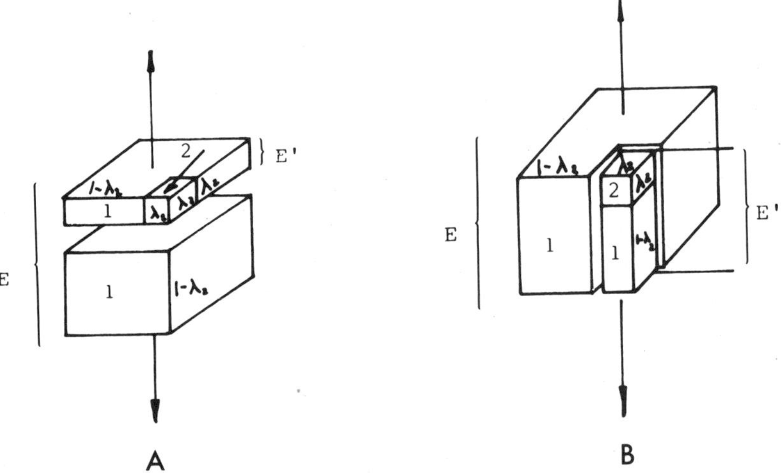

Figure 10.   Three dimensional Barentsen models a and b (partly pa-
             rallel and partly series).

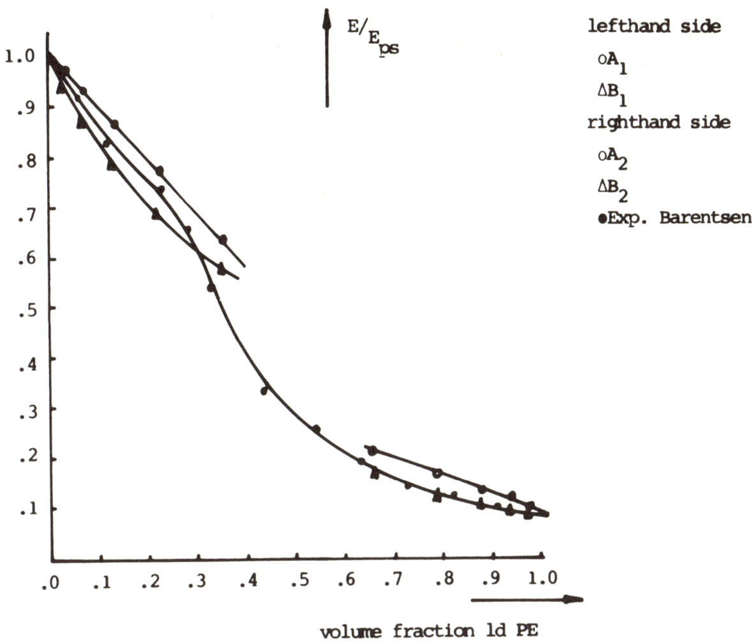

Figure 11.   Relative Young's modulus of PS/ldPE blends.
             Δ,o Models of Barentsen
              .   Experiments of Barentsen

3. MODULUS-CONCENTRATION CURVES AND THE MORPHOLOGY OF PS/PE BLENDS TO WHICH GRAFT OR BLOCK COPOLYMERS ARE ADDED

<u>Materials</u>

Graft copolymers

Hoen added graft and block copolymers of PS and PE to PS-rich PS/PE blends[8,10]. The graft copolymers were prepared as described before, by a Friedel-Craft reaction. The atactic PS used for preparation of the grafts, however, was not prepared by anionic polymerization. A PE-like polyolefin was prepared by polymerising butadiene with sec.-BuLi in toluene at 30-40°C followed by hydrogenation.

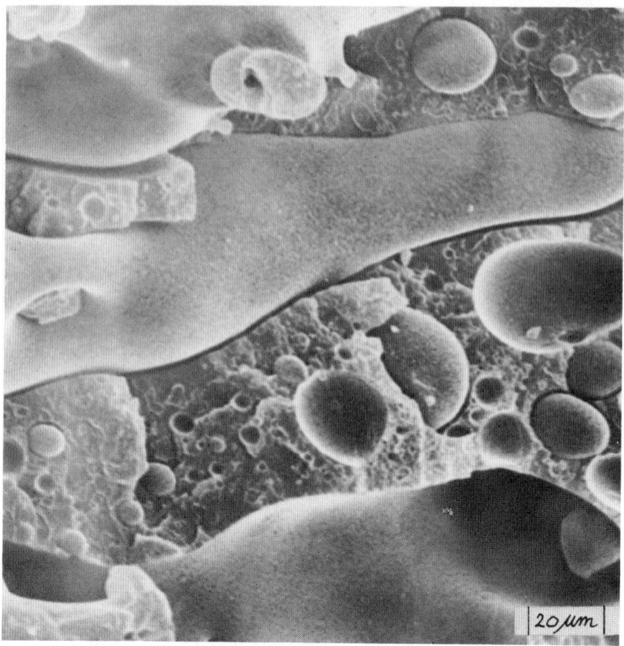

Figure 12.   Fracture surface of PS/PE blend with semi-continuous
             smooth PE phases (PS/PE, 60/40).

        The microstructure as determined by means of I.R.-spectroscopy
was:
        trans-vinylene 54.6 mole percent;
        cis-vinylene   33.8 mole percent;
        vinyl   11.6 mole percent.
The polymer was hydrogenated with the catalyst system NiDIPS/Al
(iBu)$_3$ (mol ratio 1.4) up to measured low degrees of unsaturation
between 4 and 0.5 percent. By reacting these polymers in chosen
ratios with PS and AlCl$_3$, graft copolymers of PS and "PE" of different complexity were made. By many elaborate small scale reactions,
Hoen had confirmed[11] that the reaction between PS and 1dPE was a
normal network reaction of the type discussed by Stockmayer and
Gorden[12,13]. Using this result he could calculate the composition
and complexity of the different copolymers which were prepared for
modifications of the PS/PE blends. The structures of these graft
copolymers are presented in Table 1.

TABLE 1

Structures of graft copolymers of PS and PE

Number	Type	wt % PS	M_n	M_w/M_n	gel %
GC1	Regular graft copolymer. 2/3 consists of a PE backbone (45.000) with, on the average, four PS branches (10.000)	53	110.000	1.4	0
GC2	Complex graft copolymer consists mainly of large complex finite molecules	80	450.000	8.5	0
GC3	Partly cross-linked graft copolymer. Mixture of infinite network (gel) and smaller finite molecules.	71	---	–	59
GC4	Cross-linked graft copolymer	60	---	–	100

Block copolymers

Three block copolymers were also prepared (BC1, BC2 and BC3). A PE/PS-diblock copolymer was obtained by hydrogenation of a PB/PS-diblock copolymer, that was prepared by a successive anionic polymerization (n-BuLi) of butadiene and styrene. The polybutadiene block contained 13 mole % of vinyl groups, so that the short chain branching of the hydrogenated block (with 32.5 ethylbranches/1000 C-atoms) is comparable with normal low density PE. PB/PS-diblock copolymers Solprene 1205 and Solprene 410 from Phillips Petroleum Co. were hydrogenated to the corresponding PE/PS copolymers BC2 and BC3. The diblock character of these copolymers is only partial, which means that the transition between the PS-block and the PB-block (PE-block after hydrogenation) is formed by a block of more or less random character. This specific structure is caused by a different polymerization technique. Table II contains the data for the block copolymers used.

TABLE 2

Structure of block copolymers of PS and PE

Number	Type $M_n \times 10^{-3}$	wt % PS	M_n
BC1	Diblock copolymer PS-PE 65-48	58	113.000
BC2	Partial diblock PS-(PS/PE)$_{random}$-PE 14-10-54	25	78.000
BC3	Partial diblock PS-(PS/PE)$_{random}$-PE 22-22-25	48	69.000

Modulus curves and the number of phases of copolymer modified PS/PE blends

Tensile tests were performed on PS-rich blends with concentrations of PE between 0 and 25 percent. In parallel series free PE was replaced for 5, 30, 60 or 100 percent by bound PE by blending PS with appropriate amounts of PE and graft-or block copolymer. The rate of deformation was 0.40 min^{-1}. Resulting modulus-values are plotted against total weight percentages of PE. The hypothesis was that the block- or graft copolymer molecules were situated at the interfaces between the PE particles and the PS-matrix (Figure 2). In that case it can be expected that the contribution of PE to the modulus of a blend can be calculated as the sum of the contribution of the free and bound PE. The same reasoning should be true for bound and free PS. In this way the Young modulus should be only a function of the total PE content of the PS/PE blends.

Figures 13 and 14 show that the results are in accordance with this concept for the graft copolymer modified blends that contain finite graft copolymer molecules. Some deviation occurs for PS/PE blends modified with the crosslinked graft copolymer GC4 microgel but only so when all free PE is replaced by the microgels of the graft copolymer. So for graft copolymer it appears that finite graft copolymer molecules are situated at the PS/PE interface. For block copolymer modified PS/PE blends for any series with a fixed

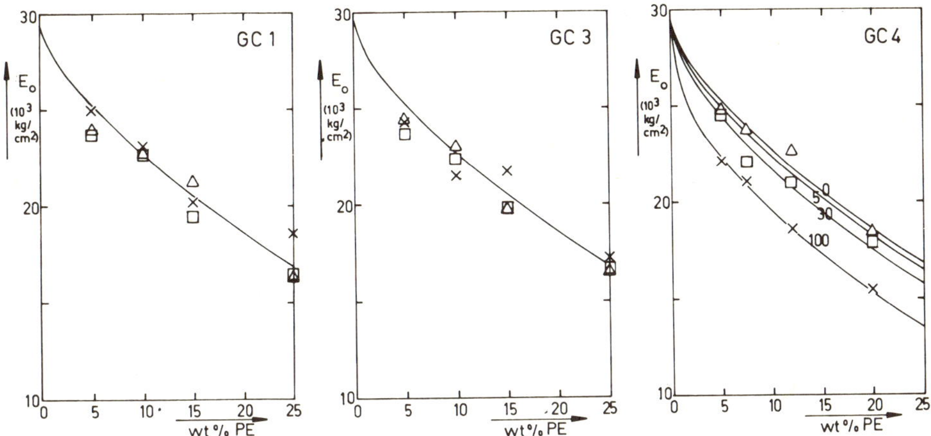

Figure 13. Young's modulus as function of total PE content in
            PS/PE blends in which 5 ($\triangle$), 30 ($\square$) or 100 (x) percent
            of free PE is replaced by PE bound in added graft
            copolymer GC1, GC3 or GC4.

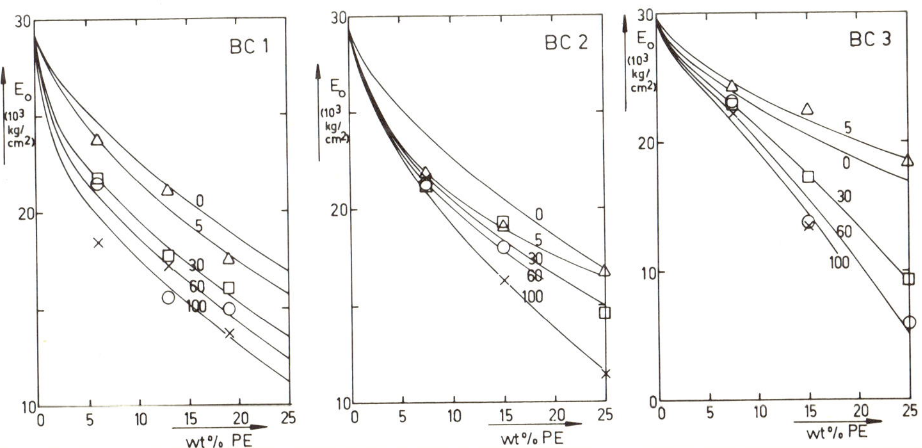

Figure 14. Young's modulus as function of total PE content for
            PS/PE blends in which 5 ($\triangle$), 30 ($\square$), 60 (o) or 100 (x)
            percent of free PE is replaced by PE bound in added
            block copolymer BC1, BC2 and BC3.

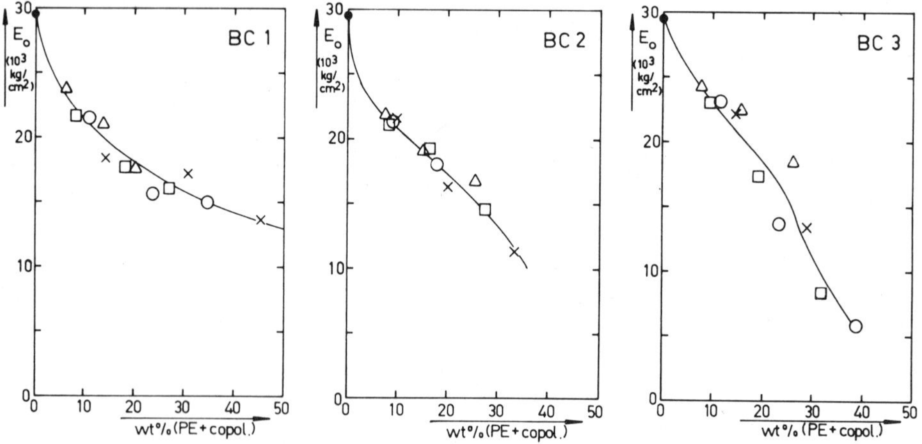

Figure 15.   Young's modulus as function of content of total
             dispersed phase (PE + block copolymer) of blends
             of Figure 14.

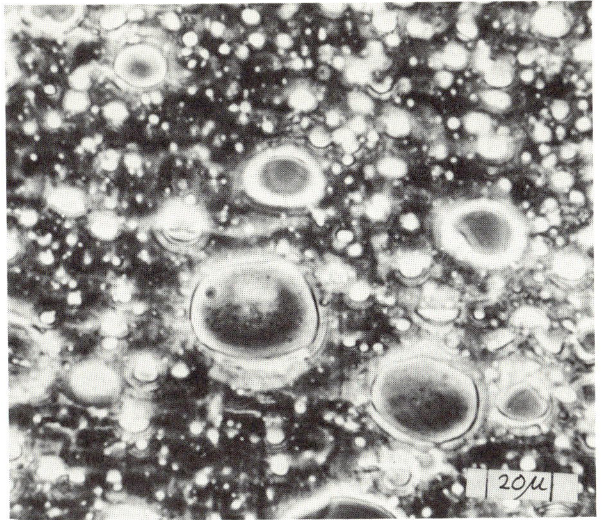

(a)

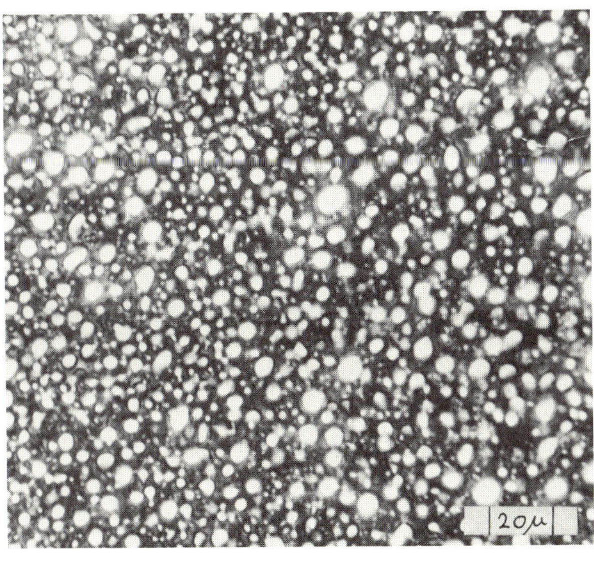

(b)

Figure 16. Influence of addition of block copolymer BC1 on PE
            particle dimensions composition.
            16a. PS/PE 75/25
            16b. PS/BC1/PE 80/2.25/17.75

percentage of replacement (5, 30, 60 or 100) of free PE by bound PE,
a new curve of modulus versus total PE concentration results. This
can be explained simply by supposing that the block copolymer molecules form a second dispersed phase of its own besides by being absorbed at the PS/PE interfaces. Experimentally it is found that the
blockcopolymers have moduli, that are much lower than those of PS
and that are only somewhat higher than that of the PE. So this hypothesis could be tested by putting the moduli of blends versus the
total amount of low modulus dispersed materiali e.g. versus the sum
of the weight percentages of PE and block copolymer. The results
are presented in Figure 15 and, as more or less single curves are
obtained, are in accordance with the hypothesis presented. Block
copolymer molecules indeed seem to form a separate phase dispersed
in PS and to be absorbed only partly at the interfaces. The tendency
to form a block copolymer phase can be understood by the ability of
these materials to form well-known stable superstructures in the 10
nm range. The fact that the block copolymers are also absorbed at
the interfaces is shown by the decrease of PE particle dimensions
in PS/PE blends if block copolymer is added to these blends (Figure
16a,b). To explain the modulus curves it is not necessary to assume
that the BC-microphases are separated from the PE particles. It is
certainly possible that a BC phase may surround PE phases in a more
or less regular manner by forming superstructures with alternating
PS-PE layers. In this way the block copolymer modified blends can
be compared with oil-water suspensions to which soap is added. The
soap will be absorbed at the oil-water interfaces but at the same
time it will form micelles containing regular ordered molecules in
the aqueous phase. As it was found that also PS-block copolymer
blends contain a visible dispersed low density phase, the block copolymer molecules seem not to be molecularly dispersed in the PS-
matrix as the soap molecules (Figure 17).

The fact that contrary to the block copolymer molecules, the
graft copolymer molecules seem to be absorbed only at the interfaces
of PE and PS, may be explained by the inability of the irregular
shaped graft copolymer molecules to form a superstructure which is
more stable than the interfacial layer. Moreover, this interfacial
layer of graft copolymer may be stabilized by absorption of PE and
PS oligomers from the homopolymers in its interfacial regions.

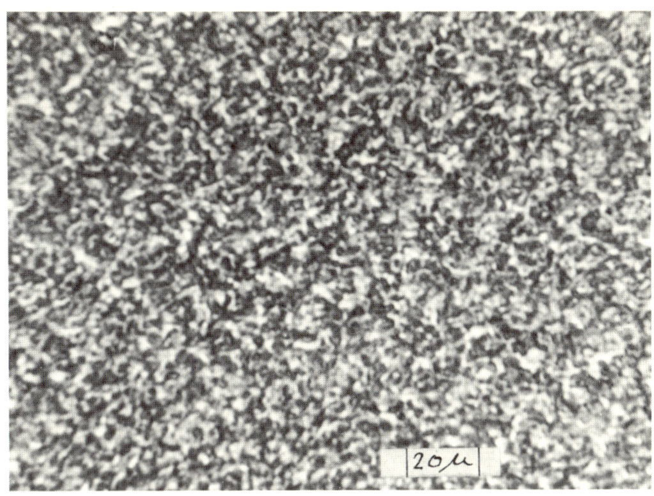

Figure 17. Micrograph of cross section of blend of PS/block copo-
           lymer BC1 (50/50). White phase of block copolymer is
           continuous.

Shape of the modulus curves and morphology

More careful analysis of the form of the modulus-concentration
curves in Figure 15 shows that in case of BC1 modified blends the
curve remains concave up to concentrations of 50 percent of dipersed
phase, indicating a particulate dispersion in that range, whereas
for BC2 and BC3 blends the curves bend downwards for concentrations
higher than 20 percent revealing the formation of semi-continuous
to continuous phases. The morphology indicated by the curve of
BC3 modified blends (Figure 15) has been confirmed by microscopic
observation of cross sections. Microscopic cross sections of blends
containing larger concentrations of BC1 showed a very small scale
morphology (< 0.1 mμ) of probably particulate character. The differences in morphology of blends modified by the different block
copolymers must be explained by the different rheological properties and molecular structure of the block copolymers. BC1 ha the
highest modulus of the three block copolymers (600 N/m^2), as compared with 400 and 200 N/m^2 for BC2 and 3 respectively and may form
a better and more cohesive superstructure than the tapered block

copolymers BC2 and 3. Thus differences of moduli but also of rheological properties during melt blending and the formation of a particulate, BC1 phase and of a semi-continuous viscous phase consisting of BC2 and of BC3 can be understood in the same way.

A general conclusion from the investigation on graft-and block
copolymer modified blends is that modulus-concentration curves are
a tool for investigating morphology and the number of phases in the
blends.

4. DEFORMATION MECHANISMS IN PS/PE BLENDS

Introduction

Samples of PS-rich PS/ldPE blends containing small amounts of
block copolymers of graft copolymers have impact fracture surfaces
indicating crazing as an important failure mechanism (Figure 18).
The same is true for PS/PE blends where a larger percentage of free
PE is replaced by PE belonging to graft copolymer. When however
in PS-rich blends 30, 60 or 100 percent of PE is replaced by bound
PE of tapered block copolymers BC2 and BC3, the fracture surfaces
show microshear (Figure 19). As these latter blends contain, as
mentioned in the foregoing section, semi-continuous or continuous
low modulus phases of block copolymer and PE, it is probable that
shearing will take place there. This, however, does not necessarily
exclude the possibility of crazing, especially not in the PS-matrix.

In order to get more quantitative informations about the contributions of elastic deformation, of crazing and of shearing to
the elongation, volume changes were measured during tensile deformation[14,15]. Measurement of volume changes to distinguish between
crazing and shearing were carried out earlier in creep experiments
by Bucknall et al[16,17,18].

Experiments

Apparatus

A dilatometer was constructed as described in reference 15.

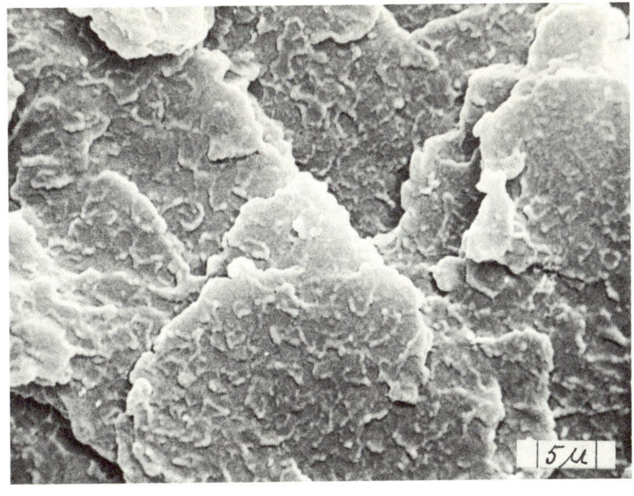

(a)

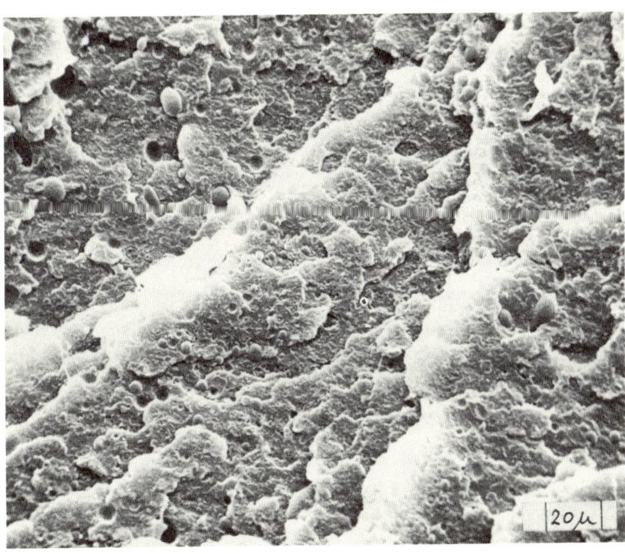

(b)

Figure 18.   Craze planes in fracture surfaces (SEM) of PS/PE blends
             with small percentages of graft (GC2) or block copoly-
             mers (BC1).
             18a. PS/GC2/PE (75/5/19)
             18b. PS/BC1/PE (80/2.75/17.75)

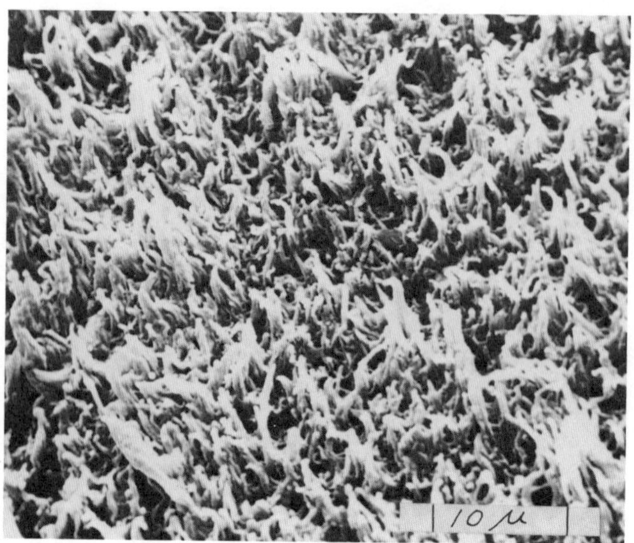

Figure 19.   Microshear in fracture surface of PS/PE blend with a
             large percentage of block copolymer BC3 PS/BC3/PE (60/
             30/10).

Tap water was used as the dilatometer liquid. Volume changes were
measured by conductometric determination of the rise of the water
in a capillary. The testbars had an oblong shape and were strained
at a rate of 0.40 min^{-1}. The elongation was measured directly from
the displacement of the moving sample clamp with a displacement
transducer whereas the load signal was taken directly from the stress
transducer of the tensile machine used. During the experiments the
signals for load and volume change were registrated on a recorder
as a function of elongation.

Data analysis

An example for a commercial high impact polystyrene, is shown
in Figure 20. Volume change and load and strain are in absolute
values. It can be seen that during the initial linear part of the
stress versus strain curve, that the volume strain versus elongational strain curve is linear as well. If the relative values of vo-
lume change $\Delta V/V_o$ plotted as a function of elongation $\varepsilon = 1/1_o$, the

initial slope of the curve is equal to $1 - 2\nu$ and so the Poisson ratio ν can be calculated from this slope. The second part of the stress-strain and volume strain-elongational curve beyond the linear part can be explained as follows. At higher stresses crazes are initiated. The crazes will thicken in the direction of the applied stress during elongation. The initiation as well as the thickening in longitudinal direction of the crazes will procede with rates that depend on the stress applied. The contribution of this thickening of crazes to the applied elongation rate increases with increasing elongation. This causes the stress to increase less than proportional with strain: the stress strain curve bends, goes through a maximum and in the end becomes practically constant. At constant load the elastic elongation will be a constant and equal to $\varepsilon_{el} = \dfrac{\sigma}{E}$, where σ is the stress and E the initial Young's modulus. In a preliminary model[19] in which deformation by crazing and elastic mechanism are acting in series the contribution to the volume strain of the elastic mechanism will be equal to

$$(\Delta V/V_o)_{el} = (1 - 2\nu)\,\sigma/E$$

In Figure 21 this elastic contribution is calculated and drawn for a hypothetical stress-strain curve. The difference between the upper and lower yield stress is exaggerated to show the features of the volume strain curve. The contribution of crazing to volume strain will be equal to

$$(\Delta V/V_o)_{cr} = \varepsilon_{cr} = \varepsilon - \varepsilon_{el} = \varepsilon - \sigma/E$$

This contribution can be drawn also (Figure 21). The total volume change is given by

$$\Delta V/V_o = (\Delta V/V_o)_{el} + (\Delta V/V_o)_{cr} = \varepsilon - 2\nu\,\frac{\sigma}{E}$$

The slope of the volume strain curve will be

$$d(\Delta V/V_o)/d\varepsilon = 1 - \frac{2\nu}{E}\,\frac{d\sigma}{d\varepsilon}$$

For the linear initial part of the stress strain curve the slope will be $1 - 2\nu$ and thus a value of the Poisson ratio is obtained. When the stress increases the stress strain curve bends and the slope $d(\Delta V/V_o)/d\varepsilon$ increases to become 1 at the upper yield stress maximum $(d\sigma/d\varepsilon = 0)$. Between the upper and lower yield stress the $d\sigma/d\varepsilon$ is

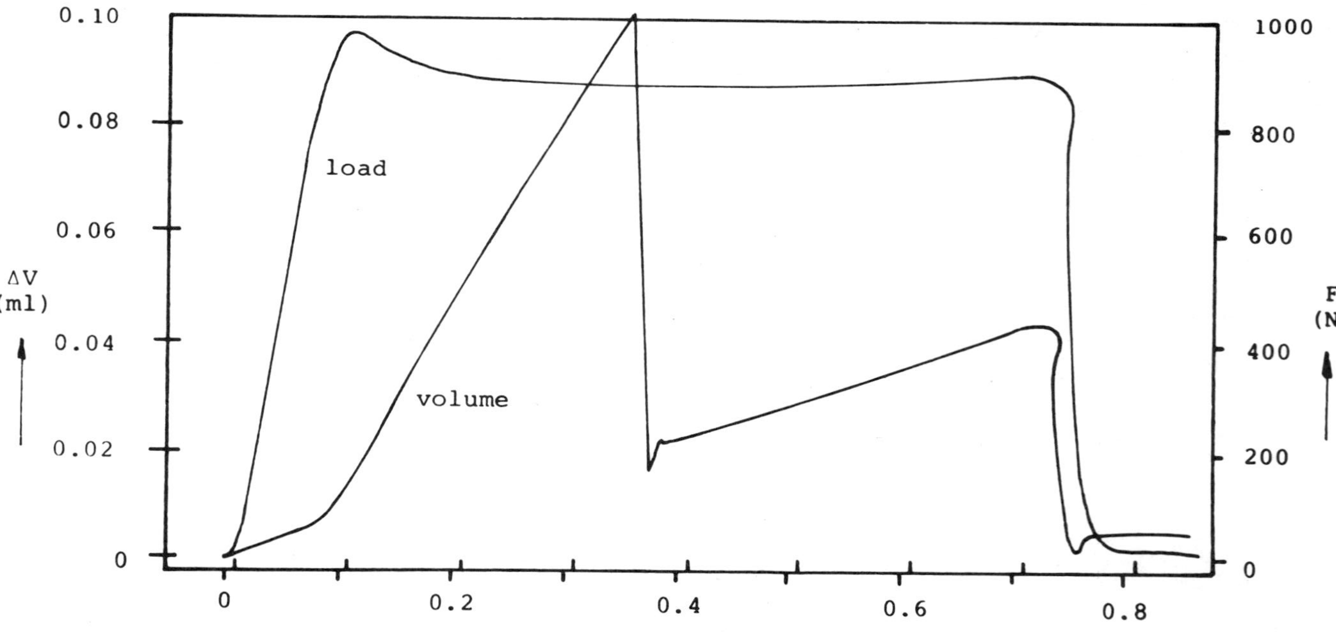

Figure 20. Stress and volume change as function of elongation of a commercial high impact polystyrene unoriented sample. At about 40 percent elongation the sensitivity of the volume change measurements was enhanced five times. After break the volume and length was largely recovered.

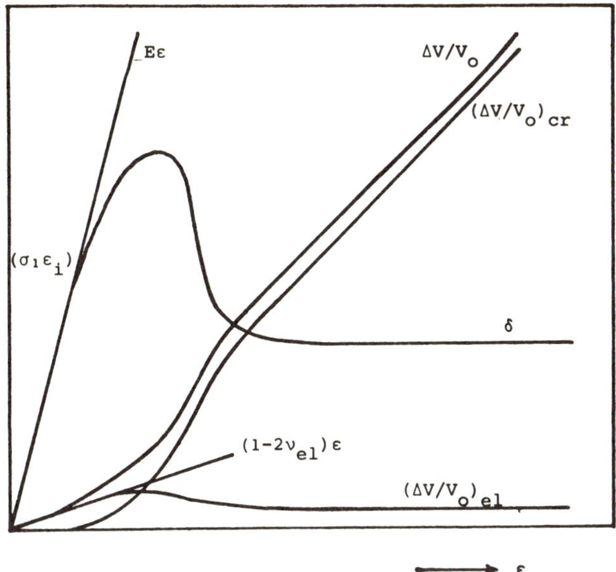

Figure 21   Volume strain and volume strain contributions of the
elastic and crazing mechanism calculated for a hypothetical stress strain curve after the model suggested in
the text

negative and the slope will be larger there than at higher elongation where the slope will be again equal to one $(d\sigma/d\varepsilon = 0)$. These
changes in slope can be observed in Figure 21 and the experimental
HIPS-curve (Figure 20). When shearing also contributes to elongation (ε_{sh}) the equations should be modified:

$$\Delta V/V_o = (1 - 2\nu)\frac{\sigma}{E} + \varepsilon_{cr} = \varepsilon - 2\nu\frac{\sigma}{E} - \varepsilon_{sh}$$

$$\varepsilon = \varepsilon_{cr} + \varepsilon_{sh} + \varepsilon_{el} = \varepsilon_{cr} + \varepsilon_{sh} + \frac{\sigma}{E}$$

Shearing will not contribute to volume changes. Differentiation
results in

$$d(\Delta V/V_o)/d\varepsilon = 1 - \frac{2\nu}{E}\ d\sigma/d\varepsilon - d\varepsilon_{sh}/d\varepsilon$$

$$d\varepsilon_{cr}/d\varepsilon = 1 - d\varepsilon_{sh}/d\varepsilon - \frac{1}{E}\ d\sigma/d\varepsilon$$

A volume strain curve is drawn for an assumed stress strain curve and presented in Figure 22. The difference between the experimental volume strain $\Delta V/V_O$ and the elastic contribution to the volume strain equals the contribution of crazing to the volume strain $\varepsilon_{cr} \cdot \varepsilon_{sh}$ can be calculated to be the difference between $\Delta V/V_O$ and $\varepsilon - 2\nu \frac{\sigma}{E}$. This last term represents the hypothetical contribution of crazing when no shearing is present (Figure 22).

For constant or nearly constant σ the slope of the second part of the volume strain curve will be equal to $1 - d\varepsilon_{sh}/d\varepsilon = d\varepsilon_{cr}/d\varepsilon$ and so the contributions of crazing and shearing are easy distinguished (e.g. for a slope of a certain elongation equal to 0.70, the contributions of shearing and crazing to the deformation are 30 and 70 percent respectively). Using this principle volume strain curves were analyzed. The initial slope gives the Poisson ratio and the slope of the second part indicates the contributions of shearing and crazing to elongation at any elongation.

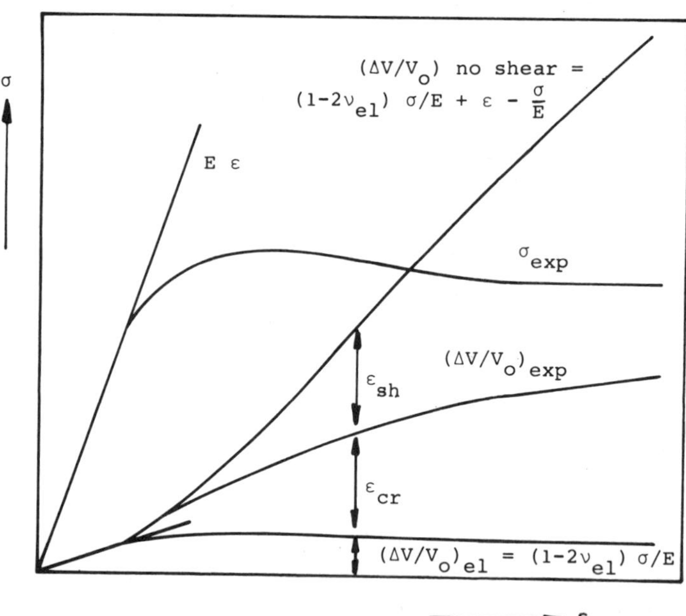

Figure 22. Volume strain contributions for the elastic and crazing mechanisms and the strain contributions for shear calculated for hypothetical stress-strain and volume strain-elongational strain curves designated: σ_{exp} and $(\Delta V/V_O)_{exp}$.

Presentation of samples investigated and of results

Samples were made by melt blending PS, ldPE and BC3. Stress strain curves (that are averaged curves from 5 samples of each composition) are presented in Figures 23, 24 and 25. Values of tensile, yield strength and of Poisson ratio are compiled in Tables 3 and 4. These Tables also give the values of the slope of the volume strain-elongational strain curve in the second part of the volume strain curve.

TABLE 3

Comparison of blends containing PS and PE and blends containing PS and block copolymer (BC) with respect to their mechanical data, dilatometric data and deformation mechanism.

Sample[a]	Dispersed phase[b] (wt %)	$\varepsilon_y/(\%)$[c]	$\varepsilon_b/(\%)$[c]	Poisson ratio[d]	Slope[e]	mechanism[f]
Blends of PS and PE						
23.0	7.5	1.8	6.0	0.33	1.13–1.04	cr.
24.0	15.0	1.8	4.0	0.32	1.04	cr.
25.0	25.0	1.0	1.0	0.31	g	cr.
Blends of PS and BC3						
23.100	14.4	1.7	1.7	0.36	g	cr.
24.100	29.0	2.55	2.60	0.41	0.68	cr.+sh.
25.100	48.0	h	18	0.47	0.04	sh.

[a] The first number of this notation to Figure 23, 24 or 25. The second number corresponds to the curve numbers in these figures and indicates the percentage of free ldPE being replaced by bound ldPE of the block copolymer;

[b] ldPE plus block copolymer;

[c] Elongational strains at yield point (ε_y) and at break (ε_b);

d
Derived from initial slope of volume strain/elongational strain
curve (elastic behaviour);

e
relates to second part of volume strain/elongational strain curve
(yielding behaviour);

f
cr. = crazing and sh. = shearing;

g
breaks just at yield point;

h
no stress maximum observed.

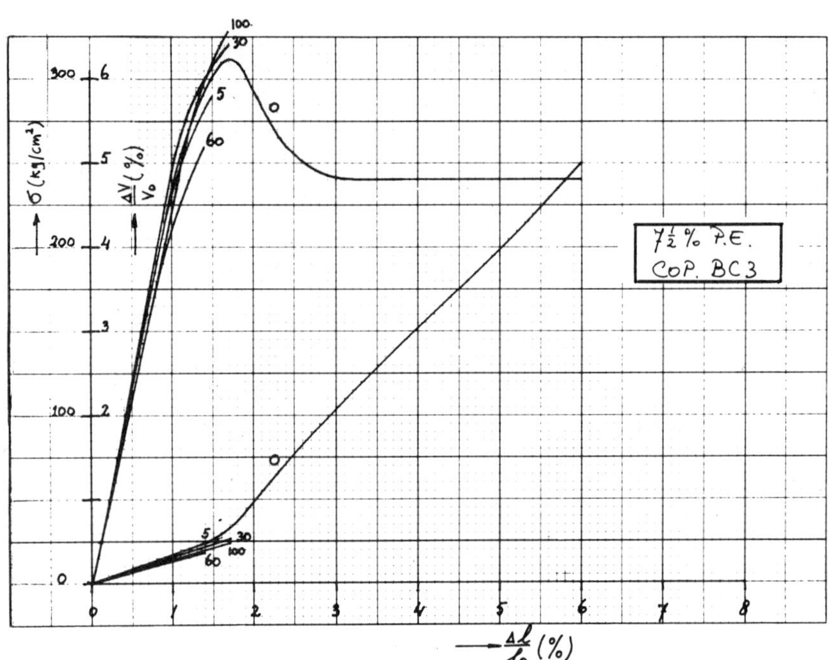

Figure 23. Experimental stress-strain and volume strain-elongation
            strain curves for blends containing 7.5 (free + bound)
            PE. Figures at curves indicate the percentage of bound
            PE present by appropriate addition of block copolymer
            BC3.

TABLE 4

Mechanical data, dilatometric data and deformation mechanism of PS/block copolymer (BC)/PE blends.

Sample[a]	E-modulus x 10^{-2} MPa	Total PE (wt %)	Dispersed phase[b] (wt %)	$\varepsilon_y/(\%)$[c]	$\varepsilon_b/(\%)$[c]	Poisson ratio[d]	Slope[e]	Mechanism[f]
23.5	29.3	7.5	7.9	g	1.5	0.33	g	cr.
23.30	23.9	7.5	9.6	g	1.7	0.34	g	cr.
23.60	22.0	7.5	11.6	g	1.4	0.37	g	cr.
24.5	19.4	15.0	15.7	2.1	7.6	0.35	0.99	cr.
24.30	19.3	15.0	19.2	2.2	2.4	0.39	1.15	cr.
24.60	13.7	15.0	23.3	2.6	2.8	0.43	0.75	cr.+sh.
25.5	15.8	25.0	26.1	2.3	7.2	0.38	0.87	cr.+sh.
25.30	11.2	25.0	32.0	4.5	6.3	0.44	0.67	cr.+sh.
25.60	7.7	25.0	38.8	h	>18	0.47	0.07	sh.

Explanation of the superior letters in this table as for Table 3.

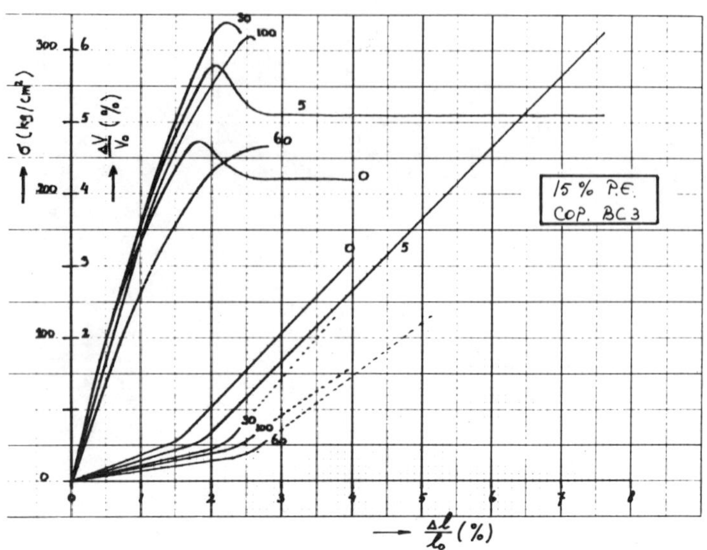

Figure 24. Experimental stress-strain and volume strain-elongation
            strain curves for blends containing 15 (free + bound) PE.
            Figures at curves indicate the percentage of bound PE
            present by appropriate addition of block copolymer BC3.

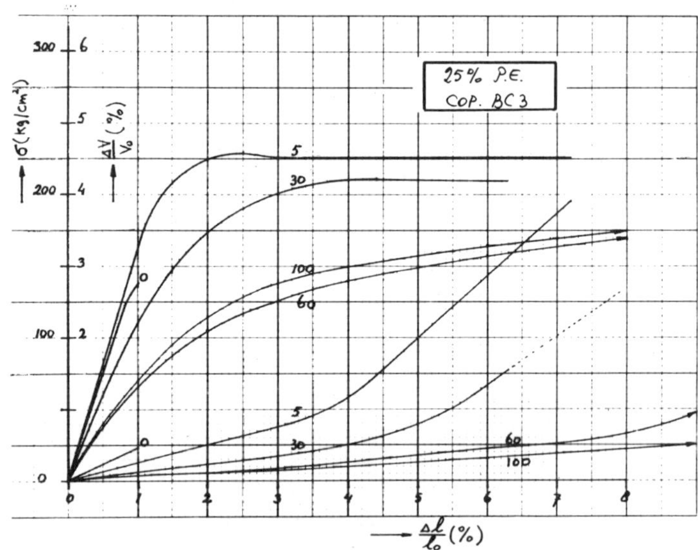

Figure 25. Experimental stress-strain and volume strain-elongation
            strain curves for blends containing 25 (free - bound)
            PE. Figures at curves indicate the percentage of bound
            PE present by appropriate addition of block copolymer BC3.

DISCUSSION

Poisson ratio of copolymer free PS/ldPE blends

Comparing the results of the samples containing only PS and PE
it is clear that the Poisson ratio decreases with increasing PE content and becomes lower than 0.33 (Table 3). As the Poisson ratio's
of PS and PE are 0.33 and 0.48 respectively this result is unexpected.
The explanation is that during cooling of PS-rich PS/PE blends after
melt blending, the PE particles contract much more than the PS matrix.
Consequently the particles of PE will have nearly no mechanical contact with the PS matrix in the sample at 20°C. Due to this lack of
mechanical contact the PS matrix will expand during elongation as
if spherical holes were dispersed in PS. In this way a Poisson ratio
with a value of 0.33 being equal to that of the matrix should be
expected. Due to stress concentration near the holes, however, the
material near the voids will be more deformed than in case of a PS
sample without voids, so a higher dilatation and a lower Poisson
ratio than that of pure PS will be measured. Calculations of Poisson
ratios of a matrix in which voids are dispersed as performed by
Sjoerdsma[9] using the aggregated grain of Kerner led to the same conclusions.

Poisson ratio of copolymer modified PS/PE blends

The Poisson ratios of PS/PE blends to which small amounts of
block copolymer BC3 are added increase with increasing PE concentration (Table 4). As certainly part of the BC3 molecules are present
at the interface between the PS and PE phase, these block copolymer
molecules will secure the mechanical contact between the phases
during cooling of the blend. This leads to an increase of the
Poisson ratios from 0.33 to higher values as the PE particles now
counteract the dilatation of the PS matrix.

Poisson ratios of PS/BC blends, increase with increasing BC
content (Table 3, samples 23.100, 24.100, 25.100). In these blends
the two materials will adhere to each other and the Poisson ratio
increases from 0.33 to 0.47. Sjoerdsma et al discussed[9] also the
form of the curve of the Poisson ratio versus concentration PS in
the blends by comparing the experimental curves with those calculated from the aggregated grain model of Kerner. In this model the

particles of the dispersed phase are allowed to aggregate in a statistical way so that a smooth phase reversal is achieved in a symmetrical fashion around 50 percent concentration. The results for Poisson ratios for concentrations below 20 percent are in accordance with the model in which mechanical contact between phases is assumed. The deviations between the results and the calculated curves at higher concentrations could be explained by the excessive tendency of PE, of BC3 and of combinations of these materials, to form a continuous, shear inducing, phase during melt blending.

Shearing and crazing of PS/PE blends

- Crazing

As already explained above the slope of the second part of the volume strain curve where the stress nearly becomes constant is equal to the fractional contribution of crazing to elongation.

By observing the results given in Table 3 it can be concluded that all PS-rich blends with PE concentrations below 20 percent show crazing as the only deformation mechanism. Blends below 20 percent of PE contain PE in particulate form only as revealed by micrographs of cross section and as can be derived by analyzing modulus concentration curves (Figure 15). Evidently the explanation of the failure must be that in these blends, the PE particles only initiate crazes. At low numbers of crazes, due to the low numbers of particles, causes the elongation at break to be small due to fatal isolated craze formation. At larger concentrations of particles and thus of craze initiation sites the elongation at break will be larger due to craze stopping mechanism as proposed by Nielsen[20] that enable deformation of the network of PS bulk material, that has come into existence. At still higher concentrations of PE particles the great number of crazes combine into cracks and elongation at break will decrease again. The overall result is that samples containing PE as dispersed phase only have elongations at break that decrease with increasing PE content. As pure PS is brittle and the data in Table 3 show a decrease in elongation at break with an increase of PE concentration beyond 10 percent, the concentration optimum for elongation will be about 10 percent of PE, which is in accordance with the results found by Barentsen (Figure 3 and reference 1 and 2).

The elongation at break of PS blends containing PE and that are modified by adding small percentages of BC3 to ensure adhesion between phases, samples 23.5 and 24.5, Table 4, increases with increasing concentration of PE. The optimum for these blends for elongation seems to be higher, it is now about 15-20 percent of PE. At still higher concentration a decrease of the elongation at break would be expected. However, at these PE concentrations above 20 percent semi-continuous low modulus phases are formed causing the elongation at break still to increase with elongation (samples 25.5, 25.30 and 25.60, Table 4 and 25.100, Table 3). This elongation is partly caused by shearing and will be discussed at the end of this section.

It is now appropriate to discuss the concentration for maximum elongation at break and thus toughness in tensile experiments for PS/PE blends in general. The optimum will be determined by the number of crazes formed during elongation and the number of crazes in turn will be determined by the number of dispersed particles. As the optimum PE concentration for BC3 modified PS/PE blends is higher than 15 percent and as this concentration is lower than 10 percent for BC3-free PS/PE blends, the conclusion must be that the PE particles covered by BC3 and adhering to the PS matrix initiate a lower number of crazes than non-adhering PE particles in BC-free blends. This effect must be rather large as a copolymer modified blend has much more PE particles than a no copolymer containing blend with the same PE content. The fact, however, that Barentsen[1,2] found for graft copolymer modified PS/PE blends an optimum in elongation at break at about 7.5 percent of PE shows that the number of crazes per particle will also be dependent on the kind of copolymer used for anchoring.

Shearing

Blends of rather high PE and block copolymer content (total dispersed phase larger than 20 percent) all show besides crazing, a contribution of shearing to elongation (Table 4, samples 24.60, 25.5, 25.30, and 25.60). As all shearing blends contain semi-continuous to continuous PE/BC phases, the conclusion can be drawn that shearing takes place in the low modulus phases whereas crazing will take place in the high modulus PS-phase.

5. GENERAL CONCLUSIONS

Some general conclusions about the deformation mechanism of the copolymer modified PS/PE blend are:

1. These blends are very useful examples to understand mechanical behaviour of blends such as high impact PS, ABS or impact PP, in which a high modulus polymer is blended with a low modulus polymer to improve impact strength.

2. The form of modulus-concentration curves of blends gives insight of the morphology of the blends.

3. The presence of particles and semi-continuous phases are of importance for the appearing of crazing and of shearing as deformation mechanisms.

4. Measurements of Poisson ratios in dependence of blend composition can give indication about the presence or absence of mechanical contact, and probably, of adhesion between phases.

5. Adhesion by the block copolymer suppresses the crazing initiation ability of the PE particles in the blend under consideration, and increases yield and fracture stress.

ACKNOWLEDGEMENTS

Important contributions to the knowledge of PS/PE blends came from Dr. ir. W. Barentsen[a] and Dr. ir. N. Hoen[b]. Their work led to their theses and to publications[1,2,3,4,8,10,11]. Research on preparation and properties of graft and block copolymers and on hydrogenation was carried out by P. Piet. Without these contributions this work had been impossible[3,10,11]. A dilatometer for measurement of volume strain was developed by Ir. J. Coumans[c] during preparation of his engineers thesis[14,15]. Mr. H. Ladan made numerous micrographs of cross sections and SEM pictures of fracture surfaces for all publications from this Laboratory.

a) AKZO Zoutchemie, Research, Hangelo
b) Central Laboratory DSM, Geleen
c) Laboratory for physical Chemistry, Eindhoven University of Technology

REFERENCES

1. W. M. Barentsen, Thesis, Eindhoven University of Technology
    (1972) (in Dutch).
2. W. M. Barentsen and D. Heikens, Polymer $\underline{14}$, 579 (1973).
3. W. M. Barentsen, D. Heikens and P. Piet, Polymer, $\underline{15}$, 119 (1974).
4. D. Heikens and W. M. Barentsen, Polymer, $\underline{18}$, 69 (1977).
5. T. Okamoto, and H. Takayanagi, J. Polym. Sci. $\underline{C23}$, 597 (1968).
6. B. Paul, Trans. AIME, $\underline{218}$, 36 (1960).
7. L. Bohn, Angew. Makromol. Chem., $\underline{29/30}$, 25 (1973).
8. D. Heikens, N. Hoen, W. Barentsen, P. Piet and H. Ladan,
    J. Polym. Sci., Polymer Symposium, $\underline{62}$, 309-341 (1978).
9. S. D. Sjoerdsma, A. C. A. M. Bleijenberg and D. Heikens,
    Polymer, accepted.
10. N. Hoen, Thesis, Eindhoven University of Technology (1977)
    (in English).
11. P. Piet, N. G. M. Hoen, J. H. G. M. Lohmeyer and D. Heikens,
    IUPAC 23th Int. Symp. on Macromolecules, Madrid, Sept. 1974,
    Preprints vol. $\underline{1}$, II 1-33.
12. L. H. Stockmayer, J. Polym. Sci., $\underline{9}$, 69 (1952).
13. M. Gordon, Proc. R. Soc. London, Ser. A $\underline{268}$, 240 (1962).
14. W. J. Coumans, D. Heikens and S. D. Sjoerdsma, Polymer, $\underline{21}$,
    103-108 (1980).
15. W. J. Coumans and D. Heikens, Polymer, accepted.
16. C B. Bucknall and D. Clayton, Nature, Phys. Sci., $\underline{231}$, 107
    (1971).
17. C. B. Bucknall and D. Clayton, J. Mater. Sci., $\underline{7}$, 202 (1972).
18. C. B. Bucknall, "Toughened Plastics", Applied Science, London,
    Chapter 7, 1977.
19. To be published elsewhere.
20. L. E. Nielsen, "Mechanical Properties of Polymers", Reinhold,
    New York, p. 133, 1963.

MECHANICAL MODELS OF HETEROGENEOUS POLYMERIC MATERIALS

T. Pakula

Centre of Molecular and Macromolecular Studies

Polish Academy of Sciences, Lodz, Poland

INTRODUCTION

The theoretical description of mechanical properties of heterogeneous materials has been a subject of investigation of several authors in last few years. The interest in this field is involved by
technological importance of composite polymeric materials like polymer blends, filled polymers and block copolymers. Blending of polymers is usually used to improve properties and processability of materials. On the other hand most of polymers that are considered to
be chemically homogeneous become physically heterogeneous when considered on a microscopic scale. An example is the coexistence of
crystalline and amorphous regions in most crystallizing polymers.

Mechanical properties of all of these materials are dependent
on composition, properties of components and on geometrical configuration and shapes of structural elements. The elastic or viscoelastic behaviour of two solid phases firmly bonded together is then
the subject of interest. Various ways of solution of the problem
have been proposed. They can be calssified into several categories
as proposed by Chamis and Sendeckyj[1]. The main of them are:
1) simple phenomenological models[2],
2) exact calculations of upper and lower bounds of the elastic constants[3,5],
3) self-consistent theories in which the environment of a filler is
   approximated by a continuum[4,6],

4) exact numerical solutions for some particular geometries[7,8].

Practically all of the solutions relate macroscopic mechanical properties of the composite with properties of components for an arbitrary composition but in most cases no restriction is made on the distributions of shapes, sizes and orientations of the structural elements. However, it becomes important when properties of materials with more complex structures have to be described. As an example blends of semicrystalline polymers can be considered in which shapes, sizes and eventual correlations of orientations of anisotropic elements can influence the local and macroscopic properties of the blend.

Recently structural models have been proposed[9] in which mechanical models are used as unit elements for representation of local behaviour of a particular structural arrangement of components. The unit elements whose properties are variable according to the fluctuations of the local properties in the material arising from distributions of orientations, structural forms and composition are built up into aggregates the properties of which can be calculated by various averaging methods.

An important aspect of such treatment is the dimensionality of the model. It was shown that not all aspects of interactions between components of the composite material can be properly described by models with reduced dimensionality as for example in the case of Takayanagi's models. A complete solution describing the mechanical properties of the three-dimensional model suitable for application as an unit element representing the local properties of two-component materials has been obtained[9,9a].

This paper will discuss different methods of averaging the properties of unit elements built up into aggregates representing complex arrangements of structural elements. A new averaging concept will be shown.

TYPES OF HETEROGENEITIES AND RELATED STRUCTURES

To introduce convenient geometries of models which could represent properties of local structure of materials one should take into account various types of heterogeneities in different composi-

te materials, their forms and mechanical nature. Three types of
materials are considered here: block copolymers, blends of polymers
and semicrystalline polymers. The latest being of special importance
when used as components in blended materials.

The relatively simple structures are observed in block copolymers. They can be classified into three groups: droplet like, fibrillar and lamellar as appear dependently on the composition of
comonomer units in the polymer. What is characteristic of these
structures is that they usually form textures of high symmetry and
uniformity of dimensions of elements. The correlation of dimensions
and positions of structural elements can be local ones when samples
are obtained from solutions[10,11] or even on a macroscopic scale if
samples are prepared during an extrusion process[12,13]. The best
examples of such structures have been demonstrated on polystyrene-
polybutadiene-polystyrene copolymers.

Similar structural forms but with much lower order and regularity can be obtained in blends of polymers if two amorphous and incompatible polymers are mixed together. The type of the structure
depends in this case on the composition, type of blending process
and on the mechanical state of the molten blend from which it was
cooled down and solidified. It was demonstrated that slow or alternatively fast cooling of the blend being extruded can lead respectively to droplet like or fibrillar structures in the solidified filament[14]. The situation becomes much more complex when the components of blends can crystallize. The possible morphologies in such
blends have been discussed by Stein et al[15]. If one component can
crystallize, crystals or spherulites can be dispersed in an amorphous
homogeneous or heterogeneous phase, dependently on compatibility
of polymers. On the other hand the noncrystallizable component may
be dispersed within the spherulites in larger domains or even included within the spherulites between lamella of crystallizable component. Such cases have been observed in blends of polycaprolactone
and polyvinylchloride[16] and blends of polyphenylenoxide with isotactic polystyrene[17]. If both components of binary blend crystallize
the following structures can be considered: 1) crystals of two
components dispersed in an amorphous matrix, 2) one or both components exhibiting spuperstructures such as spherulites, hedrites or
other types, 3) separate spherulites of two components or mixed
spherulites containing crystals of both components.

Various morphologies that might be encountered in blends of crystallizable polymers are well illustrated by Stein et al[15] and their appearance is thought to be related to the structure of blend in molten state determined by compatibility of components and to the cooling rate from molten state which determines diffusion and crystallization rates of components. The crystalline elements of crystallizing components can, moreover, assume different geometrical forms being of the lamellar or bundle types, in both cases possessing high mechanical anisotropy. This variety of morphologies and their complexness shows that an attempt of adeguate theoretical description of mechanical properties of such complicated systems can be an extremely difficult task.

From the above short review it comes out, however, that it is possible to distinguish some basic elemental structural units common for all cases discussed in all systems. These are lamellar, fibrillar or spherical inclusions of one phase inside the other. Further complications of structures are based on different correlations of positions and orientations of these elements, on details of their shapes and size distributions as well as on the nature of mechanical properties of phases. The theoretical considerations in this paper are based on the above remark. It is assumed that the mechanical properties of basic structural elements can be described by some elemental mechanical units with parameters related to details of local structure and properties of components while more complicated morphologies can be represented by aggregates of such units in which specific arrangements of the structural elements of the material are considered.

SIMPLE MECHANICAL MODELS
<u></u>

Several simple models have been developed to calculate the static elastic moduli and dynamic viscoelastic functions of polymeric composite materials. All of them are based on a combination of the so-called parallel and series models[18] which present the upper and lower bounds of the modulus of binary materials. As an example two cases considered by Takayanagi[19] are shown in Figure 1. The moduli of models A and B (Figure 1) are

$$E_A = \left[\frac{\varphi}{(1 - \lambda)E_m + \lambda E_i} + \frac{1 - \varphi}{E_m} \right]^{-1} \tag{1}$$

$$E_B = \lambda \left[\frac{\varphi}{E_i} + \frac{1 - \varphi}{E_m} \right]^{-1} + (1 - \lambda)E_m \qquad (2)$$

where E_i and E_m are the moduli of internal component and of the matrix respectively, the $\lambda\varphi$ being the surface of the rectangle (Figure 1) is the volume fraction of the dispersed phase. Different in form but in consequence analogous result is described by Kerner equation[20] which can be expressed in the following form[21] for two component system

$$\frac{E}{E_m} = \gamma \frac{v_m E_m + \beta (\alpha + v_i)E_i}{(1-\alpha v_i)E_m + \alpha\beta v_m E_i} \qquad (3)$$

where $\beta=(1+\mu_m)/(1+\mu_i)$, $\gamma=(1+\mu)/(1+\mu_m)$, v_m and v_i are volume fractions of components and μ_m and μ_i are the Poison coefficients. The α is a constant dependent on the shape of inclusions. It has been demonstrated that the two models are exactly equivalent[21,22] and the relationships between respective parameters have been given[23].

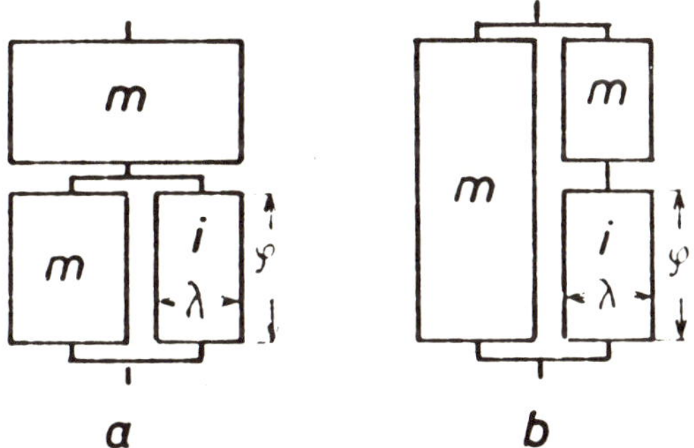

Figure 1.   Two cases (a) and (b) of Takayanagi models for two-phases system corresponding to eqs. (1) and (2) respectively.

An attempt to introduce two-dimensional models has been made by Pakula et al[2,4] and was extended later to three dimensional case[9]. The two and three-dimensional models differ from the uni-dimensional one in the way in which two components are bonded together as well they consider deformation of model elements both under normal and shear stresses. The differences between various models can be well demonstrated if one considers their application for description of lamellar structures. Only in this case the geometrical parameters of various models have the same meaning. Models of different dimensionality are shown in Figure 2. It was pointed out[9] that moduli of various models are strongly dependent on the values of Poison ratio of both components and the dimensionality of models. Figure 3 shows an example of dependences of modulus on composition for all models presented in Figure 2 deformed in two directions parallel and perpendicular to the interface (the uni-dimensional model is represented by separate parallel and series cases). As it is seen in Figure 3 considerably different dependences are predicted by various models if the Poison ratio of the component with lower modulus is close to value of 0.5 (for dependences in Figure 3 $\mu=0.45$). The de-

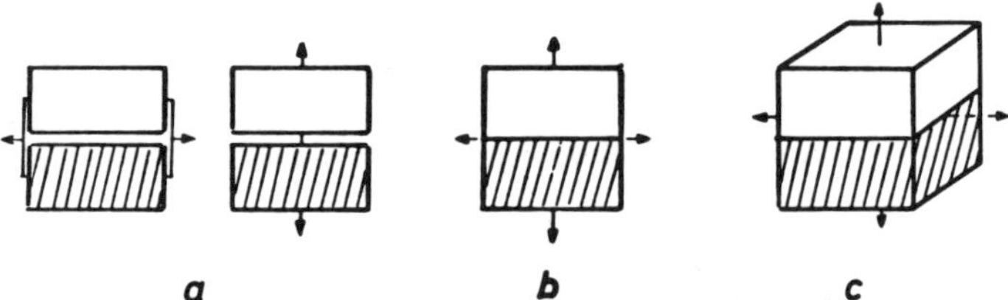

Figure 2.   Models of lamellar structure (a) according to Takayanagi, (b) two-dimensional and (c) three-dimensional.

pendences will coincide if μ is assumed to be zero. This demonstrates well the importance of dimensionality of models. As it has been discussed in the previous paper[9], only three-dimensional models can reflect all aspects of interactions of firmly bonded phases. Therefore the only model suitable for representation of local properties of various structures is the three-dimensional one.

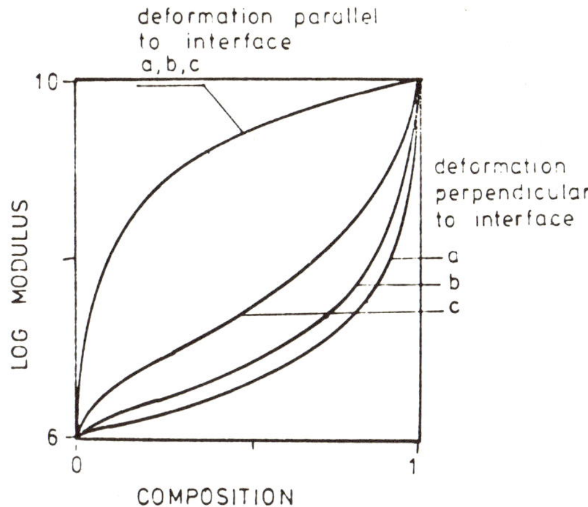

Figure 3. Dependence of modulus on composition for various models
from Figure 2 (for μ=0.45 – the Poisson ratio of the
component with lower modulus).

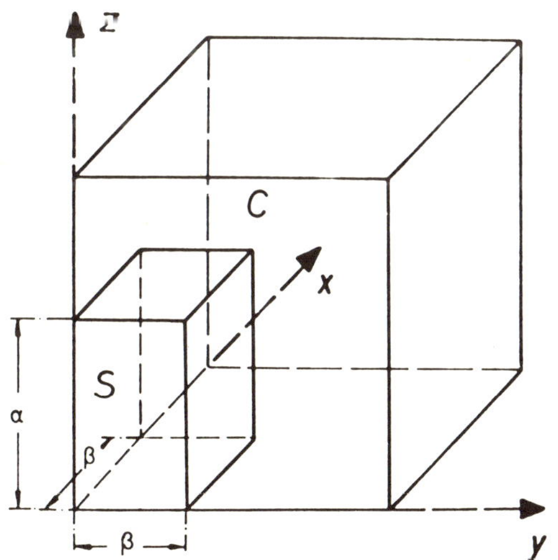

Figure 4. Geometry of the three-dimensional model.

Geometrical representation of such unit model element is shown
in Figure 4. It is assumed that the whole unit has a form of cube
with a unit side length related to the characteristic dimensions of
the structure which it should represent. For example, in the case
of lamellar structure it could be assumed that its dimensions are
related to long spacing. The cube is composed of two materials with
different mechanical properties being however in full cohesion, so
that under deformation of the cube, stresses are fully transmitted
through the interface. One of the components representing the
suspended (or internal) phase has a form of prism with height α and
square cross-section with side β. The other component representing
the continuous (or external) phase fills up the remaining volume of
the cube. The volume contribution of the internal component is
given by $V=\alpha\beta^2$ and of external component by $1-V$.

The internal part of the unit element when representing the
crystalline component is assumed to have transverse isotropy around
z-direction (Figure 4). Actual mechanical properties of polymer
crystals do not have transverse isotropy. In most polymer crystals
the elastic modulus in the direction of molecular chain is, however,
far higher than that in the transverse direction usually by two
orders of magnitude. Consequently a small anisotropy within the
transverse direction can be neglected in comparison with the extra-
ordinary anisotropy in the direction of the molecular chain. The
external component in the model is assumed to have isotropic mechanical properties.

The geometry assumed for the unit model is flexible enough to
represent the principal cases of the local morphology of semicrystalline polymers, blends and block copolymers. The lamellar thickness or the fibril length is usually very different from its lateral
dimensions. Therefore if we relate the value of α to lamellae thickness we have $\alpha\ll\beta$ as a condition for lamellar structure and if α
is related to fibril length then the internal geometry of the model
must satisfy the condition $\alpha\gg\beta$. By assuming $\alpha=\beta$ we approximate
the droplet like structures of dispersed spheres. The mechanical
transverse isotropy of internal component coincides with the geometrical symmetry of the unit element. Therefore the whole unit element exhibits also the transverse isotropy of mechanical properties.
It is obvious that such a model can be easily reduced to the description of simpler cases when material consists of two isotropic components. The mechanical properties of model components are assumed

to be fully described by a set of compliance values characteristic
for a given symmetry and type of mechanical behaviour. In the case
of elastic material these are compliance constants while in the ca-
sese of linear viscoelastic properties they can be given in the form
of complex numbers. To describe mechanical properties of such model
element one has to find a set of mechanical constants by means of
which it is possible to describe all types of deformations of this
element. The solution has been already presented both for deformation under normal stresses[9] and for shear deformation of the model[9a].

MODELING OF COMPLEX STRUCTURES AND THEIR MECHANICAL PROPERTIES

It is expected that the mechanical properties of any heterogeneous material will depend on the details of the local arrangement
of structural elements, determined mainly by the local orientations
of elements if they possess any type of anisotropy. In the case of
regular structures with uniform composition and orientation of the
structural units the properties of the material can be represented
by a single unit element equivalent to the repeat unit of the structure. To describe more complex cases such as materials with composition and orientation distributions or fluctuations the aggregate
model proposed originally by Ward[25] can be applied as a first approximation. According to this model it is proposed here that any heterogeneous material may be regarded as the aggregate of unit elements whose mechanical properties are those calculated for mechanical unit elements with geometrical parameters being correlated with
the geometry of possible structural elements in the material. The
aggregate should be built up, in such a case, from a number of unit
elements whose properties are variable according to the fluctuations
of the local properties in the material.

The average mechanical quantities for the aggregate can be obtained in two ways: 1) by assuming uniform stress throughout the
aggregate (which implies a summation of compliances) and 2) by assuming uniform strain (which implies a summation of stiffnesses).
There are the geometrical representations of the aggregate model
related to the above different averaging procedures. These are
shown in Figure 5 as a "series" and "parallel" aggregates respectively. Under the assumption that each elemental cube in the aggregate has the same mechanical properties with transverse isotropy,
the expressions for both types of averages for an aggregate with
a certain orientation distribution of the symmetry axes of the unit

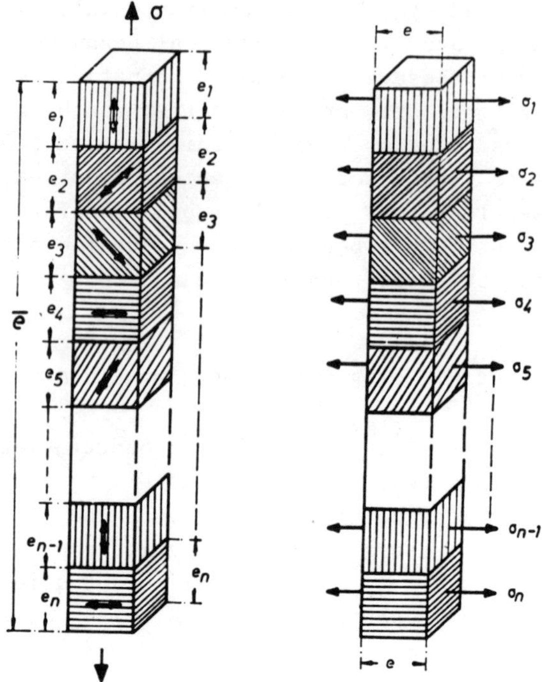

Figure 5. "Series" (a) and "parallel" (b) aggregate models.

elements have been given by Ward[25]. The expressions assume a simple
forms in the case of random orientations of unit elements, it is in
the case when one would like to express the properties of isotropic
body ba properties of an aggregate with randomly oriented anisotropic units. In such a case for series model the compliance components are found to be

$$\overline{M}_{33} = \frac{1}{15}\ (8M_{11}+3M_{33}+4M_{13}+2M_{44}) \tag{4}$$

$$\overline{M}_{13} = \frac{1}{15}\ (M_{11}+M_{33}/3M_{12}+8M_{13}-M_{44}) \tag{5}$$

and for parallel model the stiffness components are

$$\overline{T}_{33} = \frac{1}{15}\ (8T_{11}+3T_{33}+4T_{13}+8T_{44}) \tag{6}$$

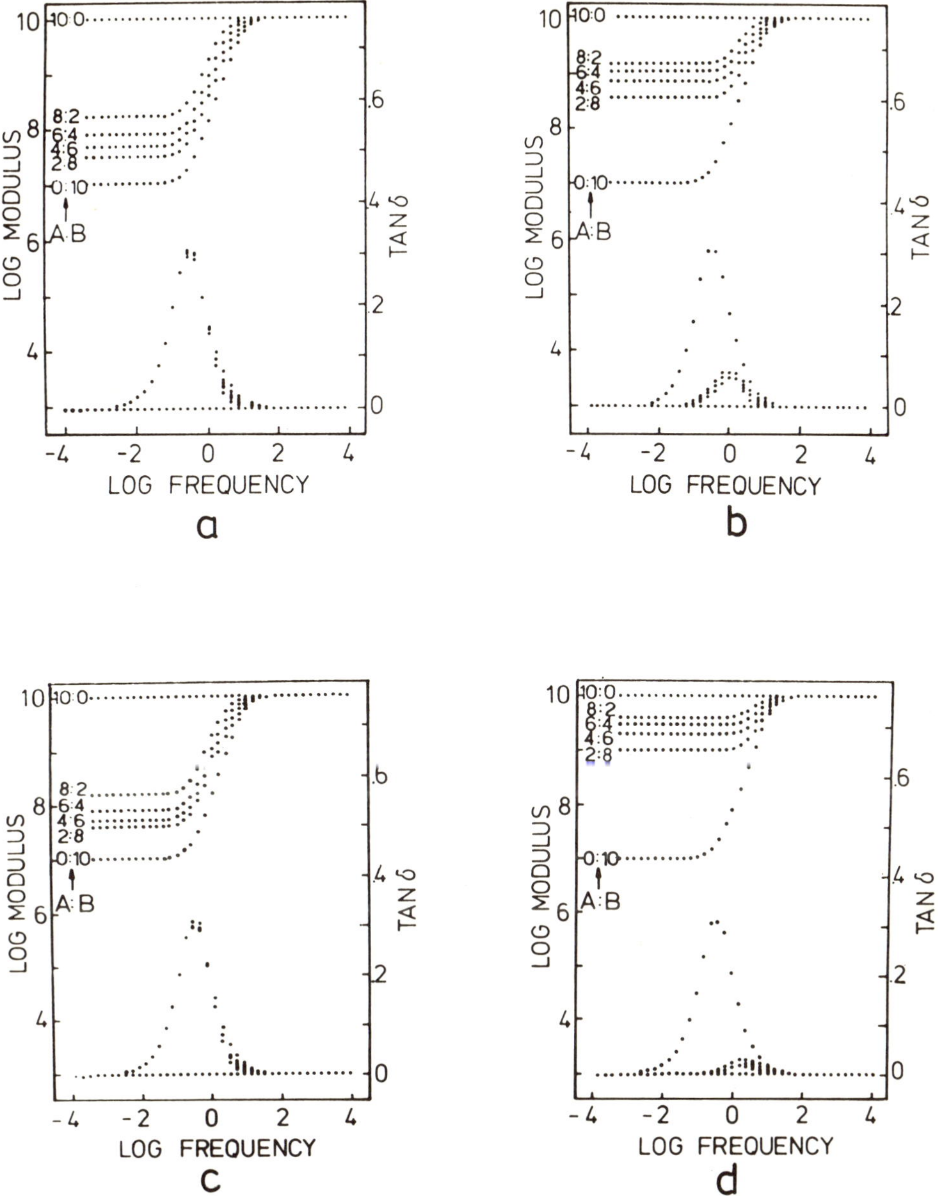

Figure 6.   Viscoelastic functions (E and tgδ vs. frequency)
            for various compositions of two-component systems
            calculated according to "series" (a,c) and "parallel"
            (b,d) models for fibrillar (a,b) and lamellar (c,d)
            structures with randomly oriented elements.

$$\overline{T}_{13} = \frac{1}{15}\,(T_{11}+T_{33}+3T_{12}+8T_{13}-4T_{44}) \tag{7}$$

where T_{ij} and M_{ij} are the stiffness and compliance components respectively for the anisotropic unit element considered in the aggregate. Both types of averages can be directly compared by calculating values of moduli for respective models

$$\overline{E}_s = \frac{1}{\overline{M}_{33}} \qquad \text{(for series model)} \tag{8}$$

$$\overline{E}_p = \overline{T}_{33} - 2\,\frac{\overline{T}_{13}^{\,2}}{\overline{T}_{33}+\overline{T}_{13}} \qquad \text{(for parallel model)} \tag{9}$$

To illustrate the effects of both averaging methods, in the case when they are applied to description of viscoelastic properties of heterogeneous material, results obtained for an aggregate of randomly oriented, two component unit elements will be presented. Two particular cases are considered for the geometry of the unit elements: $\alpha<\beta$ and $\alpha>\beta$ related to lamellar and fibrillar structures respectively. For numerical calculations both components are assumed to be isotropic buth with different mechanical behaviour: one of them is an elastic body (component A) and the other one viscoelastic (component B). Calculations have been performed for different compositions related to the case when A is internal and B external component. The properties of elastic component A are given by values of the modulus and the Poison ratio while the properties of viscoelastic component are assumed to be described by the frequency dependences of E' and tan δ according to the equations for "standard viscoelastic body"[9,25]. Results are presented as frequency dependences of E' and tan δ of the aggregates as shown in Figure 6.

It follows from the examples presented that by means of the above averaging methods extremely different properties of the aggregate can be predicted, dependent only on the averaging type. This effect is greater for aggregates with higher anisotropy of unit elements as for example for lamellar structures. It means that for such structures mechanical properties can not be described uniquely by means of these methods and that the range of uncertainty is very large. Such effects have been, however, expected because both "pa-

rallel" and "series" models of the aggregate are in principle uni-
dimensional and have the same weaknesses as was shown in relation
to dimensionality of models of the unit element. Both models are
moreover constructed artificially because no cohesive connections
between elements are assumed. This involves nonuniformity of strains
of aggregate elements in all directions which means that after deformation the aggregate has no longer the regular shape as in Figure
5. The above disadvantages of both types of averaging can be eliminated by replacing the uni-dimensional aggregate models by a
three-dimensional one. The proposition for construction of such a
model is the following:

The basic unit elements are defined in the same way as before,
which means that compliance matrix of each elemental unit in the
aggregate is calculated according to the three-dimensional model of
two-component unit element considering orientation of its symmetry
axis with respect to a direction distinguished in the aggregate
(for example direction of external extensional stress). It means
simply that compliance matrix of each unit element calculated at
first in its coordinate system has to be transformed to compliances
expressed in the coordinate system of the aggregate according to
the orientation of the unit element in the aggregate. Every eight
unit elements constitute a three-dimensional packet having a form
of a cube. The full cohesion between all elements in the packet
and in all directions is assumed. The packet built up from eight
basic unit elements will be called the first-order packet. Every
eight first-order packets constitute a new second-order packet,
constructed in the same way as the former one but with properties
of the elemental units given by the properties calculated independently for eight packets of order one. The procedure can be repeated
n times to form the final n-order packet being a three-dimensional
aggregate of 8^n unit elements. A graphical illustration of the
procedure of constructing such an aggregate is shown in Figure 7.

To calculate the properties of any packet of 8 elements a procedure similar to that applied for the unit element is proposed.

To satisfy the continuity inside each packet the equality of
strains in all directions for adjacent elements must be assumed.
This can be expressed by the following equations

$$\varepsilon_x = \left[\varepsilon_x(1,j,k) + \varepsilon_x(2,j,k)\right]/2 \qquad (10a)$$

$$\varepsilon_y = \left[\varepsilon_y(i,1,k) + \varepsilon_y(i,2,k)\right]/2 \qquad (10b)$$

$$\varepsilon_z = \left[\varepsilon_z(i,j,1) + \varepsilon_z(i,j,2)\right]/2 \qquad (10c)$$

The sum of stresses acting on all elements of the packet must be
equal to the external stress acting on the packet. This leads to
the following equations of the stress belance

$$\frac{1}{4}\sum_{jk}\sigma_x(1,j,k) = \frac{1}{4}\sum_{jk}\sigma_x(2,j,k)=\sigma_x \qquad (11a)$$

$$\frac{1}{4}\sum_{ik}\sigma_y(i,1,k)= \frac{1}{4}\sum_{ik}\sigma_y(i,2,k)=\sigma_y \qquad (11b)$$

$$\frac{1}{4}\sum_{ij}\sigma_z(i,j,1)= \frac{1}{4}\sum_{ij}\sigma_z(i,j,2)=\sigma_z \qquad (11c)$$

Properties of every element of the packet can be given by constitutive equation in the following form

$$\begin{bmatrix} \sigma_x(i,j,k) \\ \sigma_y(i,j,k) \\ \sigma_z(i,j,k) \end{bmatrix} = \begin{bmatrix} E_{mn}(i,j,k) \end{bmatrix} \cdot \begin{bmatrix} \varepsilon_x(i,j,k) \\ \varepsilon_y(i,j,k) \\ \varepsilon_z(i,j,k) \end{bmatrix} \qquad (12)$$

where only normal stress and strain components are considered and
$E_{mn}(i,j,k)$ are respective moduli of the element (i,j,k) in the packet. In the equations (10-12) the i,j and k values are integers
indicating coordinates of elements in the aggregate along x,y and
z directions respectively.

Simultaneous solution of equations (10-12) leads to the constitutive equations of the packet. The solution for calculation
of properties of the packet of 8 elements is shown in appendix.
If such solution is known repetition of calculations for all packets

of every order can give the properties of the aggregate. The method is suitable only for computer calculations

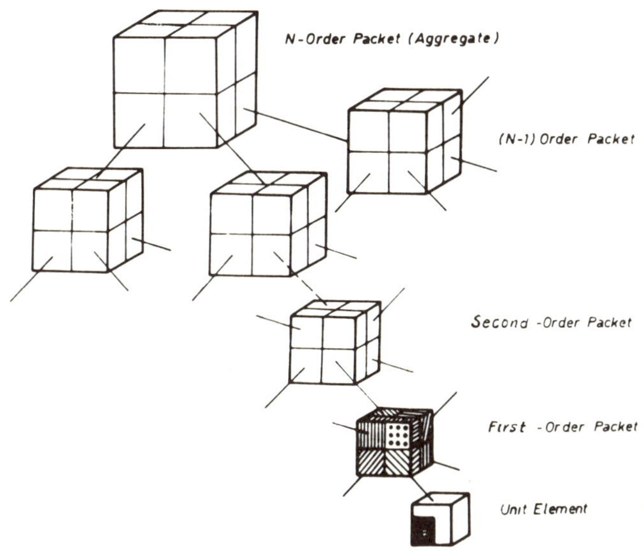

Figure 7. Schematic illustration of construction of three-dimensional aggregate ("packet" model).

EXAMPLES OF APPLICATION OF THE PACKET MODEL

The averaging method according to the packet model of an aggregate has been compared with series and parallel models for the case
shown in the former paragraph where those models were used to
describe viscoelastic properties of randomly oriented lamellar structures. Assuming the same properties of components as before and
considering an aggregate of the third order (containing 512 unit
elements) the frequency dependences of E' and $\tan\delta$ have been calculated for various compositions of the system. The results are
shown in Figure 8. They should be compared with dependences from
Figures 6c and 6d related to parallel and series models respectively.
It is seen that the results obtained using the packet model lie
between extremes predicted by series and parallel models.

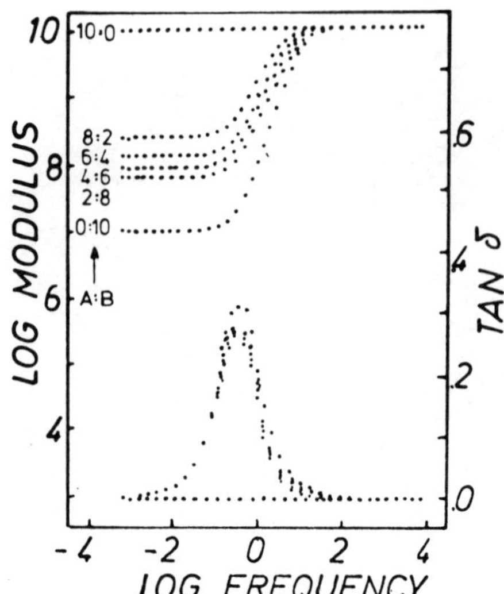

Figure 8. Viscoelastic functions calculated according to "packet" aggregate model with randomly oriented lamellar elements (compare with Figure 6c, d).

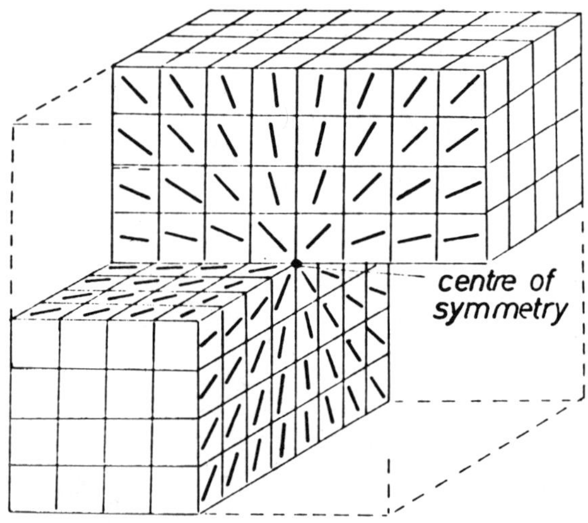

Figure 9. Representation of a spherulite by the aggregate model.

Another comparison in shown in Figure 10 where the static elastic moduli are calculated according to various models and for different structures of unit elements. Additionally a model in which orientational correlation of unit elements has been assumed is considered. Schematic diagram of the aggregate in which the elements have the anisotropy axes oriented radially with respect to the symmetry centre of the aggregate is shown in Figure 9. Such a model can be regarded as representation of the spherulitic structure. The results presented in Figure 10 show that in the case when the spherulitic aggregate is built up from lamellar elements the moduli calculated are lower than those for aggregates with randomly oriented elements. This result can be however well justified by the presence of correlated bands in the spherulitic structure (oriented at 45° w.r.t. external stress) which have the low resistance to shear deformation expecially in the case when the unit elements in the aggregate have the structure of the lamellar type. It is also seen from the above comparison of various models that in the case when aggregates are built up from elements with isotropic properties all models coincide.

The packet model has been also used to estimate the experimental data determined for blends of styrene-acrylonitrile copolymer (SAN) and a graft copolymer made by grafting SAN on polybutadiene (PBD graft)[26]. The temperature dependences of E' and tanδ for pure components and blends with various compositions are shown in Figure 11. Aggregates with unit elements related to various structures have been considered to calculate properties of blends assuming experimentally measured properties of components. The best agreement between experimental data and functions predicted by models has been obtained in the case when the aggregate was constructed as a mixture of randomly oriented unit elements with various structures lamellar, fibrillar and droplet like and when elements of both types Ain B and Bin A with respective compositions were considered. Results obtained in this way are plotted in Figure 11. Examination of structure of blends by means of electron microscopy for blends of this type shows that it is difficult to asign a particular structure type to the observed one. This shows that good estimation of experimental data by a model with a mosaic structure created as described above can be regarded as a satisfactory result.

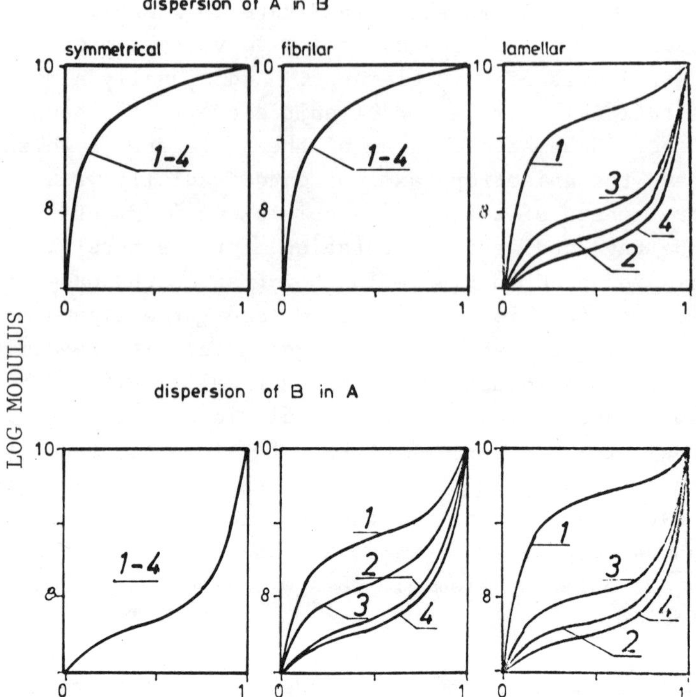

Figure 10.   Modulus – composition functions for various structures
calculated according to various aggregate models:
(1) "parallel" model
(2) "series" model   random orientation
(3) "packet" model   of elements
(4) "packet" model of the spherulite

CONCLUSIONS

Several advantages of the proposed averaging procedure for
three-dimensional aggregates (the packet model) can be noticed:
- Equivalence of strains is only assumed for the nearest neighbours
  in the aggregate and is not necessary for the elements far away
  from each other.
- Any regular or statistical arrangements of structural elements

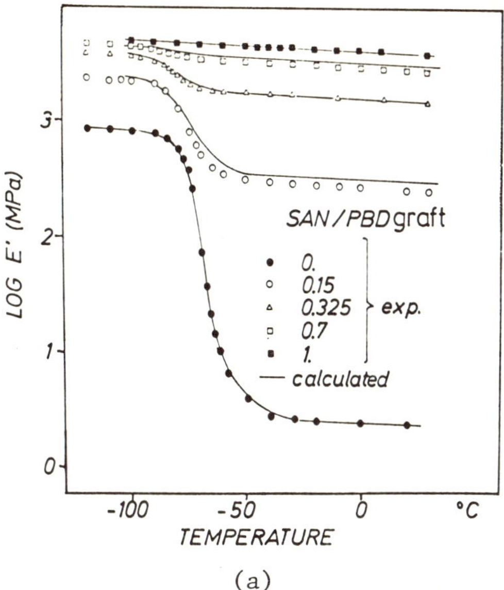

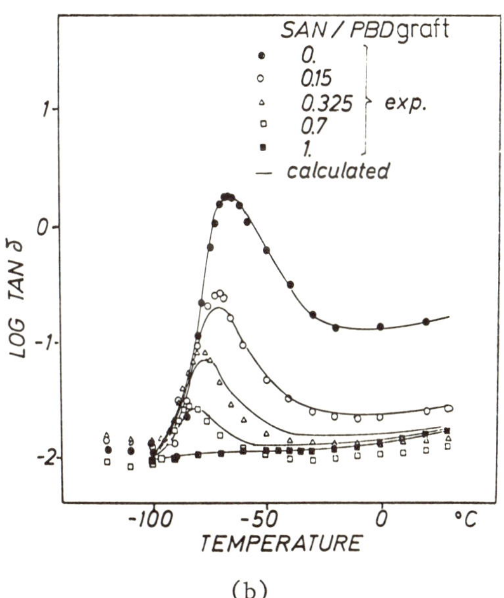

Figure 11. Estimation of experimental viscoelastic functions for blends of graft polybutadiene and styrene-acrylonitrile copolymer by functions calculated according to "packet" model.

can be introduced into the aggregate both in local and macrosco-
pical scale.
- An additional parameter, size of structural elements is introduced indirectly and is determined by the order (n) of the aggregate.
In this way even relationships between structures of different
size levels can be analysed.
- An unique equivalence between particular detailed arrangements
of unit elements in the aggregate and properties of the aggregate
is obtained.

Further detailed comparison of the model with experimental
results is needed expecially for samples with well characterized
structures.

ACKNOWLEDGMENTS

The author is very grateful to dr. G. Locati for his giving
an access to experimental results presented in this work.

APPENDIX

Calculation of moduli for the "packet" model

Introducing asignments

$$
\left.\begin{array}{l}
\varepsilon_x(1,j,k) = \varepsilon_1 \\
\varepsilon_x(2,j,k) = \varepsilon_2 \\
\varepsilon_y(i,1,k) = \varepsilon_3 \\
\varepsilon_y(i,2,k) = \varepsilon_4 \\
\varepsilon_z(i,j,1) = \varepsilon_5 \\
\varepsilon_z(i,j,2) = \varepsilon_6
\end{array}\right\} \qquad \text{(for } i,j \text{ and } k = 1,\ 2) \qquad\qquad (I)
$$

and combining eqs. (11) and (12) (see the text) we can write

$$
\sum_n P_{1n}\varepsilon_n = 4\sigma_x \qquad\qquad (IIa)
$$

$$
\sum_n P_{2n}\varepsilon_n = 4\sigma_x \qquad\qquad (IIb)
$$

$$
\sum_n P_{3n}\varepsilon_n = 4\sigma_y \qquad\qquad (IIc)
$$

$$
\sum_n P_{4n}\varepsilon_n = 4\sigma_y \qquad\qquad (IId)
$$

$$
\sum_n P_{5n}\varepsilon_n = 4\sigma_z \qquad\qquad (IIe)
$$

$$
\sum_n P_{6n}\varepsilon_n = 4\sigma_z \qquad\qquad (IIf)
$$

where the matrix of P_{mn} coefficients has the following form

$$
\begin{bmatrix}
\sum_{jk} E_{11}(1jk), & 0 & , \sum_k E_{12}(11k), & \sum_k E_{12}(12k), & \sum_j E_{13}(1j1), & \sum_j E_{13}(1j2) \\
0 & , \sum_{jk} E_{11}(2jk), & \sum_k E_{12}(21k), & \sum_k E_{12}(22k), & \sum_j E_{13}(2j1), & \sum_j E_{13}(2j2) \\
\sum_k E_{21}(11k), & \sum_k E_{21}(21k), & \sum_{ki} E_{22}(i1k), & 0 & \sum_i E_{23}(i11), & \sum_i E_{23}(i12) \\
\sum_k E_{21}(12k), & \sum_k E_{21}(22k), & 0 & , \sum_{ki} E_{22}(i2k), & \sum_i E_{23}(i21), & \sum_i E_{23}(i22) \\
\sum_j E_{31}(1j1), & \sum_j E_{31}(2j1), & \sum_i E_{32}(i11), & \sum_i E_{32}(i21), & \sum_{ij} E_{33}(ij1), & 0 \\
\sum_j E_{31}(1j2), & \sum_j E_{31}(2j2), & \sum_i E_{32}(i12), & \sum_i E_{32}(i22), & 0 & , \sum_{ij} E_{33}(ij2)
\end{bmatrix}
$$

where E_{mn} are modulus constants of respective packet elements with coordinates ijk in directions x,y and z respectively.
From equations (IIa), (IIc) and (IIe) we can calculate the strain components in the following form

$$\varepsilon_1 = (S_{11}\varepsilon_2 + S_{12}\varepsilon_4 + S_{13}\varepsilon_6 + S_{14}\sigma_x + S_{15}\sigma_y + S_{16}\sigma_z)/S$$
$$\varepsilon_3 = (S_{21}\varepsilon_2 + S_{22}\varepsilon_4 + S_{23}\varepsilon_6 + S_{24}\sigma_x + S_{25}\sigma_y + S_{26}\sigma_z)/S \qquad (III)$$
$$\varepsilon_5 = (S_{31}\varepsilon_2 + S_{32}\varepsilon_4 + S_{33}\varepsilon_6 + S_{34}\sigma_x + S_{35}\sigma_y + S_{36}\sigma_z)/S$$

where coefficients S_{mn} are given by determinants

$$S = \begin{vmatrix} P_{11} & P_{13} & P_{15} \\ P_{31} & P_{33} & P_{35} \\ P_{51} & P_{53} & P_{55} \end{vmatrix}$$

$$S_{11} = \begin{vmatrix} -P_{12} & P_{13} & P_{15} \\ -P_{32} & P_{33} & P_{35} \\ -P_{52} & P_{53} & P_{55} \end{vmatrix}, S_{12} = \begin{vmatrix} -P_{14} & P_{13} & P_{15} \\ -P_{34} & P_{33} & P_{35} \\ -P_{54} & P_{53} & P_{55} \end{vmatrix}, S_{13} = \begin{vmatrix} -P_{16} & P_{13} & P_{15} \\ -P_{36} & P_{33} & P_{35} \\ -P_{56} & P_{53} & P_{55} \end{vmatrix}$$

$$S_{21} = \begin{vmatrix} P_{11} & -P_{12} & P_{15} \\ P_{31} & -P_{32} & P_{35} \\ P_{51} & -P_{52} & P_{55} \end{vmatrix}, S_{22} = \begin{vmatrix} P_{11} & -P_{14} & P_{15} \\ P_{31} & -P_{34} & P_{35} \\ P_{51} & -P_{54} & P_{55} \end{vmatrix}, S_{23} = \begin{vmatrix} P_{11} & -P_{14} & P_{15} \\ P_{31} & -P_{36} & P_{35} \\ P_{51} & -P_{56} & P_{55} \end{vmatrix}$$

$$S_{31} = \begin{vmatrix} P_{11} & P_{13} & -P_{12} \\ P_{31} & P_{33} & -P_{32} \\ P_{51} & P_{53} & -P_{52} \end{vmatrix}, S_{32} = \begin{vmatrix} P_{11} & P_{13} & -P_{14} \\ P_{31} & P_{33} & -P_{34} \\ P_{51} & P_{53} & -P_{54} \end{vmatrix}, S_{33} = \begin{vmatrix} P_{11} & P_{13} & -P_{16} \\ P_{31} & P_{33} & -P_{36} \\ P_{51} & P_{53} & -P_{56} \end{vmatrix}$$

$$S_{14} = \begin{vmatrix} P_{33} & P_{35} \\ P_{53} & P_{55} \end{vmatrix} \quad S_{15} = \begin{vmatrix} P_{13} & P_{15} \\ P_{53} & P_{55} \end{vmatrix} \quad S_{16} = \begin{vmatrix} P_{13} & P_{15} \\ P_{33} & P_{35} \end{vmatrix}$$

$$S_{24} = \begin{vmatrix} P_{31} & P_{35} \\ P_{51} & P_{55} \end{vmatrix} \quad S_{25} = \begin{vmatrix} P_{11} & P_{15} \\ P_{51} & P_{55} \end{vmatrix} \quad S_{26} = \begin{vmatrix} P_{11} & P_{15} \\ P_{31} & P_{35} \end{vmatrix}$$

$$S_{34} = \begin{vmatrix} P_{31} & P_{33} \\ P_{51} & P_{53} \end{vmatrix} \quad S_{35} = \begin{vmatrix} P_{11} & P_{13} \\ P_{51} & P_{53} \end{vmatrix} \quad S_{36} = \begin{vmatrix} P_{11} & P_{13} \\ P_{31} & P_{33} \end{vmatrix}$$

If the values of ε given by eqs. (III) are introduced to eqs. (IIb), (IId) and (IIf) the latest can be rearanged in the following form

$$T_{11}\ \varepsilon_2 + T_{12}\ \varepsilon_4 + T_{13}\ \varepsilon_6 = T_{14}\ \sigma_x + T_{15}\ \sigma_y\ T_{16}\ \sigma_z$$

$$T_{21}\ \varepsilon_2 + T_{22}\ \varepsilon_4 + T_{23}\ \varepsilon_6 = T_{24}\ \sigma_x + T_{25}\ \sigma_y + T_{26}\ \sigma_z \qquad\text{(IV)}$$

$$T_{31}\ \varepsilon_2 + T_{32}\ \varepsilon_4 + T_{33}\ \varepsilon_6 = T_{34}\ \sigma_x + T_{35}\ \sigma_y + T_{36}\ \sigma_z$$

where

$$T_{11}=P_{22}+(P_{23}S_{21}+P_{25}S_{31})/S \qquad T_{14}=1-(P_{23}S_{24}+P_{25}S_{34})/S$$

$$T_{12}=P_{24}+(P_{23}S_{22}+P_{25}S_{32})/S \qquad T_{15}=-(P_{23}S_{25}+P_{25}S_{35})/S$$

$$T_{13}=P_{26}+(P_{23}S_{23}+P_{25}S_{33})/S \qquad T_{16}=-(P_{23}S_{26}+P_{25}S_{36})/S$$

$$T_{21}=P_{42}+(P_{41}S_{11}+P_{45}S_{31})/S \qquad T_{24}=-(P_{41}S_{14}+P_{43}S_{34})/S$$

$$T_{22}=P_{44}+(P_{42}S_{12}+P_{43}S_{32})/S \qquad T_{25}=1-(P_{41}S_{15}+P_{43}S_{35})/S$$

$$T_{23}=P_{46}+(P_{42}S_{13}+P_{43}S_{33})/S \qquad T_{26}=-(P_{41}S_{16}+P_{43}S_{36})/S$$

$$T_{31}=P_{62}+(P_{61}S_{11}+P_{63}S_{21})/S \qquad T_{34}=-(P_{61}S_{14}+P_{63}S_{24})/S$$

$$T_{32}=P_{64}+(P_{61}S_{12}+P_{63}S_{22})/S \qquad T_{35}=-(P_{61}S_{15}+P_{63}S_{25})/S$$

$$T_{33}=P_{66}+(P_{61}S_{13}+P_{63}S_{23})/S \qquad T_{36}=1-(P_{61}S_{16}+P_{63}S_{26})/S$$

From eqs. (IV) one can calculate the strain components

$$\varepsilon_2 = A_{41}\ \sigma_x + A_{42}\ \sigma_y + A_{43}\ \sigma_z$$

$$\varepsilon_4 = A_{51}\ \sigma_x + A_{52}\ \sigma_y + A_{53}\ \sigma_z \qquad\text{(V)}$$

$$\varepsilon_6 = A_{61}\ \sigma_x + A_{62}\ \sigma_y + A_{63}\ \sigma_z$$

where

$$A_{41} = \frac{1}{A} \begin{vmatrix} T_{14} & T_{12} & T_{13} \\ T_{24} & T_{22} & T_{23} \\ T_{34} & T_{32} & T_{33} \end{vmatrix} \quad A_{12} = \frac{1}{A} \begin{vmatrix} T_{15} & T_{12} & T_{13} \\ T_{25} & T_{22} & T_{23} \\ T_{34} & T_{32} & T_{33} \end{vmatrix} \quad A_{43} = \frac{1}{A} \begin{vmatrix} T_{16} & T_{12} & T_{13} \\ T_{26} & T_{22} & T_{23} \\ T_{36} & T_{32} & T_{33} \end{vmatrix}$$

$$A_{51} = \frac{1}{A} \begin{vmatrix} T_{11} & T_{14} & T_{13} \\ T_{21} & T_{24} & T_{23} \\ T_{31} & T_{34} & T_{33} \end{vmatrix} \quad A_{52} = \frac{1}{A} \begin{vmatrix} T_{11} & T_{15} & T_{13} \\ T_{21} & T_{25} & T_{23} \\ T_{31} & T_{35} & T_{33} \end{vmatrix} \quad A_{53} = \frac{1}{A} \begin{vmatrix} T_{11} & T_{16} & T_{13} \\ T_{21} & T_{26} & T_{23} \\ T_{31} & T_{36} & T_{33} \end{vmatrix}$$

$$A_{61} = \frac{1}{A} \begin{vmatrix} T_{11} & T_{12} & T_{14} \\ T_{21} & T_{22} & T_{24} \\ T_{31} & T_{32} & T_{34} \end{vmatrix} \quad A_{62} = \frac{1}{A} \begin{vmatrix} T_{11} & T_{12} & T_{15} \\ T_{21} & T_{22} & T_{25} \\ T_{31} & T_{32} & T_{35} \end{vmatrix} \quad A_{63} = \frac{1}{A} \begin{vmatrix} T_{11} & T_{12} & T_{16} \\ T_{21} & T_{22} & T_{26} \\ T_{31} & T_{32} & T_{36} \end{vmatrix}$$

$$A = \begin{vmatrix} T_{11} & T_{12} & T_{13} \\ T_{21} & T_{22} & T_{23} \\ T_{31} & T_{32} & T_{33} \end{vmatrix}$$

and the other strain components according to (III) and (V) are

$$\varepsilon_1 = A_{11}\sigma_x + A_{12}\sigma_y + A_{13}\sigma_z$$
$$\varepsilon_3 = A_{21}\sigma_x + A_{22}\sigma_y + A_{23}\sigma_z \qquad\qquad (VI)$$
$$\varepsilon_5 = A_{31}\sigma_x + A_{32}\sigma_y + A_{33}\sigma_z$$

where

$$A_{11} = (S_{14} + S_{11}A_{41} + S_{12}A_{51} + S_{13}A_{61})/S$$
$$A_{12} = (S_{15} + S_{11}A_{42} + S_{12}A_{52} + S_{13}A_{62})/S$$
$$A_{13} = (S_{16} + S_{11}A_{43} + S_{12}A_{53} + S_{13}A_{63})/S$$

$$A_{21}=(S_{24}+S_{21}A_{41}+S_{22}A_{51}+S_{23}A_{61})/S$$

$$A_{22}=(S_{25}+S_{21}A_{42}+S_{22}A_{52}+S_{23}A_{62})/S$$

$$A_{23}=(S_{26}+S_{21}A_{43}+S_{22}A_{53}+S_{23}A_{63})/S$$

$$A_{31}=(S_{34}+S_{31}A_{41}+S_{32}A_{51}+S_{33}A_{61})/S$$

$$A_{32}=(S_{35}+S_{31}A_{42}+S_{32}A_{52}+S_{33}A_{62})/S$$

$$A_{33}=(S_{36}+S_{31}A_{43}+S_{32}A_{53}+S_{33}A_{63})/S$$

Then according to equation (10) in the text of this paper we have

$$\varepsilon_x = \tfrac{1}{2}(A_{11}+A_{41})\sigma_x + \tfrac{1}{2}(A_{12}+A_{42})\sigma_y + \tfrac{1}{2}(A_{13}+A_{43})\sigma_z$$

$$\varepsilon_y = \tfrac{1}{2}(A_{21}+A_{51})\sigma_x + \tfrac{1}{2}(A_{22}+A_{52})\sigma_y + \tfrac{1}{2}(A_{23}+A_{53})\sigma_z \qquad \text{(VII)}$$

$$\varepsilon_z = \tfrac{1}{2}(A_{31}+A_{61})\sigma_x + \tfrac{1}{2}(A_{32}+A_{62})\sigma_y + \tfrac{1}{2}(A_{33}+A_{63})\sigma_z$$

which gives a constitutive equation of the packet in the form

$$
\begin{bmatrix} \varepsilon_x \\ \varepsilon_y \\ \varepsilon_z \end{bmatrix}
=
\begin{bmatrix} C_{11} & C_{12} & C_{13} \\ C_{21} & C_{22} & C_{23} \\ C_{31} & C_{32} & C_{33} \end{bmatrix}
\begin{bmatrix} \sigma_x \\ \sigma_y \\ \sigma_z \end{bmatrix}
\qquad \text{(VIII)}
$$

where C_{ij} are the compliances of the whole packet expressed by respective coefficients in equations (VII). Thus the equation (VIII) can be treated as the final solution.

REFERENCES

1. C. C. Chamis and G. P. Sendecky, J. Compos. Mat., $\underline{2}$, 332 (1968).
2. M. Takayanagi, H. Harima and Y. Iwata, Mem. Fac. Engng., Kyushu
    Univ., $\underline{23}$, 1 (1963).
3. R. Hill, J. Mech. Phys. Solids, $\underline{12}$, 199 (1964).
4. L. J. Walpole, J. Mech. Phys. Solids, $\underline{17}$, 235 (1969).
5. Z. Hashin, J. Mech. Phys. Solids, $\underline{13}$, 119 (1965).
6. R. Hill, J. Mech. Phys. Solids, $\underline{13}$, 189 (1965).
7. G. A. Van Fo Fy and G. N. Savin, Mechanika Polimerov, $\underline{1}$, 131,
    (1963).
8. C. H. Chen and S. Cheng, J. Compos. Mat., $\underline{1}$, 30 (1967).
9. T. Pakula and M. Kryszewski, Plaste u.Kautschuk, $\underline{24}$, 761 (1977)
9a. T. Pakula, paper submitted for publication.
10. H. Handus, K. Illers and E. Ropte, Kolloid Z.u.Z. Polymere,
    $\underline{216}$, 110 (1967).
11. M. Matsuo, Japan Plast., $\underline{2}$, 6 (1968).
12. A. Keller, E. Pedemonte and F.M. Willmouth, Nature, $\underline{225}$, 538
    (1970).
13. J. Dlugosz, A. Keller and E. Pedemonte, Kolloid Z.u.Z. Polymere,
    $\underline{242}$, 1125 (1970).
14. T. Pakula, J. Grebowicz and M. Kryszewski, paper presented on
    V° Discussion Conference on Phases and Interphases in Macromo-
    lecular Systems, Prag 1976.
15. R. S. Stein, F. B. Khambatta, F. P. Warner, T. Russel,
    A. Escula and E. Balizer, J. Polym. Sci. Polym. Symp., $\underline{63}$, 313
    (1978).
16. C. J. Ong and F. P. Price, J. Polym. Sci. Polym. Symp., $\underline{63}$, 45
    (1978).
17. W. Wenig, F. E. Karasz and W. J. MacKnight, J. Appl. Phys., $\underline{46}$,
    4194 (1975).
18. B. Paul, Trans. AIME, $\underline{218}$, 36 (1960).
19. T. Okamoto and M. Takayanagi, J. Polym. Sci., $\underline{C23}$, 597 (1968).
20. E.H. Kerner, Proc. Phys. Soc., $\underline{69B}$, 808 (1956).
21. R. A. Dickie, J. Appl. Polym. Sci., $\underline{17}$, 45 (1973).
22. D. Kaplin and N. W. Tschoegl, Polym. Ing. Sci., $\underline{14}$, 43 (1974).
23. S. P. Hersh, J. Appl. Polym. Sci. Appl. Polym. Symp., $\underline{31}$, 37
    (1977).
24. T. Pakula, M. Kryszewski, J. Grebowicz and A. Galeski, Polymer
    J., $\underline{6}$, 94 (1974).
25. I. M. Ward, "Mechanical Properties of Solid Polymers", Wiley
    Interscience, (Ch. 10), 1971.
26. G. Locati, unpublished results.

MECHANICAL PROPERTIES OF MULTICOMPONENT POLYMERIC MATERIALS:

A PREDICTIVE CLASSIFICATION OF BINARY SYSTEMS

R. Greco

Istituto di Ricerche su Tecnologia dei Polimeri e
Reologia del C.N.R., Arco Felice, Napoli, Italy

INTRODUCTION

Multicomponent polymeric materials, including polymer blends,
block, graft and random copolymers, interpenetrating and semi-inter-
penetrating networks, gradient polymers or combinations of the above
listed systems, are becoming, recently more and more interesting
from both practical and scientific points of view[1-3]. In fact, it
is possible to obtain on the one hand new plastic materials (oriented
towards well definite end-uses) just optimizing specific properties
coming from different polymers. This result can be obtained by
various processing procedures: 1) melt-mixing two or more homopolymers or copolymers (blends); 2) polymerizing two or more monomers
(copolymers)[1]; 3) forming one or two networks (semi or interpretating
networks or gradient polymers). On the other hand, the concept of
the compatibility of two or more high molecular weight substances
is a very attractive scientific subject. In fact, even for the
simplest case of two amorphous rubbery components in a blend, the
segmental compatibility between the two species is a rare event,
since generally the entropy gain in the mixing is very low[4].
However a few couples of polymers, exhibiting strong attractive
interactions can show a good compatibility or at least a good homogeneity (very fine domains of the dispersed phase). The problem
becomes more and more complicated if also crystallizable or glassy
components are mixed in a blend. In these cases, indeed, many
other phenomena must be taken into account (different chain mobility
of the diverse phases, cocrystallization effects, etc.). For the

purpose of the present work, the assumption of a substantial incompatibility of the two components will be made in an attempt to classify binary polymeric systems, treating as exceptions of this scheme the few "compatible" couples known in the literature[1-4].

BASICAL CLASSIFICATION PRINCIPLES

The binary systems to be classified may involve, as previously mentioned, various types of components: homopolymers, copolymers, networks, etc.. The basical concepts, on which the proposed classification will be based, are the following ones: crystallizability and "glassiness" of the two single components considered at the end-use temperature T_a of the obtained bicomponent material[5].

The crystallizability is an intrinsic stereospecific property of the macromolecular chains but what is really important from practical and scientific points of view, is the actual fractional crystallinity X_c and the melting points, T_m of the used polymers (both depending on the used processing conditions). The "glassiness"[6] of a material, which can be considered to be inversely dependent on the total free volume amount existing in the material at a given temperature and pressure, is more difficult to define. As it is well known[7,8], any polymeric substance undergoes at certain conditions, to a transition from a glassy to a rubbery consistency. Such transition is a multivariable phenomenon which depends on temperature, pressure, time of responses or frequency, presence of diluents and on several other parameters which can influence the changes in the free volume. However from a practical point of view the most important variable, keeping all the other parameters constant, is the temperature, (generally it is the case when a hand-made article is working at its end-use conditions). Following Ferry[8] and Boyer[9] it is possible to make the assumption that at $T=T_g$, where T_g is the glass transition temperature, all the polymers are in iso-free-volume state, that is they contain the same amount of free volume in their amorphous regions. This is true, of course, if one neglects the secondary transition mechanism contributions occurring at $T<T_g$[10] and the dependence of the T_g on the amount of crystallinity. With regards to this last aspect multiple T_g's has been invoked by Boyer[11] to take into account the mobility of the different amorphous phases existing in a semicrystalline polymer. In fact it is possible to recognize at least two rubbery phases: a

pure amorphous region having a higher chain mobility and a more
constrained amorphous phase interconnected with the polymer crystallites (chain loops, chain ends or tie molecules) and therefore two
diverse T_g's must be considered. The former independent of fractional crystallinity X_c and the latter strongly depending on it[11].
However any furhter detailed discussion on this subject is beyond
the scope of this paper. Therefore at a very first approximation
one can choose $\Delta T = T_a - T_g$ as a gross measure or best as a qualitative indication of the "glassiness" of a polymer. A more refined
definition, still on a semiempirical basis, could be referred to
certain criteria proposed by Boyer to define the true nature of a
polymer glass transition phenomenon[12]. Also a third important
criterion in the proposed classification is the role played by the
composition of the examined binary system. In other words, it is
the component which constitutes the matrix of the material to
determine, at a greater extent, the mechanical behaviour of the
polymeric systems. Of course there are also other characteristics
that may influence the overall material features, such as the rubber
cross-linking of one or both the components, the secondary transition mechanisms existing a $T < T_g$ and so on. But, at first approximation the effects of such material conditions can be neglected
in this classification, in which only a very qualitative and gross
mechanical behaviour will be forecasted on the basis of the three
main concepts above mentioned: crystallizability, "glassiness" of
the two components and composition of the binary system.

BINARY POLYMERIC SYSTEM CLASSIFICATION

In Table 1 the various bicomponent polymeric systems have been
divided in three main classes with respect to the crystallizability
of the two constituents α and β.
- Class 1: both α and β are crystallizable.
- Class 2: α is crystallizable, β is not.
- Class 3: both α and β are not crystallizable.
In the third column a subdivision of these three classis has been
qualitatively made with respect to the glassy or rubbery consistency
of α and β comparing their T_g's with the end-use temperature T_a
(irrespectively of their effective "glassiness"). In the fourth
column a further subdivision is made taking into accounts the composition effects. In fact, when the two components have almost the
same consistency (glassy or rubbery) the mechanical behaviour is,

TABLE 1. <u>Binary polymeric system classification</u>

Class	Crystallizability	"Glassiness"	Composition	Qualitative mech. prediction	Binary systems
1	α,β both amorphous	1.1. $T_{g\alpha}$, $T_{g\beta} < T_a$	any	rubber	BR–PIB EPDM–EPM
		1.2. $T_{g\alpha} < T_a < T_{g\beta}$	1.2.1. $C_\alpha >> C_\beta$	reinforced rubber	SBS SIS
			1.2.2. $C_\beta >> C_\alpha$	toughened glass	HIPS
		1.3. $T_{g\alpha}$, $T_{g\beta} > T_a$	any	glass	PS–PMMA
2	α crystallizable β amorphous	2.1. $T_{g\alpha}$, $T_{g\beta} > T_a$	2.1.1. $C_\alpha >> C_\beta$	toughened plastics	HDPE–PIB(EPM)
			2.1.2. $C_\beta >> C_\alpha$	reinforced rubber	PIB–HDPE
		2.2. $T_{g\alpha} < T_a < T_{g\beta}$	2.2.1. $C_\alpha >> C_\beta$	reinforced plastics	HDPE–aPS
			2.2.2. $C_\beta >> C_\alpha$	quasi glass	aPS–iPP
		2.3. $T_{g\beta} < T_a < T_{g\alpha}$	2.3.1 $C_\alpha >> C_\beta$	toughened plastics	P4MP–PIB Ny6–EPM iPP–EPM
			2.3.2. $C_\beta >> C_\alpha$	reinforced rubber	PIB–P4MP EPM–Ny
		2.4. $T_{g\alpha}$, $T_{g\beta} > T_a$	any	glass	aPS–P4MP
3	α,β both crystallizable	3.1. $T_{g\alpha}$, $T_{g\beta} < T_a$	3.1.1. any	plastics	HDPE–iPP
		3.2. $T_{g\alpha} < T_a < T_{g\beta}$	3.2.1. C C	reinforced plastics	HDPE–P4MP
			$C_\beta >> C_\alpha$	quasi glass	P4PM–HDPE
		3.3. $T_{g\alpha}$, $T_{g\beta} > T_a$	any	glass	P4MP–Ny

with minor differences from case to case, almost independent of
composition; when, on the other hand, one of the two polymers has
a very different consistency from the other, it is of the utmost
importance to know which one constitutes the matrix and which one
the dispersed phases of the system. Of course there are also
intermediate behaviour in the central range of composition where
both polymers may form continuous interconnected phases and where
an inversion of matrix can take place. Finally in the fifth column
the qualitative mechanical behaviour is predicted on the basis of
the three basical concepts used in this classification. In the
sixth (last) column a few examples of polymer systems are indicated;
and their mechanical properties are briefly discussed in the
remainder of this work. They have been arbitrarily chosen to give
only a practical explanation of the above mentioned classification
through some examples, without any pretence to make an exhaustive
review of the mechanical behaviour of binary polymeric systems.
The following abbreviations have been used throughout the text:

a) <u>Homopolymers</u>:

BR = Butyl rubber
PIB = Polyisobutylene
aPS = Atactic polystyrene
PVC = Poly(vinyl chloride)
B = Polybutadiene
HDPE = High density polyethylene
LDPE = Low density polyethylene
iPP = Isotactic polypropylene
Ny 6 = Nylon 6
P4MP = Poly(4 methyl pentene 1)
PC = Polycarbonate
PET = Polyethylene terephtalate

b) <u>Copolymers</u>

EPDM = Ethylene-propylene-diene random terpolymer
EPM = Ethylene-propylene random copolymer
SBS = Poly(styrene-co-butadiene-co-styrene)
SIS = Poly(styrene-co-isoprene-co-styrene)
SAN = Poly(styrene-co-acrylonitrile)
S-co-B = Poly(styrene-co-butadiene)
B-g-SAN = Polybutene-g-Poly(styrene-co-acrylonitrile)
NR = (Nitrile rubber)butadiene-acrylonitrile random copolymer
ABS = Poly(acrylonitrile-co-butadiene-co-styrene)

MBS = Poly(methylmetacrylate-co-butadiene-co-styrene)

Class 1.1 $(T_{g\alpha}, T_{g\beta}) < T_a$

    In this system both α and β are rubbery polymers and form an overall rubber-like system. In the chosen example α (BR or EPDM) is a crosslinkable component containing in the average a double bond every a few hundreds of C atoms, whereas β (PIB or EPM) is a saturated component of the same chemical nature. After crosslinking such a system is constituted by an α network containing a certain amount of β unattached reptating molecules (see Figure 1). The α network is the supporting materials which reacts to the equilibrium conditions; the β molecules instead can only respond in transient phenomena like creep, stress relaxation or dynamic mechanical tests.

    The resulting material, rubbery in nature, is able to damp vibration or noises at frequencies which are proportional to about the third power of the β weight average molecular weight[13-15] at an extent depending on the relative β amount. It is possible through this way to design elastomers with damping peaks at certain given frequencies for many practice purposes.

    The mechanical dynamic data in agreement with the De Gennes[16] and Edwards[17] theories and with Doi's calculations[18] are reported elsewhere[13-15]. This example is also very interesting because it can give a diverse view of the compatibility concept. In fact, chemically speaking, due to the nature of the two components, a segmental compatibility is assured between α and β macromolecules. However each component mantains in the blend its own individuality, due to crosslinking of α and to the high molecular weight of the β unattached macromolecules, as shown by the dynamical mechanical spectra. Therefore in such a case the overall entropy depends remarkably also on the way in which the two constituents are interconnected on a large scale and not only on the local segmental compatibility (which is on the other hand a necessary condition to avoid macro or microphase separations with very different relaxation mechanisms).

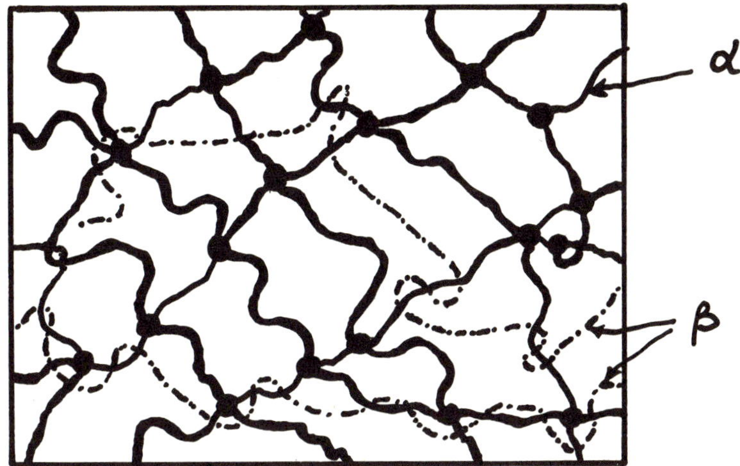

Figure 1. Schematic representation of a binary system consisting
of an α network with β unattached reptating molecules.

Class 1.2 ($T_{g\alpha} < T_a < T_{g\beta}$)

Since α is rubbery and β is glassy it is very crucial which
one of them is the matrix of the system, therefore subcases must be
considered:

Class 1.2.1 $C_\alpha \gg C_\beta$

The chosen example is the SBS (or SIS) block copolymer with a
prevailing amount of polybutadiene (β) reinforced by spherical particles of polystyrene (α) (see Figure 2). These systems, known as
"thermoplastic elastomers", are an outstanding example of improved
processing procedures directed towards definite end-use materials.
In fact, they can help to avoid the inconvenience of compounding
rubber with carbon black and of thermosetting it permanently in slow
cumbersome curing cycles[2]. This has been accomplished by synthesizing αβα sandwich block copolymers, in which the very large central
soft block (M_w=6x10^5) is surrounded by two shorter "hard" segments
M_w=1x10^5) agglomerated in firm spherical clusters in finished

products. The hard domains act both as reinforcing fillers and
thermoplastic multiple "crosslinks"[2].

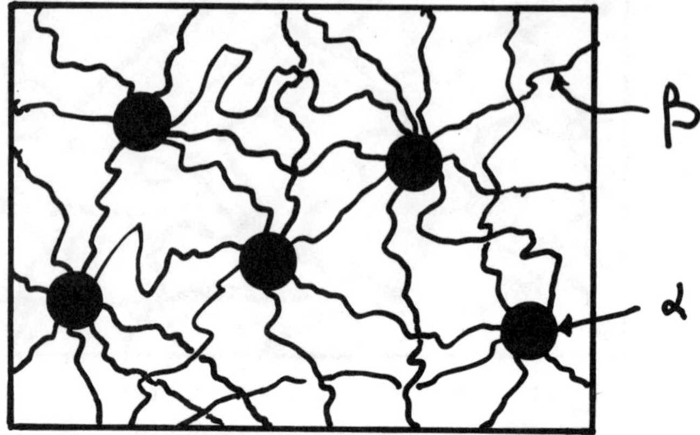

Figure 2.   Schematic representation of an $\alpha\beta\alpha$ thermoplastic elastomer.

Class 1.2.2 $C_\beta \gg C_\alpha$

This is the reversed system with respect to the previously
discussed one. In fact the matrix is now a glassy polymers (β)
whereas the dispersed particles are rubbery (α). Also these kind of
systems have become of the utmost commercial importance as "toughened
polymers". In Table 2 are reported many examples of rigid brittle
polymers (first column) like aPS, SAN, PVC to which a rubbery impact
modifier (second column) has been added. The strong modification due
to such an additive brings to a series of features (third column)
as yielding and stress whitening and to a great improvement of some
mechanical properties such as elongation at break and impact strength,
with some losses in the modulus and in the tensile strength. Such
characteristics are completely absent in the original very brittle
materials.

For any further discussions on this subject, which is far beyond
the scope of this work, the reader is referred to more specific
papers[19-20].

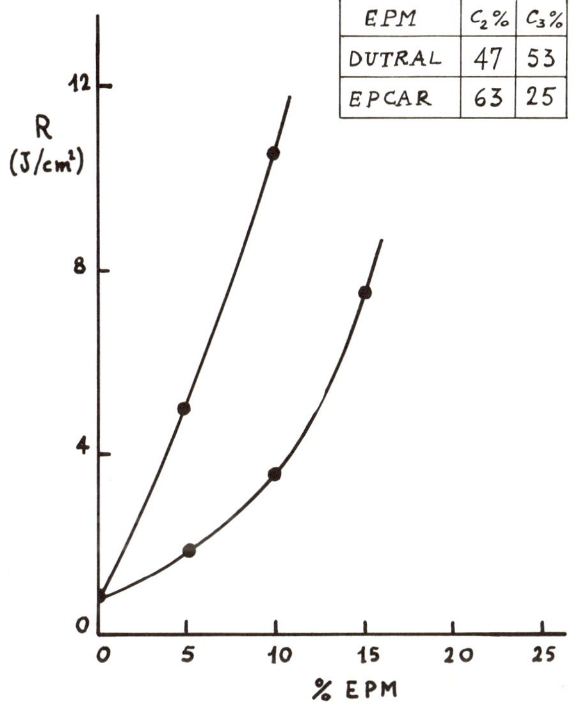

Figure 3. Impact strength R versus percentage of EPM. The two
curves are relative to two different commercial EPM
copolymers (Dutral and Epcar) having ethylene percentage
in weight as indicated).

TABLE 2

Toughened glassy polymers

Glassy matrix	Rubbery additives	Modified polymer features
aPS	S-co-B B	High elongation at break
SAN	B-g-SAN ABS NR	Yielding Stress whitening
PVC	NR ABS MBS	High impact strength

<u>Class 1.3</u> $(T_\alpha, T_{g\beta} > T_a)$

Both the components are glassy therefore such a system is not of a very great interest since in any case a brittle mechanical behaviour is to be expected (see for example aPS-PMMA blends).

<u>Class 2.1</u> $(T_{g\alpha}, T_{g\beta} < T_a)$

In this case there is no symmetry since α is crystallizable and β is not, therefore two subcases must be taken into account:
<u>Class 2.1.1</u> $C_\alpha \gg C_\beta$ as an example the couple HDPE + rubber (EPM or PIB) can be considered. The mechanical behaviour is a even more toughened plastics as shown in Figure 3 in which the Izod impact strength R is reported as a function of two commercial EPM copolymer weight content[21].
<u>Class 2.1.2</u> $C_\beta \gg C_\alpha$ one can consider inversely rubber + HDPE, whose mechanical behaviour can yield a slightly reinforced rubber.

__Class 2.2__ $(T_{g\alpha} < T_a < T_{g\beta})$

An example of such class can be given by PO-aPS blends. Such systems have been extensively studied in our laboratory and the results have been presented in previous papers and are also discussed in this book. The processing conditions, the thermal swelling and mechanical properties, have been studied over the entire range of composition and related to the resulting morphology observed by optical and electron microscopy[22-25]. Of course there are two extreme conditions:

__Class 2.2.1__ $C_\alpha \gg C_\beta$ where the mechanical behaviour is that of a reinforced semicrystalline matrix, like i.e. HDPE-aPS, with aPS dispersed phase.

__Class 2.2.2__ $C_\beta \gg C_\alpha$ whose behaviour is very similar to a glassy aPS matrix minor different features.

__Class 2.3__ $(T_{g\beta} < T_a < T_{g\alpha})$

This class is just the reversed case, in which a glassy semicrystalline polymer like Nylon or P4MP, is mixed with a rubber polymer (like EPM or PIB). Also in this case the extreme conditions in the composition are given by:

__Class 2.3.1__ $C_\alpha \gg C_\beta$ Yielding a toughened plastics. An example is treated in a paper of this book, in which an EPM copolymer (Dutral of Montedison) has been added to the Ny6[26].

__Class 2.3.2__ $C_\beta \gg C_\alpha$. Yielding a reinforced rubber.

__Class 2.4__ $(T_{g\alpha}, T_{g\beta} > T_a)$

Both the semicrystalline and the amorphous component are glassy in nature. No particular features different from a semibrittle behaviour are to be expected by such a system (aPS-Ny).

__Class 3.1__ $(T_{g\alpha}, T_{g\beta} < T_a)$

Here, both the components are semicrystalline polymers with amorphous rubbery regions. Therefore the overall behaviour of such blends is substantially ductile or semiductile as that of the two homopolymers. HDPE-iPP as prototype of this class has been studied in few papers in our laboratory encompassing the entire range of composition[27,28].

Class 3.2 ($T_{g\alpha} < T_a < T_{g\beta}$)

Here, one semicrystalline polymers is rubbery whereas the
another one is glassy in its amorphous regions (i.e. HDPE-P4MP).
Two subcases can be given with respect to the composition effect:
$C_\alpha \gg C_\beta$ yielding a reinforced plastics like HDPE-P4MP or $C_\beta \gg C_\alpha$
yielding a more rigid material like P4MP-HDPE.

Class 3.3 ($T_{g\alpha}, T_{g\beta} > T_a$)

In all the composition this system is substantially glassy
and therefore of a little practical interest.

It must be noted that in all the classes above mentioned where
glassy components are involved, it is of a certain value the
existence, for some of these polymers, of secondary relaxation
mechanisms below T_g, just as in the case of PC and nylon 6. In
fact in these cases the toughness is higher than in others, like
aPS, where these phenomena are totally absent.

TERNARY POLYMERIC SYSTEMS

On the basis of the proposed classification it is possible
to study ternary systems where a suitable third component is added
in relatively small amount to the previously considered bicomponent
materials to increase the compatibility of their constituents or
at least to improve the homogeneity of the material. Generally
these additives are some copolymers, whose chemical sequences are
made by the same monomers of the two polymers of the primary
bicomponent material. They can be distinguished with respect to
their chemical-physics characteristics in two different species:
a) interfacial agents, like _graft_ or _block_ polymers that can act
as emulsifiers, namely interfacial agents in a system, reducing
the surface tension of the two polymers and increasing their interfacial adhesion. The effect on the morphology of the material is
to yield smaller size domains and a good interfacial adhesion with
strong mechanical behaviour improvements. One of these systems,
is the one studied by Barentsen et al. namely LDPE-aPS blends to

which a graft LDPE-g-aPS copolymer was added[29-31].

A second example are the Ny-EPM blends to which a Ny-g-EPM
has be added. It is also possible, as shown elsewhere in this book,
to obtain the graft formation during the melt mixing of the material[26].

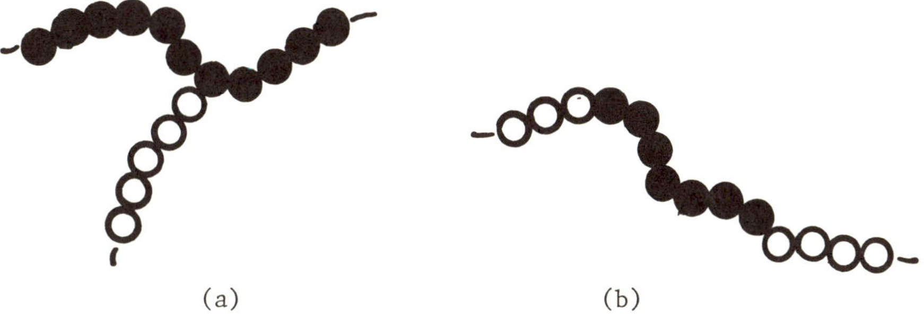

(a)   (b)

Figure 4. a) graft copolymer; b) block copolymer.

b) <u>A random copolymer</u> (Figure 5) has quite different properties
because it exhibits intermediate properties between those of the
corresponding homopolymers depending on composition (i.e. $T_g=T_g(\alpha/\beta)$)
and therefore it may act as a compatibilizer making easier the miscibility of the amorphous regions of the two components.

Figure 5.   Random copolymer.

An interesting example of this kind is that of random EPM's added
to HDPE-iPP blend[21].

In Figure 6 are reported the strength σ_r of HDPE-iPP blends to
which different quantities of a EPM (Epcar) has been added.

Also the Izod impact strength R is greatly improved passing
from the binary HDPE-iPP blends to ternary blends in which different
Epcar weight contents has been added (see Figure 7). Any further
details can be found in the original work[21].

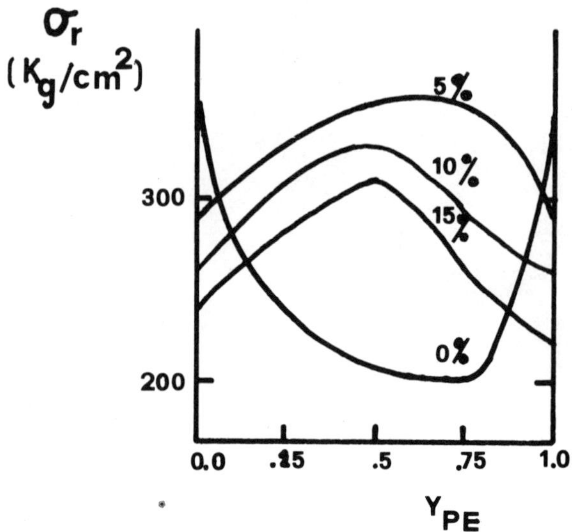

Figure 6.   Tensile strength σ_r as a function of $Y_{PE}=W_{HDPE}/W_{HDPE}+W_{iPP}$ (weight HDPE binary fraction) for ternary blends containing different weight percentages of Epcar as indicated.

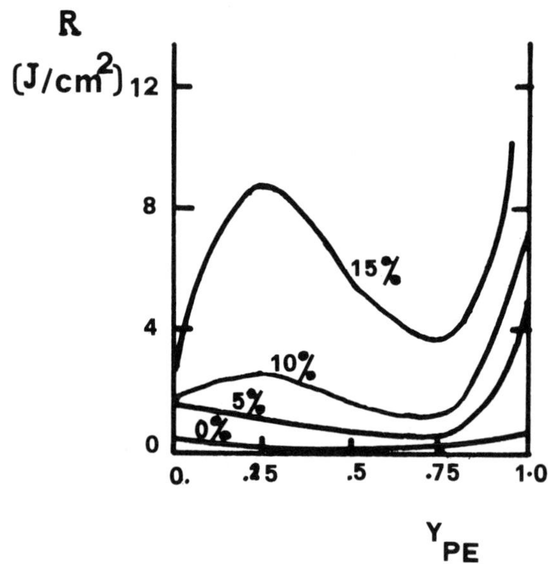

Figure 7.   Izod impact strength R as a function of Y_{PE} for ternary blends containing different weight percentages of Epcar as indicated.

CONCLUSION

In this paper an attempt has been made to classify binary polymeric systems with respect to their mechanical behaviour by the
knowledge of the characteristics of the single components at the end-
use temperature T_a. The prediction made on the basis of crystallizability, "glassiness" and composition conditions can be very useful
in choosing a given system both for practice pupuses, or to carry
on further and thourough studies on their properties taking into
accounts many other important features as crosslinkability, secondary
on mechanisms, processing conditions and so on.

The addition of additives such as suitable block, graft or
random copolymers can greatly improve the mechanical features of
binary systems.

REFERENCES

1. W. J. MacKnight, J. Stoelting and F. E. Karansk, "Multicomponent
    polymeric systems" Advances in Chemistry Series 99, Ed.
    R. F. Gould, Am. Chem. Soc., 1971.
2. H. Sperling "Recent advances in polymer blends, Graft and Blocks"
    Plenum Press, New York, 1974.
3. J. A. Manson and L. H. Sperling, "Polymer Blends and Composites"
    Plenum Press, New York, 1976.
4. R. Paul and S. Newman "Polymer Blends", Academic Press, London,
    1978.
5. R. Greco and E. Martuscelli, La Chimica & l'Industria, $\underline{61(4)}$,
    298 (1979).
6. R. Greco, Tecnopolimeri e Resine, $\underline{5}$, 33 (1979).
7. M. C. Shen and A. Eisenberg, Progr. Solide State Chem., $\underline{3}$, 407,
    1966.
8. J. D. Ferry "Viscoelastic Properties of Polymers" 2nd Ed.
    Wiley, New York, 1970.
9. R. F. Boyer, Rubber Chem. Techn., $\underline{36}$, 1303 (1963).
10. J. Heijbour, Intern. J. Polymeric Mater., $\underline{6}$, 11 (1977).
11. R. F. Boyer, J. Macromol. Sci. Phys., $\underline{B8}$, 503 (1973).
12. ibidem "Macromolecules", $\underline{6}$, 288 (1973).
13. O. Kramer, R. Greco, R. A. Neira and J. D. Ferry, J. Polym. Sci.,
    Polym. Phys. Ed., $\underline{12}$, 2361 (1974).

14. O. Kramer, R. Greco, J. D. Ferry and E. T. McDonel, ibidem 13,
     1675 (1975).

15. R. Greco, C. R. Taylor, O. Kramer and J. D. Ferry, ibidem, 13,
     1687 (1975).

15. P. G. De Gennes, J. Chem. Phys., 55, 572 (1971).

17. S. F. Edwards and J. W. V. Grant, J. Phys., A6, 1169 (1973).

18. M. Doi, Polym. J. (Japan), 5, 288 (1973).

19. C. B. Bucknall "Toughened Plastics" Appl. Sci. Publ., London,
     1977.

20. G. Ragosta and R. Greco, Tecnopolimeri e Resine, 2, 35 (1980).

21. V. Costantini and C. Quagliariello "Effect of the addition of
     random ethylene-propylene copolymers to HDPE-iPP blends"
     PhD. Thesis ITPR, Naples (1979).

22. R. Greco, H. B. Hopfenberg, E. Martuscelli, G. Ragosta and
     G. Demma, Polym. Eng. Sci., 18, 654 (1978).

23. M. Costagliola, R. Greco, E. Martuscelli and G. Ragosta,
     J. Mat. Sci., 14, 1152 (1979).

24. R. Greco, G. Ragosta, E. Martuscelli and C. Silvestre
     "Properties of Polystyrene-Polyolefine Alloys: I Processing and
     Mechanical Properties" (present book).

25. E. Martuscelli, C. Silvestre, R. Greco and G. Ragosta
     "Properties of Polystyrene-Polyolefine Alloys: II Processing
     Morphology Relationship" (present book".

26. M. Avella, R. Greco, N. Lanzetta, G. Maglio, M. Malinconico,
     E. Martuscelli, R. Palumbo and G. Ragosta "Synthesis of
     Interfacial agents and their use in Nylon 6/Rubber Alloys"
     (present book).

27. R. Greco, G. Mucciariello, G. Ragosta and E. Martuscelli,
     J. Mat. Sci., 15, 845 (1980).

28. R. Greco, G. Ragosta, E. Martuscelli and G. Mucciariello,
     J. Mat. Sci., in print.

29. W. M. Barentsen and D. Heikens, 14, 579 (1973).

30. W. M. Barentsen, D. Heikens and P. Piet, Polym.,15, 119 (1974).

31. W. M. Barentsen, P. J. Hejdenrijk, D. Heikens and P. Piet,
     Brit. Polym. J., 10, 17 (1978).

PROPERTIES OF POLYSTYRENE-POLYOLEFIN ALLOYS:

I. PROCESSING AND MECHANICAL PROPERTIES

R. Greco, G. Ragosta, E. Martuscelli and
C. Silvestre

Istituto di Ricerche su Tecnologia dei Polimeri e
Reologia, C.N.R., ARCO FELICE (Napoli), Italy

One of the most interesting perspective in the field
of science and technology of plastics is to obtain new
materials by blending two or more polymers. However,
even though many types of blends have been studied, only a
few comprehensive analysis have been presented in literature. In the present work the results of a systematic
study on binary polystyrene-polyolefin alloys are reported, as a prototype of a class of binary blends consisting
of a glassy amorphous polymer at R.T. and of a semicrystalline one[1]. Therefore the main purpose of this work is to
characterize the alloys from a thermal, swelling and mechanical point of view, and successively to correlate such
properties with the resulting morphology of the extruded
specimens, as shown in the present book.

EXPERIMENTAL

Materials

The homopolymers used are listed in table I in which from left
to right, the code, the molecular parameters, the density, the melt
flow index and the sources are indicated.

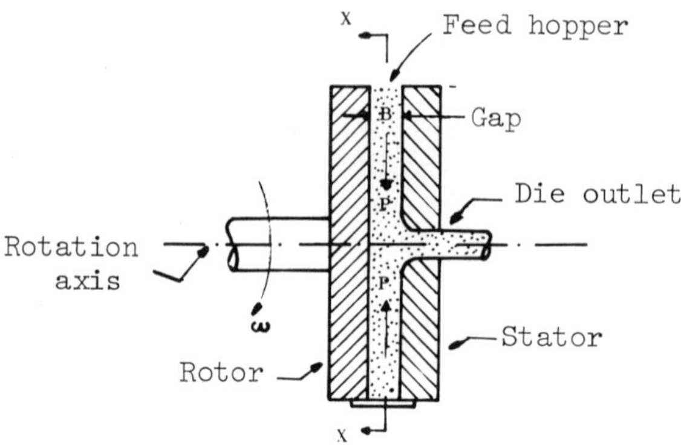

Side view (cross section)

Figure 1. Simplified version of the "elastic melt" extruder.

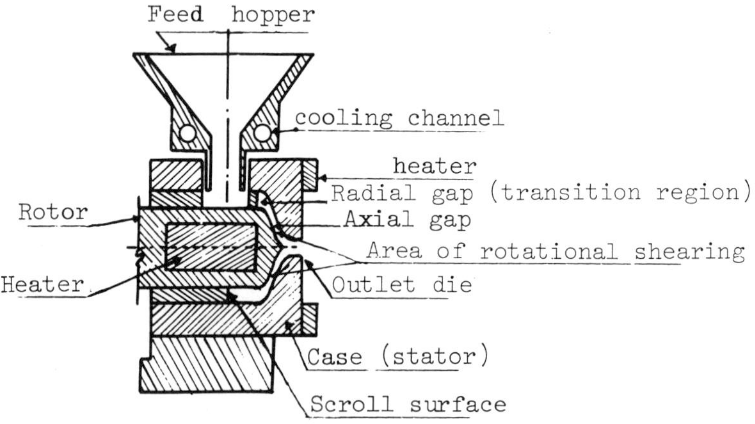

Figure 2. Cross-section of the mixing-extruder.

TABLE I

Homopolymers characterization

Polymers	Code	Molecular Parameters			Density at 30°C (g/cc)	MFI (gr/10min)	Source
		$\overline{M}n$	$\overline{M}w$	$\overline{M}w/\overline{M}n$			
Atactic polystyrene	aPS	--	10^5	--	1.075	--	BDH Chemicals
High density polyethylene	HDPE	$8 \cdot 10^3$	$92 \cdot 10^3$	11.6	0.943	3.7	Rapra
Isotactic polypropylene	PP	$15.6 \cdot 10^3$	$307 \cdot 10^3$	20	0.890	3.9	Rapra
Isotactic polybutene-1	PB	--	--	--	0.895	--	Petrotex
Poly-4-methylpentene-1	P4MP	--	--	--	0.820	--	ICI TPX RTI8
Low density polyethylene	LDPE1	$26.6 \cdot 10^3$	$215 \cdot 10^3$	8.7	0.906	0.2	Rapra
" "	LDPE3	$16.6 \cdot 10^3$	$96 \cdot 10^3$	5.5	0.916	3.3	Rapra
" "	LDPE5	$11.1 \cdot 10^3$	$77 \cdot 10^3$	7.0	0.906	24.6	Rapra

Blend and specimen preparation

The alloys were obtained by melt mixing at 230°C in a laboratory miniextruder[3,4] (Custom Scientific Instrument Inc.) aPS with each of the above mentioned polyolefins (see Table I).

A simplified version of the miniextruder[5], used for making the cylindrical specimens is shown in Figure 1. The pelletized polymers introduced in the hopper (A), are conveyed to a mixing zone between a rotor (D) and a stator (C), both kept at 230°C. In such a zone the material elements are sheared and hence elongated along a tangential path. The stored elastic tension in the mixed material generates a centripetal pressure which pushes the blend toward the rotation axis favouring its extrusion through the orifice (F). The real extruder used in this work is shown in Figure 2, in detail. The extruded continuous filament, cut in suitable lengths will provide for the cylindrical specimens.

Techniques

The blends have been characterized thermally by differential scanning calorimetry to detect the glass transition temperature Tg of aPS, the melting point, Tm, and the fractional crystallinity, X_c, of the relative polyolefin. Tensile mechanical tests, to determine the Young modulus (E), the yield points and the ultimate properties as a function of the composition, have been performed on the specimens by an Instron Machine at room temperature and at a cross-head speed of 10 mm/min. Also volumetric and axial swelling measurements, yielding indirect informations on the overall morphology, have been made on the same specimens. The processing conditions and the above mentioned properties have been correlated successively with the resulting morphology of some of these blends, as reported elsewhere[2].

RESULTS AND DISCUSSION

Thermal properties

The glass transition temperature, Tg, of aPS and the melting point, Tm, of the crystallizable component in all blends do not vary appreciably with the composition. The fractional crystallinity, X_c, as function of the aPS percentage for all blends, is reported in Figure 3. The very slight decrease in crystallinity as well as that of the melting point, with increasing the aPS content, could be attributed to kinetic limitations to crystal growth probably due to the sudden vitrification of aPS after extrusion. In addition, the same processing conditions described above, imposed on all blends, can produce across the filament a distorted radial temperature distribution with respect to that of the pure polyolefins. Such effect is due to the separation of the aPS and of the companion polyolefin in two distinct phases, and to the fact that the morphology of the blends changes as the aPS content varies[2-4].

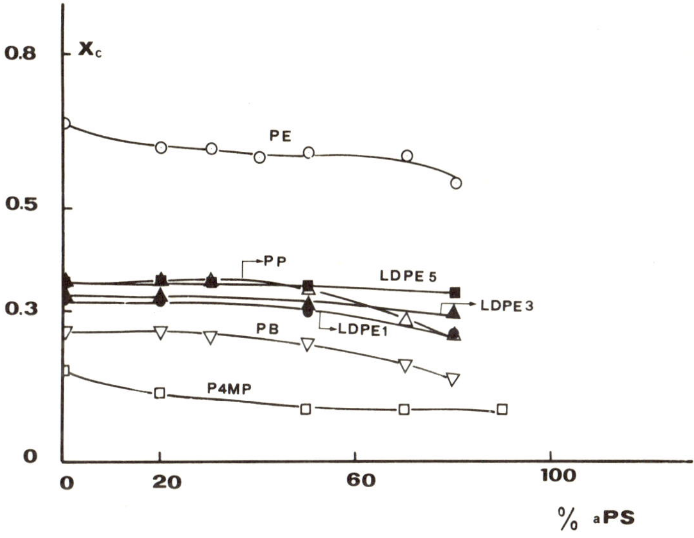

Figure 3. The effect of blend composition on fractional crystallinity, X_c, for all the blends as indicated (X_c is calculated on the basis of the polyolefin content).

Swelling properties

The apparent equilibrium volumetric absorption $(\frac{V-V_0}{V_0})$eq., in n-hexane relative to the dry specimen volume V_0, increases monotonically with increasing the total amorphous content (see Figure 4). Furthermore all the polymers absorb almost the same amount of solvent[3,4] as shown by the fact that, in the limits of the experimental errors, all the points lie on a single master curve.

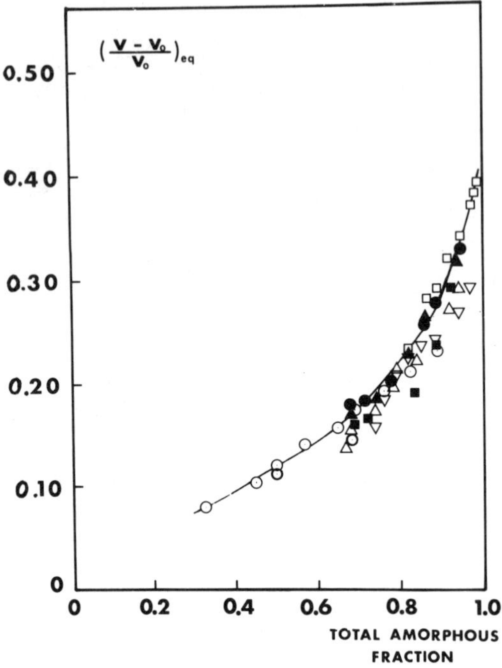

Figure 4. Master curve of apparent equilibrium volumetric swelling, $(\frac{V-V_o}{V_o})$ eq. as a function of total amorphous fraction of the following blends: o PE-aPS; $\triangle$ PP-aPS; ∇ PB-aPS; $\square$ P4MP-aPS; $\blacktriangle$ $LDPE_3$-aPS; $\bullet$ LDPE1-aPS; $\blacksquare$ LDPE5-aPS.

The corresponding axial swelling, $(\frac{L-L_o}{L_o})$ eq. relative to the
dry specimen length L_o, shows surprisingly an opposite trend with
increasing total amorphous content, as shown in Figure 5. In addition to this feature no master curve is obtainable in such a case.
The distinct curves tend to a zero value increasing the total amorphous content, whose maximum value (unity) corresponds to a pure
aPS specimen. These effects, which reveals a strong anisotropy of
the overall morphology can be due mainly to the presence in the
blend of both glassy aPS domains and rubbery polyolefin ones. The
former swells by Fickian mode[6] and the latter by the so called
"case II" kinetics[7].

The "case II" model reflects a step concentration profile,
which divides any aPS domain in an outer swollen region and in an
inner still umpenetrated glassy core. Therefore, as explained in
detail elsewhere[3,4], there is an increasing tendency of the cylindrical specimens to expand radially with increasing the aPS content.
The fact that all the polymers absorb the same n-hexane amount in
a pseudo equilibrium swelling is possible only if the glassy aPS
content left after absorption is small with respect to the total
amorphous fraction of the blend. On the other hand, the glassy aPS
cores still present in the sample domains, could explain the anisotropy of the specimens.

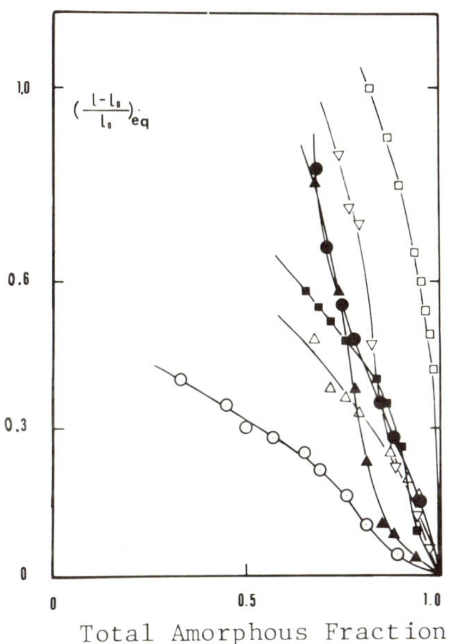

Figure 5. Curves of apparent equilibrium axial swelling $\frac{(L-L_0)}{L_0}$ eq.
as a function of total amorphous content for the
various blends: o PE-aPS; Δ PP-aPS; ∇ PB-aPS;
□ P4MP-aPS; ▲ LDPE3-aPS; ● LDPE1-aPS; ■ LDPE5-aPS.

Mechanical Properties

Typical stress-strain curves for aPS and most of the pure polyolefins used in the present work are indicated in Figure 6. The
initial slope of the curves represents the Young modulus, E, which

at a fixed cross-head speed and at vanishing elongation value is a
material property. Since in all the polyolefins the starting morphology of the extruded filaments is predominantly spherulitic, with
no appreciable orientation in the crystalline regions (as shown by
wide-angle X-ray diffraction analysis) the modulus, E, of the pure
components mainly depends on the fractional crystallinity and on
the "glassiness" of the materials. The "glassiness" can be considered to be inversely dependent on the total free volume present in
an amorphous polymer, and therefore operationally defined in a rough
way (neglecting the secondary transitions below T_g) by the difference
$\Delta T = T_g - Ta$ (where Ta is the end-use temperature). Such behaviour
is shown in Figure 7, where E increases both with the crystallinity
content and/or with the "glassiness" of the materials as in the
cases of P4MP($\Delta T = 0 \div 5°C$) and of aPS ($\Delta T = 77°C$).

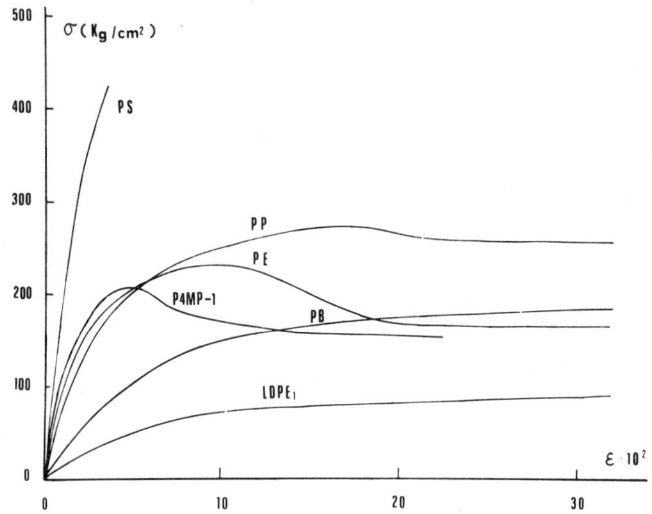

Figure 6. Stress-strain curves for the different homopolymers as
           indicated.

At very high deformation values (Figure 6), a semicrystalline
polymer can have different kinds of stress-strain curves (brittle,
abruptly yielding ductile, smoothly yielding ductile). The shape
of the curves will depend upon the rate of deformation, the temperature of testing, or upon a combination of both. However, the
cross-head speed and the temperature have been kept constant in the
present work for all the samples. Therefore, the brittle or the

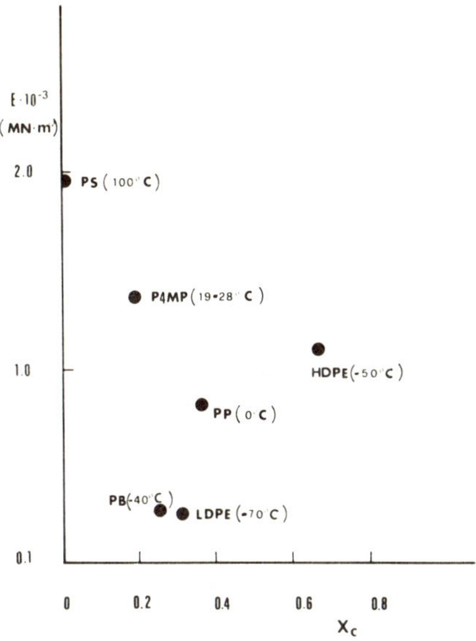

Figure 7. Young's modulus, E, as a function of X_c for different
polymers (T_g in parenthesis) as indicated.

ductile behaviour will mainly depend on the starting morphology
and/or on the crystallinity of the materials[18]. In fact, PE shows
a marked plastic behaviour with easy neeking formation. PP has a
lower crystallinity content, but it shows higher yield strength and
elongation values. PB shows instead a smoothly yielding ductile
curve without neeking, this fact depends on the level of cristalli-
nity and probably on molecular weight, as discussed previously by
Rakus et al.[9]. LDPE samples show the same mechanical behaviour as
PB again depending on their levels of crystallinity, and in this
case, also on the effect of branching. On the other hand, P4MP in
spite of its low value of crystallinity, shows an abruptly yielding
ductile stress-strain curve. Such behaviour can be understood if
one takes into account that T_g of P4MP is very close to room temperature, ranging from 15 to 29°C depending on crystallinity[10,11].
Therefore P4MP can show yielding behaviour especially if one assumes
that some further relaxation mechanism below T_g, is present at room
temperature.

Stress-strain curves for PE-aPS blends at different aPS content are reported in Figure 8. The behaviour gradually goes from

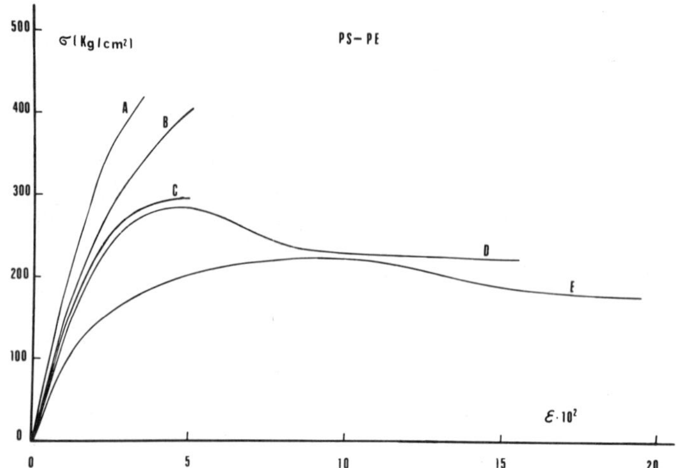

Figure 8.   Typical stress-strain curves for PE-aPS blends at different aPS percentages: A, 100%; B, 65%; C, 40%; D, 30%;
E, 0%.

brittle fracture of pure polystyrene (curve A) to the very ductile
one of the pure polyethylene (curve B). All the other blends show
analogous patterns with minor differences due to the peculiar characteristics of the different polyolefins. In any case between 30÷50%
of aPS content an inversion of matrix takes place in the blends,
which gives reason of the brittle-ductile transition. The Young
modulus E of the blends is reported in Figure 9 as a function of
aPS content. All the alloys show a monotonic increase with increasing the aPS content, of course the "glassiness" of each polyolefin
influences the whole curve (for instance P4MP-aPS blend shows higher
values than those relative to all the other blends on the entire
range of composition).

CONCLUSIONS

Polyolefins and polystyrene are very incompatible materials
in the molten state but various morphologies can be obtained according to their different viscosities at the used processing conditions, and to their relative amounts[12-15]. Even at the very low
strain rates used in their preparation[2-4], with no orientation in

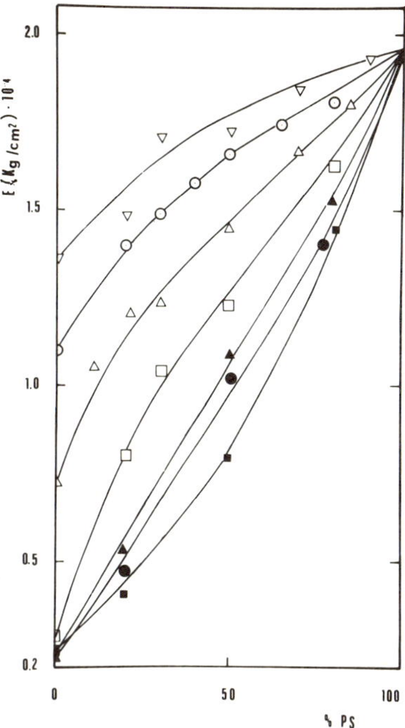

Figure 9. Young's modulus E, as a function of aPS percentage for
the blends: ∇ P4MP-aPS; o PE-aPS; △ PP-aPS; □ PB-aPS;
● LDPE$_1$-aPS; ▲ LDPE$_3$-aPS; ■ LDPE$_5$-aPS.

the crystalline regions as detected by X-ray diffraction technique,
the overall morphology of the two distinct domains can give a
marked anisotropy in the specimens, as revealed by swelling measurements.

At low aPS contents, the anisotropy can be attributed to the
presence of aPS domains elongated along the extrusion direction as
shown also by optical and scanning electron microscopy. Such micro-
fibrils are presumably obtained by squeezing the aPS particles
throughout the die and by the subsequent rapid cooling of the filaments below the aPS glass transition temperature. Therefore up to
a given aPS concentration the blend consist of a polyolefin matrix
with aPS dispersed domains oriented along the extrusion direction.
This mode of dispersion of the aPS particles gives a reinforcement

effect on the polyolefin, whose moduli and yield values increase
(see curve D of Figure 8), even though the behaviour is still ductile.

At higher concentration the aPS tends to become matrix itself
and this determines the brittleness of the blend. In this case,
the anisotropy is due to the cylindrical shape of the whole specimens, which can only expand radially up to the equilibration time.

Taking into account the master curve of Figure 4, one can say
that swelling data refer to a quasi-equilibrium state in which the
small amount of the remaining glassy aPS is still effective to give
the observed anisotropy. Moreover, the quenching effect, which the
materials undergoes (due to the high T_g of aPS) out of the orifice,
is probably greater in the latter case (aPS matrix).

The indirect morphological informations obtained by swelling
and mechanical tests have been confirmed by direct optical and
electron microscopic observations, shown in the second paper of
this series. Of course many further details are described there
such as the modes of dispersion, the sizes of the particles and
their distribution and also a particular feature, the "sheat core"
formation in the middle of the specimens in some given processing
conditions[2].

REFERENCES

1. R. Greco "Mechanical properties of multicomponent polymer mate-
    rials: a predictive classification of binary systems", in the
    present book.
2. E. Martuscelli, C. Silvestre, R. Greco and G. Ragosta "Proper-
    ties of polystyrene-polyolefin alloys: II processing-morphology
    relationship", in the present book.
3. R. Greco, H. B. Hopfenberg, E. Martuscelli, G. Ragosta and
    G. Demma, Polym. Eng. Sci., 18, 654 (1978).
4. M. Costagliola, R. Greco, E. Martuscelli and G. Ragosta,
    J. Mat. Sci., 14, 1152 (1979).
5. B. Maxwell, SPE J., 26, 48, June 1970.
6. V. T. Stannett, "Diffusion in Polymers", eds. J. Crank and
    G. S. Park, Academic Press, N. Y., 1968 (Ch. 2).
7. H. B. Hopfenberg and A. L. Frish, J. Polym. Sci., B7, 405 (1969)
8. C. A. Sperati, A. A. Franta and H. W. Starkweather jr., J. Am.
    Chem. Soc., 75, 6127 (1953).

9.   J. P. Rakus, C. D. Mason and R. J. Schaffhauser, Polym. Letters, $\underline{7}$, 590 (1969).

10. B. G. Ranby, K. S. Chan and H. Brumberger, J. Polym. Sci., $\underline{58}$, 545 (1962).

11. F. L. Saunders, Polym. Letters, $\underline{2}$, 755 (1964).

12. C. D. Han and T. C. Yu, J. Appl. Polym. Sci., $\underline{15}$, 1163 (1971).

13. C. D. Han, J. Appl. Polym. Sci., $\underline{15}$, 2579 (1971).

14. C. D. Han and Y. W. Kim, J. Appl. Polym. Sci., $\underline{18}$, 2589 (1974).

15. C. D. Han and Y. W. Kim, J. Appl. Polym. Sci., $\underline{19}$, 2831 (1975).

PROPERTIES OF POLYSTYRENE-POLYOLEFIN ALLOYS:

II. PROCESSING MORPHOLOGY RELATIONSHIP

E. Martuscelli, C. Silvestre, R. Greco and G. Ragosta

Istituto di Ricerche su Tecnologia dei Polimeri e
Reologia, C.N.R., Arco Felice (Napoli) Italy

INTRODUCTION

With the aim of formulating new plastic materials, polymer
blends are replacing homopolymers[1], more and more.

The morphological studies are very important in order to establish how the processing condition may govern the morphological
features, and hence the properties of these blends[2].

The overall morphology, i.e., the modes and the state of dispersion of different phases in an extrudate mixture of incompatible
molten polymers depends on the blend compositions, on the starting
particle sizes, molecular weight and distribution of individual
components on the type of extruder used, on the melt extrusion
temperature, and finally on the eventual premixing devices. Essentially the following basic modes of dispersion may be found:
1)   Ribbons and lamellae (stratified morphology)
2)   Rods and fribrils
3)   Droplets
The formation of ribbons, rods or droplets of the disperse phase in
a polymer blend could be predicted on the basis of the VanOene
theory[3], which predicts, for a binary blend whose component particles
are larger than 1μ, a droplet or fibril-like morphology of the disperse phase if the molecular weight and the deformability of the disperse phase are larger than those of the matrix. On the contrary

in the case of less deformability and low molecular weight of the
disperse phase, stratified morphology should be observed. Therefore
according to the theory, the modes of dispersion depend only on the
deformability (elastic term), and molecular weight of the disperse
phase, and on the initial particle size of the individual components.
All the other parameters listed above should influence the homogeneity
of the blend, i.e. dimensions and distribution of the domains of the
disperse phase.

The main goal of the present paper is to review the most important morphological features of extrudate sample of atactic polystyrene (aPS)/high density polyethylene (HDPE) and atactic polystyrene/
/isotactic polypropylene (iPP) blends and to compare the results
found in the literature on such blends processed by usual apparatus
(screw extruder, capillary rheometer) with those obtained by us on
extrudate samples, processed by using a new special laboratory
screw-less mini extruder[4].

REVIEW OF LITERATURE DATA ON MORPHOLOGY/PROCESSING AND COMPOSITION
RELATIONSHIP IN aPS/HDPE, aPS/iPP ALLOYS

Atactic polystyrene/high density polyethylene blends (aPS/HDPE)

The morphology of such blends has been studied by VanOene[3] and
Han et al.[5] on extrudate samples. According to VanOene's results
the HDPE stratifies along concentric rings in extrudates of aPS/HDPE
blends, with high aPS content. On the contrary in alloys with HDPE
matrix, the aPS disperse phase forms droplets. The materials used
by VanOene were: Polystyrene - Styron 666 manufactured by the Dow
Chemical Co. and High Density Polyethylene - Marlex 6009 manufactured
by Phillips Petroleum. The blends were extruded at 170°C by using
a capillary rheometer. The results obtained seem to confirm
VanOene's theory and as confirmed by Han et al.[5] the aPS is more
elastic than the HDPE. No marked effect on extrudate morphology
was observed if the conditions of extrusion, i.e. length of the
capillary, shear rate and melt temperature, were varied. The influence of starting particle sizes of the individual components was
investigated by preblending the powder mixture, before extrusion,
with a Kenics mixer. Extrusion of the preblended aPS/HDPE (30/70)
mixture at 190°C and 225°C gives a droplet and fibril-like morphology respectively of the aPS disperse phase. This droplet-fibril
transition is not readily explained by the author.

Han et al. studied the influence, on the same blends of the extrusion temperature on the state of dispersion[5]. The material used by the authors were Polystyrene - Styron 686 manufactured by the Dow Chemical Co. and High Density Polyethylene - DMDJ 4309 manufactured by Union Carbide. A vortex formed mainly by the HDPE was found at the centre of the extrudate filament, extruded at 200°C by means of a capillary melt rheometer in the case of 20/80 and 50/50 aPS/HDPE blends. The formation of such a vortex appears to depend mainly on the melt temperature, in fact as the temperature is increased (220°C) the vortex disappears for all blend compositions. Optical microscopy, by reflected polarized light, was used as method of analysis. The aPS was dissolved by toluene or xylene, in this way the HDPE remains to provide positive identification of the two phases.

Atactic polystyrene/isotactic polypropylene blends (aPS/iPP)

Such blends have been investigated by Han et al.[6,7] and Krasnikova et al.[8]. Different mixing devices (single screw extruder, single screw extruder plus static mixer and twin screw extruder) have a drastic influence on the state of dispersion of one polymer in another[6]. The blends were observed with an optical microscope, after dissolution of the polystyrene phase with toluene. In all cases, the particles of the disperse phase are of well defined shape - fibrils. For the 20/80 and 50/50 blends, the resultant dispersion is finer in the case of extrudate obtained by using a single screw extruder equipped with the static mixer[6] than in the cases of extrudates obtained by using a single screw extruder alone. According to Han et al. the improvement in mixing may be attributed to the breakup of the long fibrils of the disperse phase (aPS) into smaller particles as the molten polymer passes through the stationary mixing element. The twin screw extruder does not appear to give a better dispersion than the single screw extruder with the static mixer.

The effect of different types of mixing devices on the morphology of injection molded samples of 30/70 aPS/iPP blend was also investigated[7]. The following results were found:
1) A plunger machine alone gives a poor job of mixing
2) By using a plunger machine equipped with a static mixer, or a single screw extruder the degree of dispersion is increased.

3) Very high dispersion is achieved when the aPS/iPP blend was
   injection molded by using a single screw/plunger/static mixer
   system.

Krasnikova et al.[8] studied the morphology of aPS/iPP 70/30
blend by using optical and electron microscopy. The blends were
obtained by means of a constant pressure capillary viscosimeter.
The micrographies of such blends show that the iPP is dispersed in
an aPS matrix. After removal of aPS from the extruded sample by
extracting it with toluene, the samples have the form of long fibrils.
The fibrils have a spherulitic structure whereas they are well
oriented in the direction of the flow. Stereometric analysis shows
that on passing from the centre to the edges of the samples the
volume content of the disperse phase, the number of iPP fibrils and
the asymmetry of their cross section increase. The authors found
that when the conditions of formation of samples provide fewer
possibilities of relaxation (low ambient temperature at the outflow
of the melt from the capillary, increase in the shear strain and
decrease in the length of the capillary) thinner iPP fibrils arranged
in concentric layers in the aPS matrix are observed.

EXPERIMENTAL

The molecular characteristics of the polymer used for the present study and the blend compositions explored are reported in
Table 1 and 2. Cylindrical samples of aPS/HDPE, aPS/iPP blends and
of pure homopolymers were obtained by using a laboratory mini screw-
less extruder manufactured by Custom Scientific Instrument Inc..
Three extrusion temperatures were used namely 180°, 200°, and 220°C.
The materials were isothermally processed at four different extrusion
conditions, by varying the speed of the rotor (the following speed
values were used: 40, 80, 146 and 272 RPM).

In order to investigate the influence of the premixing conditions on the blend morphology, the polymers were premixed in some
cases before the extrusion process, by using a Brabender like apparatus. The details of the processing and of the apparatus used have
been described in a previous paper[4].

Thin sections (about 10 μ) of aPS/HDPE and aPS/iPP blends and
of homopolymers were cut parallel and perpendicular to the filament

axis by using a LKB Ultramicrotome at room temperature.

TABLE 1

Homopolymers characterization

		aPS	HDPE	iPP
Molecular Parameters:	M_n	–	8×10^3	15.6×10^3
	M_w	10^5	92×10^3	307×10^3
	M_w/M_n	–	11.6	20
Density at $30°C(g/cc)$		1.075	0.943	0.890
MFI (gr/10min)			3.7	3.9
Source		BDH Chemicals	Rapra	Rapra

TABLE 2

Blend compositions (wt % / wt %) investigated.

iPP	
aPS/iPP	20/80
aPS/iPP	50/50
aPS/iPP	80/20
aPS	
aPS/HDPE	70/30
aPS/HDPE	50/50
aPS/HDPE	70/30
HDPE	

The overall morphology of the blends was investigated by optical and scanning electron microscopy. Optical micrographies were taken by using a Zeiss Ultraphot microscope on sections previously immersed in microscope oil (refractive index 1.518) in order to reduce reflection from their surface. Scanning electron microscopy

was carried out by using a Cambridge Stereoscan S-4 10 apparatus.
For blends with low aPS content the morphology was also studied on
sections etched at room temperature with toluene (toluene is a good
solvent at R.T. for aPS and a non solvent for iPP and HDPE).

RESULTS AND DISCUSSION

<u>aPS/iPP blends</u>

The addition of aPS to iPP generally causes a drastic reduction
in the dimensions of iPP spherulites. This effect is clearly shown
by the comparison of the optical micrographies of Figures 1a, 1b.
This observation suggests that the aPS may act as a nucleant agent.
Karger et al.[9] and Martuscelli and al.[10] found an analogous result
by adding a rubbery impact modifier to isotactic polypropylene.
This effect is very interesting from a practical point of view, as
the superstructure strongly influences the ultimate properties, in
particular the impact strength of polymeric materials[11,12].

Optical analysis of the longitudinal and transversal sections
of the extruded aPS/iPP filament with iPP matrix shows that the
overall morphology which occurs near the edges is different from
that near the centre of the filament. In fact, as shown in Figures
2a and 2b, at the edges of the filament fibrils of iPP and aPS
oriented along the extrusion direction are observed. The iPP fibrils
have a spherulitic structure analogously to the results found by
Krasnikova et al. in the case of extrudates of 70/30 aPS/iPP blend
obtained by means of a constant pressure capillary viscosimeter[8].
The central part of the filament is, on the contrary, characterized
by a droplet-like morphology of the disperse phase i.e. aPS. In
fact iPP spherulites and aPS domains of almost regular circular
shape are visible in both transversal and longitudinal sections
(see Figures 2, 3). This could be due to the flow conditions both
in the capillary and in the orifice of the apparatus and to the
freezing of the system outside of the extruder. The micrograph of
a cross section of the filament of the aPS/iPP 20/80 blend, observed
with parallel polars, shows a dark decentralized zone of almost
elliptical shape (see Figure 3). This region contains polystyrene
prevalently. In fact at temperatures higher than 100°C (the glass
transition temperature of the polystyrene) the material in this
zone begins to flow and it can be dissolved completely with toluene

(a)

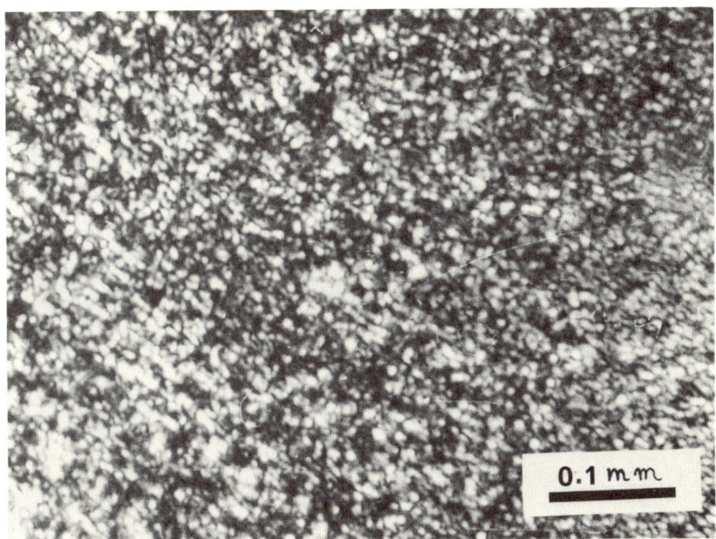

(b)

Figure 1. Optical micrographs obtained with crossed polars of extrudate cross section of pure iPP (a) 20/80 aPS/iPP blend (b).

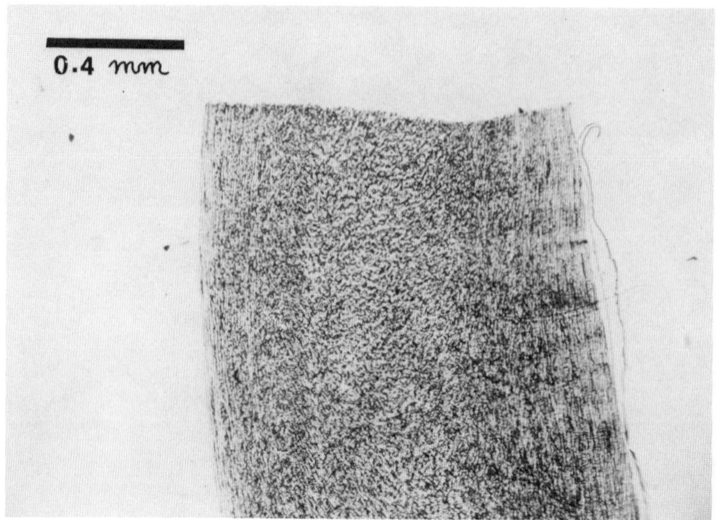

(a)

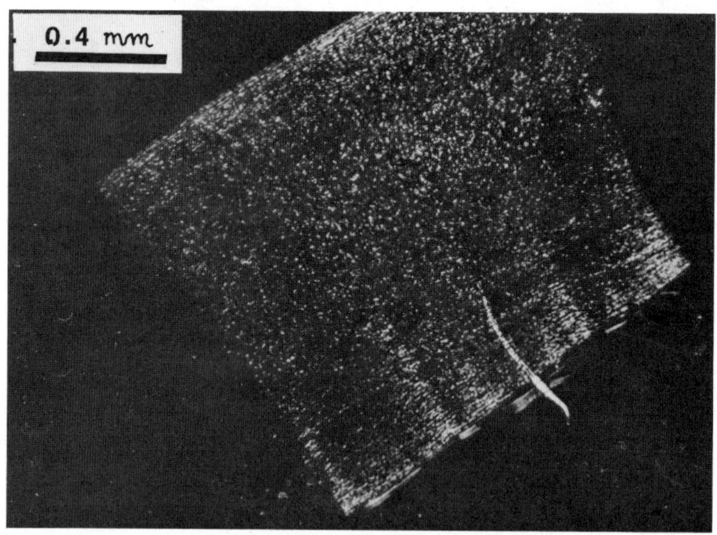

(b)

Figure 2. Optical photomicrograph of extrudate longitudinal section
           of 20/80 aPS/iPP blend: a) parallel polars; b) crossed
           polars.

at room temperature. The photomicrograph of the same section of
aPS/iPP 20/80 blend, after etching with toluene, obtained with
crossed polars, shows a hole left by the dissolved polystyrene (see
Figure 4).

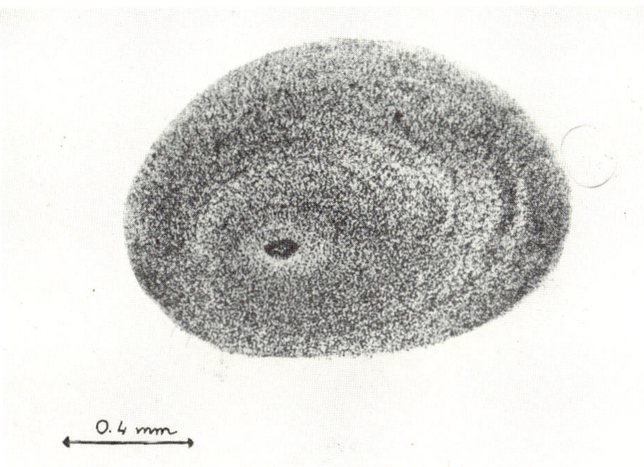

Figure 3. Optical micrograph, with parallel polars, of extrudate
           cross section of 20/80 aPS/iPP blend.

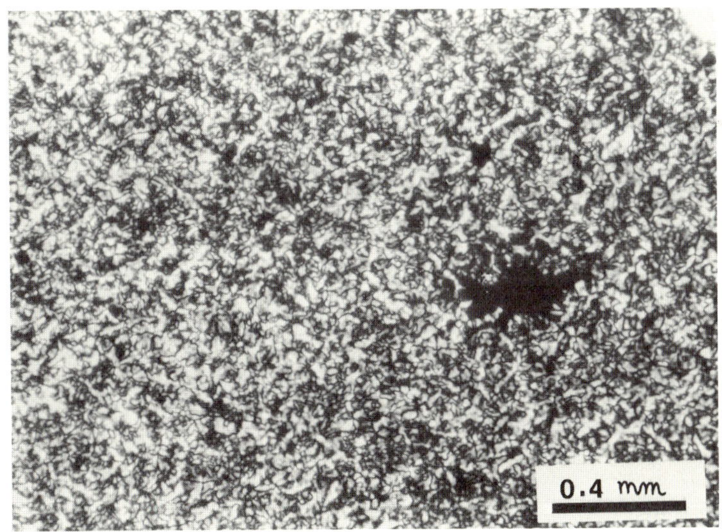

Figure 4. Optical micrograph, with crossed polars, of extrudate
           cross section of 20/80 aPS/iPP blend after etching with
           toluene.

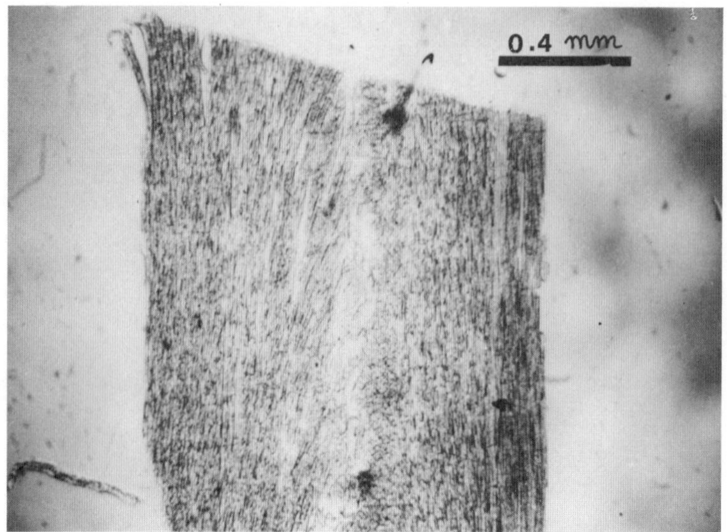

(a)

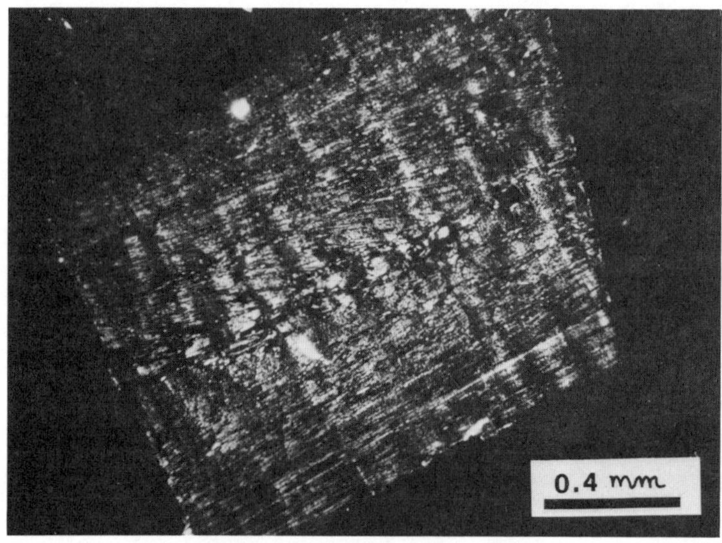

(b)

Figure 5. Optical micrographs of extrudate longitudinal section of 80/20 aPS/iPP blend: a) parallel polars; b) crossed polars.

In the case of aPS/iPP (80/20) blends with polystyrene matrix,
a prevalently fibril-like morphology is observed with the fibrils
oriented along the flow direction. The iPP fibrils have, even for
such a polystyrene matrix blend, spherulite structure (see Figure 5).
Only in a narrow region close to the filament axis the two components
form interpenetrated semicontinuous microphases. The optical micrograph of the cross section of this blend shows the presence of lamellar domains, which stratifies along concentric rings, and of a
cricular decentralized region, whose morphology is composite (see
Figure 6).

The optical microscope detailed analysis, with crossed and
parallel polars, on this section etched with toluene, brings out
that in the middle of this zone a prevalently polypropylene phase
is present surrounded by a circular high content polystyrene region.

Optical micrographs of thin sections of aPS/iPP 50/50 blends
are shown in Figures 7 and 8. From the analysis of these Figures
and also from Figure 9a (scanning electron micrograph) it emerges
that the aPS is present as disperse phase in a continuous spherulites
iPP matrix. On the other hand one cannot exclude that the aPS domains are connected since only a few sections of the specimen have
been analyzed.

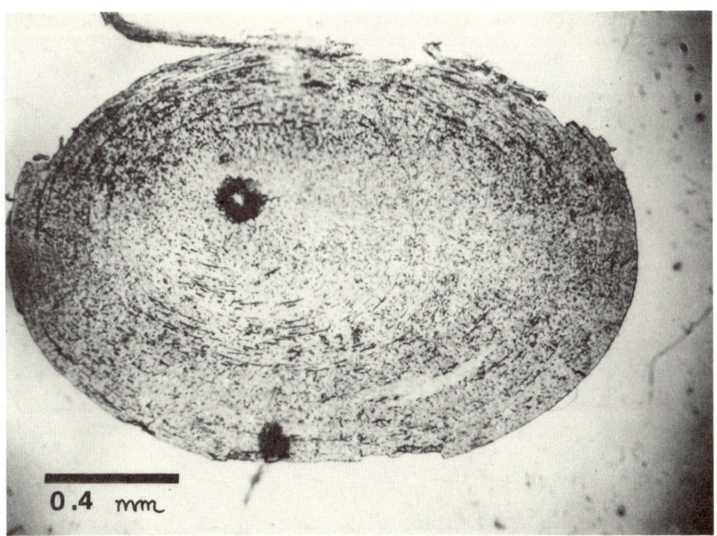

Figure 6.   Optical micrograph obtained with parallel polars of a
           extrudate cross section of 80/20 aPS/iPP blend.

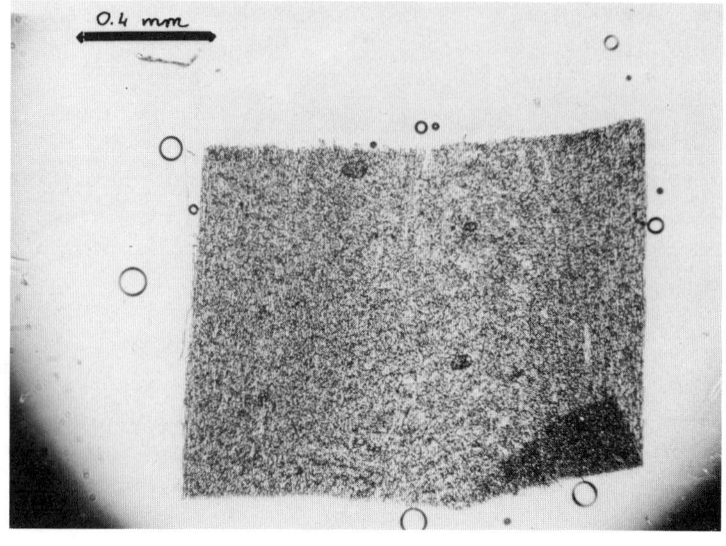

(a)

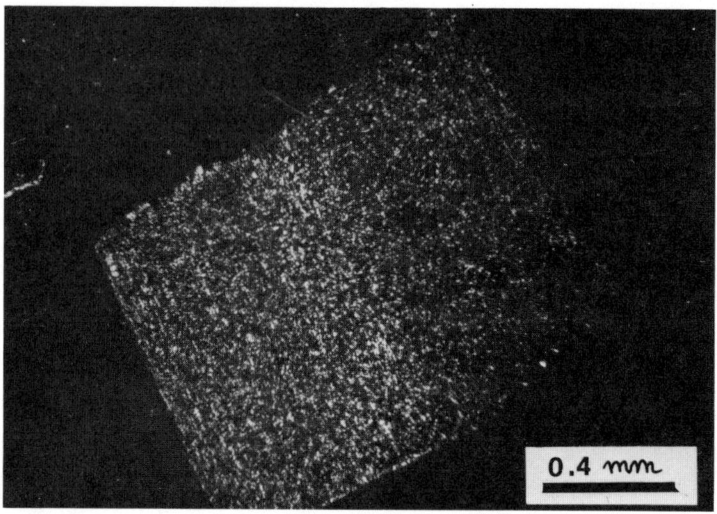

(b)

Figure 7. Optical micrographs of   extrudate longitudinal section of 50/50 aPS/iPP blend: a) parallel polars; b) crossed polars.

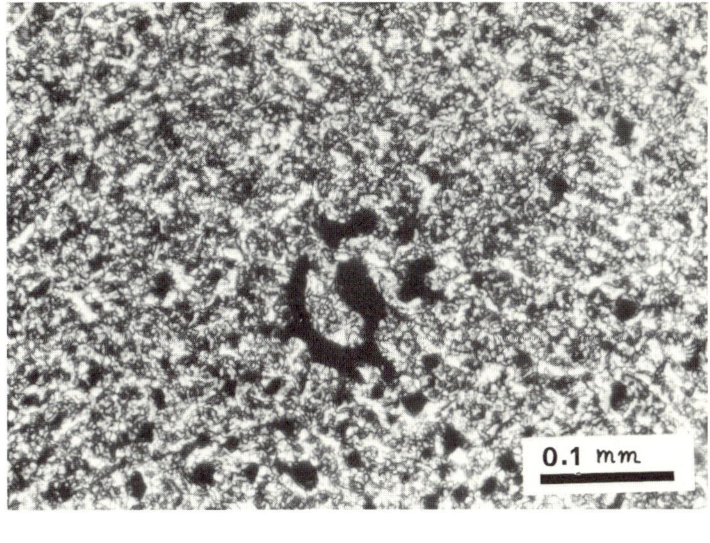

(a)

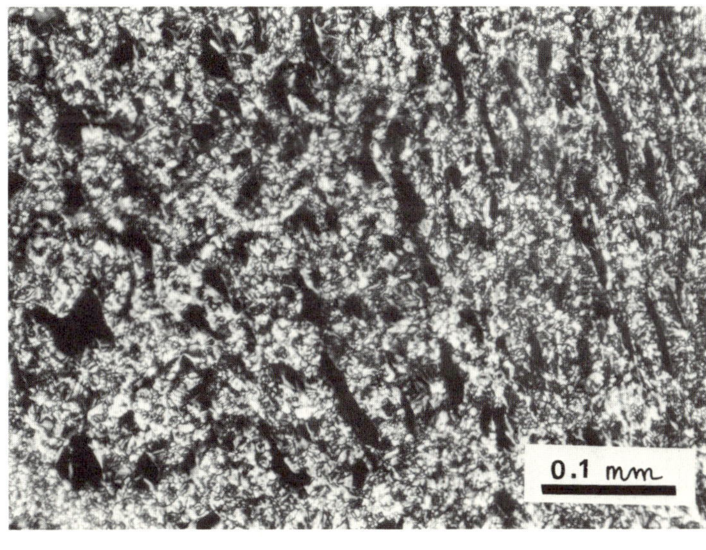

(b)

Figure 8. Optical micrographs obtained with crossed polars of extrudate sections of 50/50 aPS/iPP blends: a) cross section; b) longitudinal section.

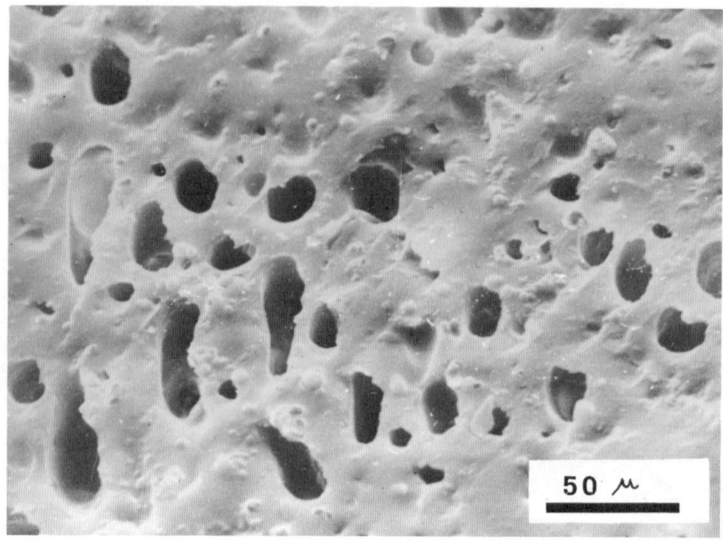

(a)

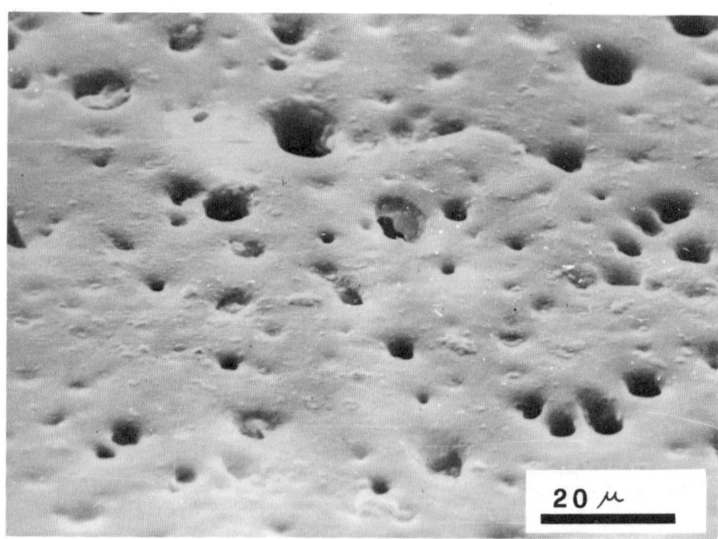

(b)

Figure 9.   Scanning electron micrographs of extrudate cross sections
            of aPS/iPP blends: a) 50/50 aPS/iPP; b) 20/80 aPS iPP.

Furthermore the mechanical behaviour which shows a brittle feature[4] in this range of composition seems to suggest on the contrary that the matrix of such a blend consists mainly of aPS. The aPS domains seem to be elongated along the axis only at the edges of the filmanent (see Figure 8b). Only few fibrils are observed in the longitudinal section of such a blend. No segregation region is observed in aPS/iPP 50/50 alloy. The shape and the dimensions of the domains of aPS were also analyzed by SEM technique on sections etched with toluene at room temperature. The application of this technique was restricted to blends with aPS content less than 50%. In fact, when the aPS is an excess the images are distorted as the iPP particles are freed by solvent action and their positions altered. Typical scanning electron micrographies of aPS/iPP blends are shown in Figure 9a,b. The average dimensions of the aPS domains, measured on the cross sections of the filaments are of 15 μ and 3 μ for the 50/50 aPS/iPP and 20/80 aPS/iPP system respectively. A decrease in the dimension of aPS domains with an increase in the iPP concentration is observed.

aPS/HDPE blends

In the case of 70/30 aPS/HDPE blend, the analysis of the cross and longitudinal sections of the filament obtained by extrusion at 180°C by means of optical microscopy with crossed and parallel polars gives the following morphological results:
1) At the edge of the extruded filament the domain shape of the individual component is fibril-like. The fibrils are oriented in the extrusion direction (see Figure 10a).
2) In the decentralized centre, a highly bright rod-shaped of about 75 μ in diameter is present due to the segregation of HDPE during the extrusion process (Figures 10b, 11). The low birefringence value, (+ 1.28 x 10^{-3}) of such a central region, calculated by a Babinet compensator, shows that the relative low orientation is superficial and only due to the cutting. To confirm this hypothesis two longitudinal sections were cut parallel and perpendicular to the extrusion direction with the same thickness. The two sections were observed under the optical microscope with a quartz plate. It is possible to note that in order to have same colour pattern, the two sections have to be rotated 90° one with respect to the other. After etching with toluene, the central zone loses the superficial birefringence because of the swelling action of the solvent, and it is possible to observe the pre-

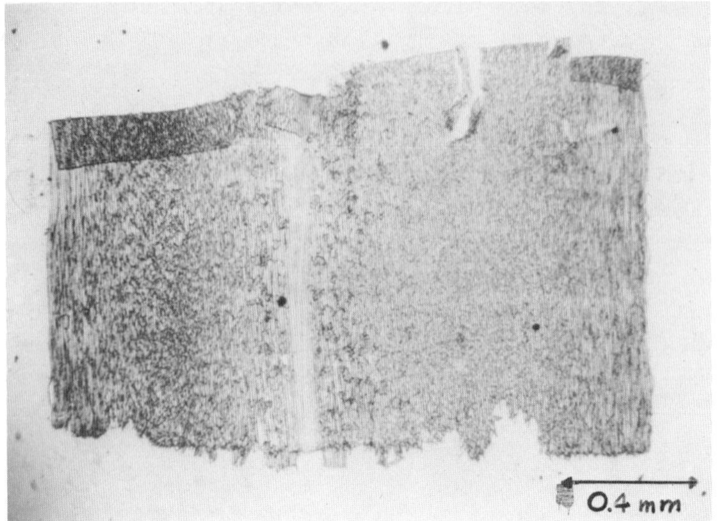

(a)

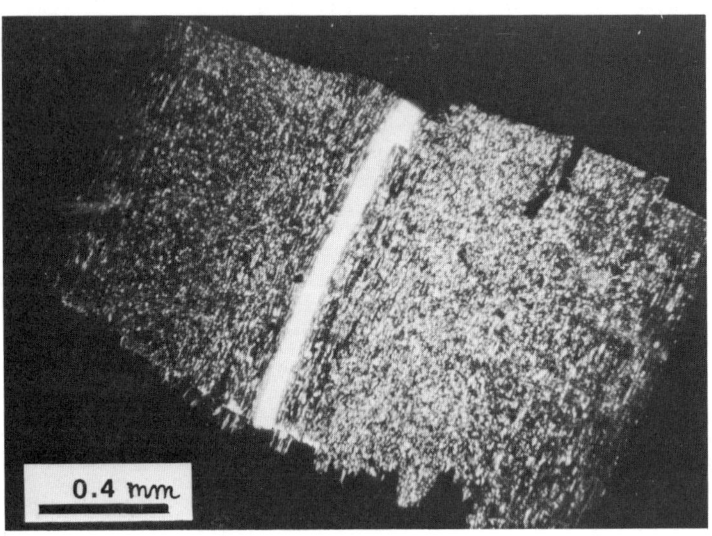

(b)

Figure 10. Optical micrographs of extruded longitudinal sections of 70/30 aPS/HDPE: a) parallel polars; b) crossed polars.

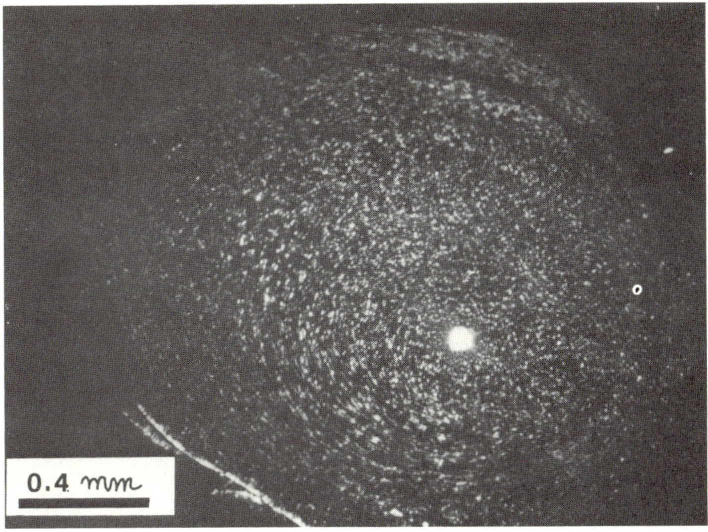

Figure 11. Optical micrographs obtained with crossed polars of
extrudate cross section of 70/30 aPS/HDPE blend.

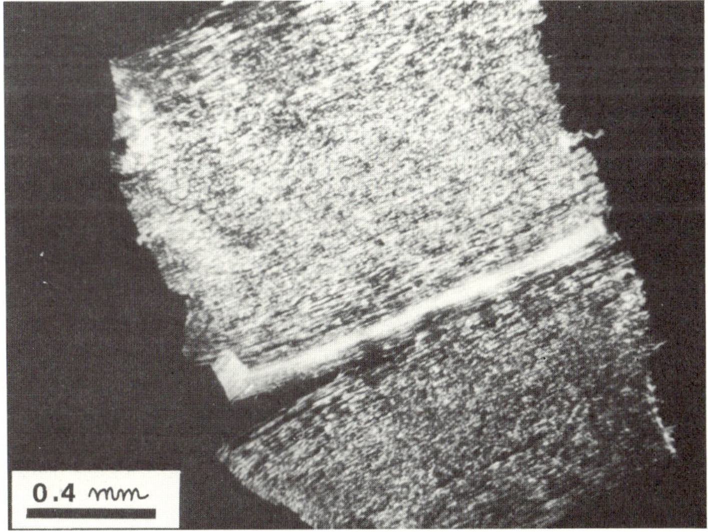

Figure 12. Optical micrograph of extrudate longitudinal section
of 70/30 aPS/HDPE blend after dissolution of the aPS
with toluene.

sence of a bundle of long thin fibrils of HDPE oriented along
the extrusion direction with a spherulitic morphology (see Figure
12).
3) As shown in the optical micrographs of the cross sections, the
   aPS and HDPE domains are oriented along concentric rings (see
   Figure 11).

In order to investigate how the processing conditions may influence the blend morphology and particularly the HDPE segregation
in 70/30 aPS/HDPE blend, the extrusion conditions and the premixing
procedure were varied. The results show that at low extrusion temperature (180°C), the observed phenomenon of segregation of the minor
component in the middle of the filament, is irrespective of rotor
speed and premixing procedure used. In fact the central HDPE rod-
like phase is also present in samples extruded at T = 180°C with
different rotor speed (40, 80, 146 and 272 RPM) and in some cases
after a premixing of the two polymers in a Brabender-like apparatus.
On the contrary, the overall state of dispersion of HDPE outside
the central HDPE phase, in blend with high aPS, depends on the preblending conditions; the preblended extruded mixtures show a coarser
dispersion of the minor component with respect to the blend obtained
only by extrusion. This result may be due to the "annealing" of
the material inside the preblending apparatus which brings about a
substantial increase of the domain sizes of the disperse phase.
This conclusion is in accord with that drawn by Danesi and Porter[13]
on polypropylene/ethylene-propylene copolymer blends. As the temperature of extrusion is increased (200° and 220°C), the formation of
a central rod is not longer observed in blends with high aPS content.

Samples of 30/70 aPS/HDPE blend, obtained by melt extrusion at
180°C, consist of aPS particles of drolet-like shape, dispersed in
the HDPE matrix (Figures 13 and 14). The morphological analysis
shown that on passing from the centre to the edges of the extruded
filament, the dispersion is finer; some fibrils are also observed
in this region. A rather darker region, near the centre of the filament, may be seen in the cross section when observed with crossed
polars (see Figure 14a). This zone is almost completely dissolved
with toluene at R.T., as shown in Figure 14b. Thus, in this blend,
a segregation phenomenon of the disperse phase (aPS) is also observed
in analogy to what is noted in the case of aPS/iPP blend with high
iPP content.

Optical micrographs of 50/50 aPS/HDPE blend are shown in

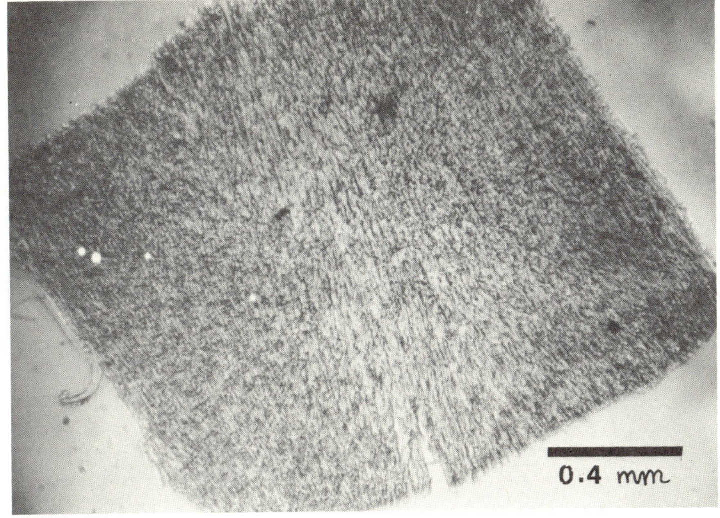

(a)

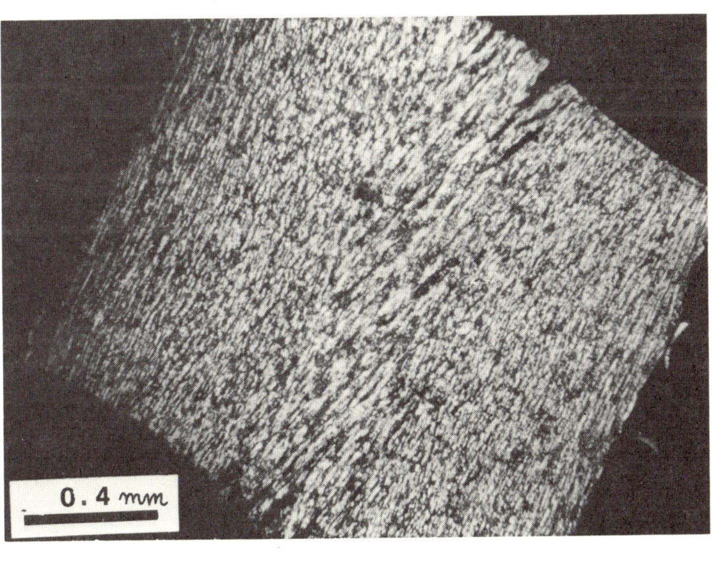

(b)

Figure 13. Optical micrographs of extrudate longitudinal section of the 30/70 aPS/HDPE blend: a) parallel polars b) crossed polars.

                     E. MARTUSCELLI ET AL.

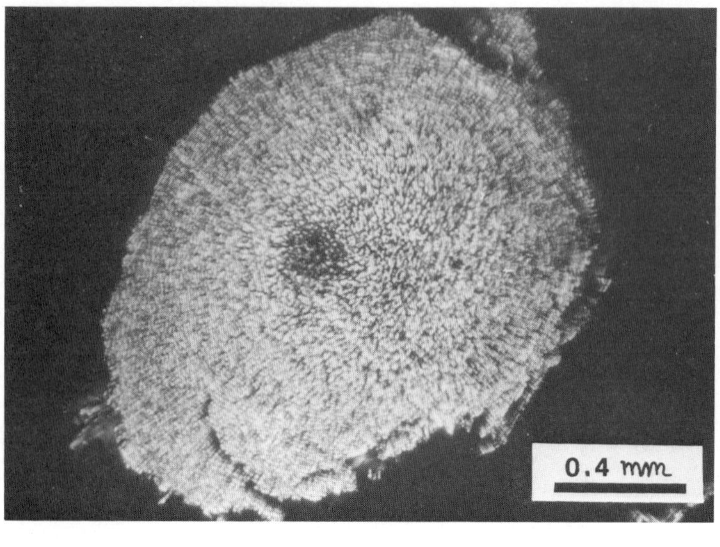

(a)

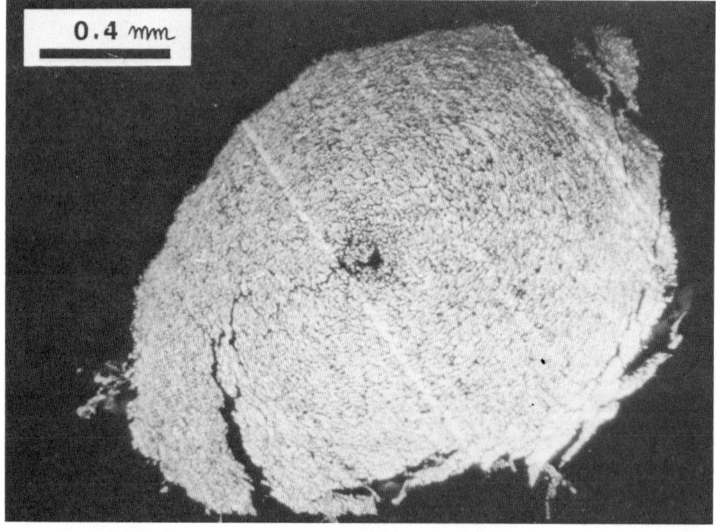

(b)

Figure 14.   Optical micrographs obtained with crossed polars of the extrudate cross sections of 30/70 aPS/HDPE blend before (a) and after (b) etching with toluene.

(a)

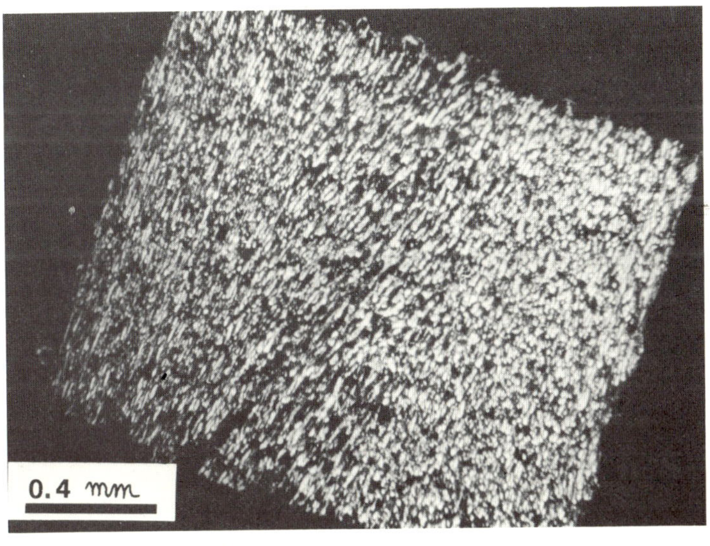

(b)

Figure 15. Optical micrographs of extrudate longitudinal section
of the 50/50 HDPE/aPS: a) parallel polars; b) crossed
polars.

Figures 15, 16. It is interesting to observe the presence of a double segregation effect in the centre of the filament (see Figure 16). In fact a region of higher HDPE content seems to be surrounded by a narrow circular shaped region with higher aPS content.

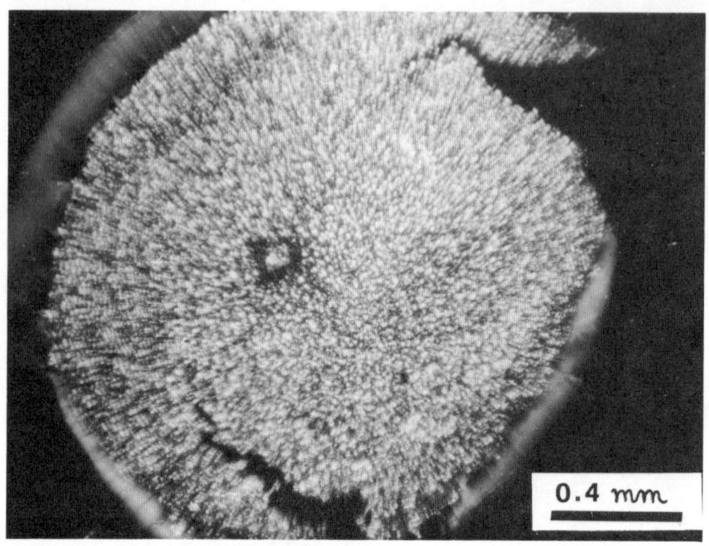

Figure 16.   Optical micrograph obtained with crossed polars of extrudate cross section of the 50/50 HDPE/aPS blend.

The experimental results of the present paper show that two polymers, such as aPS and iPP or HDPE, when they are blended, result in a heterophase system. This is in agreement with literature data reported above. The corresponding overall morphology depends on the processing history and on molecular characteristics of the components.

The following observations can be drawn in reference to the morphological analysis:
In all the cases the modes of dispersion change with the radial position in the cylindrical extrudates. This general feature is due to the flow conditions in the processing apparatus which exhibits higher shearing stress near the wall than in the central region.
In fact, at the edge of the longitudinal section, mainly fibrils can be observed. In the central zone the morphology is much more

complex depending on the local flow conditions (low shear stresses
corresponding to high velocity values). It is interesting to point
out, moreover, that in analogy to the VanOne theory and results,
our experimental observations show that the probability of finding
stratified morphologies increases increasing the aPS percent in the
blend. Only in our case is there a tendency for the disperse phase
to segregate in the middle of the extruded filament.

Such a phenomenon of segregation of the disperse phase is
certainly due to the particular forces that act on the material
when it is extruded by a screw-less apparatus used to obtain the
sample in the present study. It is likely that after the shearing
zone of the extruder, the disperse particles tend to become spherical
under the action of the interfacial forces. Then, through the die,
they tend to be squeezed and elongated along the extrusion direction;
during such an elongation a coalescence effect becomes very likely
allowing for the formation of a continuous core in the middle of
the extruded cylindrical filament. The results of the present investigation lead to the conclusion that the overall morphology of
extrudates of aPS/iPP and aPS/HDPE blends, obtained by using a screw-
less extruder, seem to be quite different from that observed in
extrudates obtained with more usual apparatus (screw extruder and
capillary rheometer). Particularly interesting, especially from
the practical point of view, is the effect of segregation of the
disperse phase in the centre of the filament observed for some compositions and extrusion temperatures.

The morphology of these filaments is very similar to that of
"sheath core" fibers. These fibers, obtained by the rather complicated process of coextrusion, are characterized by the fact that
one component is enclosed (core) by the other (sheat). The filament
with the central segregate disperse phase could be used, at least
in principle, when it is desirable to have fiber with surface properties different from those of the central core.

ACKNOWLEDGMENT

The authors wish to thank Prof. A. Keller and his collaborators
of Bristol University, for their cooperation and for the preparation
of samples to be analysed by optical and electron microscope tecni-
ques.

REFERENCES

1. R. Greco and E. Martuscelli, La Chimica & l'Industria, $\underline{4}$, 298 (1979).
    E. Martuscelli, Tecnopolimeri e Resine, $\underline{1}$, 19 (1980).
2. E. Martuscelli, Tecnopolimeri e Resine, $\underline{2}$, 14 (1979).
3. H. VanOene, J. Colloid and Interface Sci., $\underline{40}$, 3, 448 (1972).
4. R. Greco, G. Ragosta, E. Martuscelli and C. Silvestre "Properties of polystyrene/polyolefin alloys: I. Mechanical properties" in the present book.
5. C. D. Han and T. C. Yu, J. Appl. Polym. Sci., $\underline{15}$, 1163 (1971).
6. C. D. Han and Y. W. Kim, J. Appl. Polym. Sci., $\underline{19}$, 2831 (1975).
7. C. D. Han, C. A. Villamizar, Y. W. Kim and S. J. Chen, J. Appl. Polym. Sci., $\underline{21}$, 353 (1977).
8. N. P. Krasnikova, E. V. Kotova, G. V. Vinogradov and Z. Pelzbauer J. Appl. Polym. Sci., $\underline{22}$, 2081 (1978).
9. J. Karger-Kocsis, A. Kallo, A. Szafner, G. Bodor and Z. Senyel Polymer, $\underline{20}$, 37 (1979).
10. E. Martuscelli, C. Silvestre, G. C. Abate, "Crystallization, thermal behaviour and morphology of isotactic polypropylene/ /elastomer blends", in preparation.
11. C. B. Bucknall, "Toughened Plastics" Appl. Sci. Publishers Ltd. London, 1977.
12. V. Costantino, P. Quagliariello, Thesis, Istituto di Ricerche su Tecnologia dei Polimeri e Reologia, University of Naples, Italy, 1979.
13. S. Danesi and R. S. Porter, Polymer, $\underline{19}$, 448 (1978).
14. D. R. Paul, "Polymer blends" vol. 2., Edited by D.R. Paul and S. Newman, Academic Press, New York, 1978.

MORPHOLOGY AND MECHANICAL PROPERTIES OF

SXS-POLYSTYRENE BLENDS

E. Pedemonte, G.C. Alfonso, G. Siccardi

Centro Studi Chimico-Fisici di Macromolecole
Sintetiche e Naturali, Istituto di Chimica
Industriale dell'Università, Genova, Italy

INTRODUCTION

The SXS three block copolymers (where S is a polystyrenic block
and X a polyisoprenic or a polybutadienic one) form a new class of
polymeric materials at present widely employed as thermoplastic elastomers particularly at not too high polystyrene content. Their
peculiar properties are due to the thermodynamic incompatibility of
the two components in the solid state that leads to a microphase
separation. Owing to this phenomenon the material can be considered
as an elastomeric matrix physically vulcanized by the aggregation
of the polystyrene end blocks in spherical or cylindrical domains
which act also as reinforcing hard elements.

Several additives can be mixed to the copolymers with the purpose to modify some of their properties. Homopolystyrene is one of
the additives employed to improve mechanical properties and abrasion
resistance of the material; the behaviour of the resulting blend
strongly depends on the molecular characteristics of the added polystyrene.

Several authors[1-3] have deeply studied the morphology of the
SXS/PS blends. Recently Shen et al.[4,5] published a correlation
between mechanical properties and morphologies for a series of blends
of a poly(styrene-b-butadiene-b-styrene) copolymer and polystyrene
fractions of different molecular weights, cast from mutual solvents

and covering a rather wide range of compositions.

The aim of this paper is to compare the mechanical behaviour and the macrostructural characteristics of SXS/PS blends with different morphologies in order to obtain additional support to a topological model which has been assumed to explain the stress-strain curves of the pure copolymers[6].

EXPERIMENTAL

Materials

The three block copolymers used in our experiments were commercial samples of Kraton, produced by Shell Co. and labelled K 1102 and K 1107. K 1102 is a poly(styrene-b-butadiene-b-styrene) polymer while K 1107 has an isoprenic rubbery phase. The molecular characteristics of these polymers are summarized in Table I[7].

Standard fractions of polystyrene (PS), whose molecular characteristics are reported in Table II, were blended with the original copolymers.

Samples preparation

Blends were prepared in the form of thin films by casting dilute solutions of copolymer added with various amounts of polystyrene homopolymer. The evaporation was performed at T = 35°C, at very low rate (about 0.5 cm^3/h) and on a mercury surface (in order to have an homogeneous final thickness of about 0.2-0.4 mm). Films were completely dried under vacuum at 50°C, before examination and testing.

In order to obtain proper morphologies we use cyclohexane to dissolve K 1107 and its blends with polystyrene fractions and a mixture of tetrahydrofurane/methyl ethylketone (THF/MEK 90:10 by volume) for K 1102-PS blends. Previous work has revealed that using the aforementioned solvents and low evaporation rate the morphology of K 1107 is characterized by spherical polystyrene domains[8,9] as shown in Figure 1a while K 1102 exhibits the lamellar morphology reported in Figure 1b.

TABLE I

Molecular characteristics of K 1102 and K 1107 copolymers

Properties	K 1102	K 1107
Molecular weight; $\overline{M}_w$	75×10^3	1.5×10^5
Polystyrene, %	26	10
Molecular weight of each polystyrene block	10×10^3	7.5×10^3
Molecular weight of the central block	55×10^3	1.35×10^5
Molecular weight of the elastically efficient segment, M_c	14×10^3	30×10^3

TABLE II

Molecular characteristics of the polystyrene fractions

Fraction	Molecular weight, $\overline{M}_w$	$\overline{M}_w/\overline{M}_n$
PS 19	19.8×10^3	< 1.06
PS 51	51.0×10^3	< 1.06
PS 110	110.0×10^3	1.06
PS 223	223.6×10^3	1.06
PS 411	411.0×10^3	< 1.06

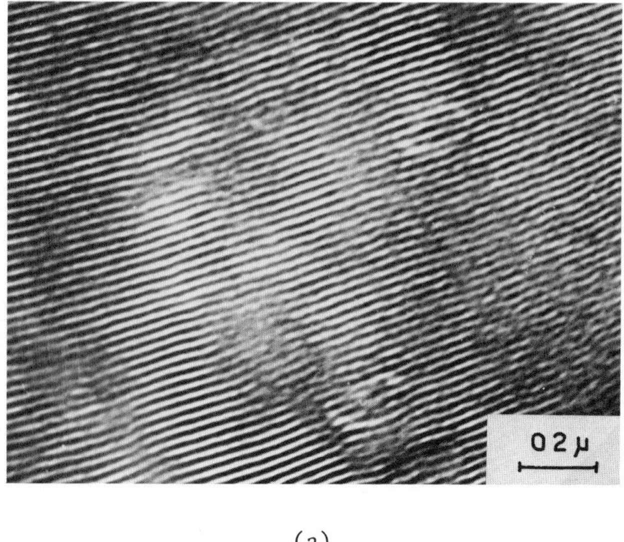

(a)

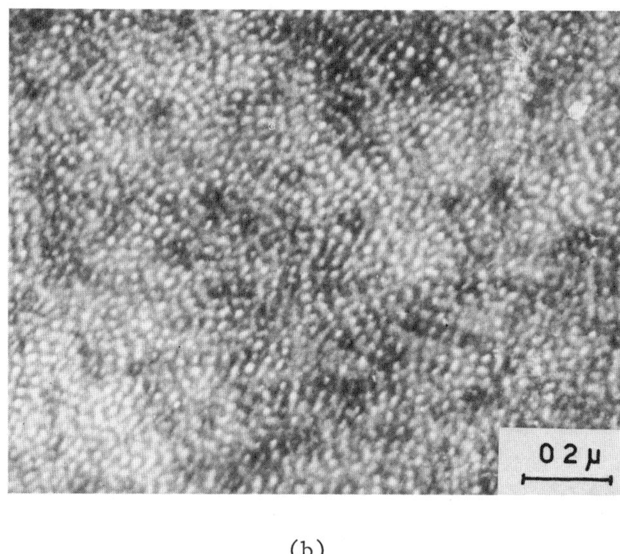

(b)

Figure 1. Electron micrographs of ultrathin sections stained with
           osmium tetroxide. a) K 1107 cast from cyclohexane;
           b) K 1102 casted from a mixture THF/MEK 90:10.

Methods

The morphological analysis was performed by electron microscopy on ultrathin sections cut by low temperature ultramicrotomy[10]. The staining was obtained by exposing the sections to vapours of aqueous osmium tetroxide solutions at room temperature; since OsO_4 is fixed by double bonds, in the electron micrographs the rubbery phase will appear dark and the polystyrene bright.

The stress-strain measurements were performed at room temperature, at a deformation rate of about 50 mm/min. The stress σ (Kg/cm^2) is defined by the ratio f/S_i, f being the load applied and S_i the initial cross sectional area; the extension ratio α is defined by $(\Delta L + L_i)/L_i$ where L_i is the initial lenght of the sample and ΔL its deformation. The actual specimen length was measured between two fiduciary lines across the strip[11].

RESULTS AND DISCUSSION

Correlation between modulus and amount of added polystyrene

The variation of the stress-strain curves of K 1107 due to the addition of different amounts of the homopolystyrene fractions PS 19 are shown in Figure 2. It results that blending the copolymer with a polystyrene fraction whose molecular weight is similar to that of the hard block of the thermoplastic elastomer leads to higher values of the elastic modulus while the general characteristics of the tensile behaviour are preserved. For istance no stress-softening effect comes out at low deformations when a second tensile cycle is performed immediately after the first one[12]. However at large deformations a progressively more important hardening effect is present on increasing the amount of added homopolymer: anyhow the loads necessary to deform the samples are higher for the original copolymer and its blends than for a chemically vulcanized polyisoprenic elastomer.

The behaviour can be easily correlated with the morphologies of the blends. Electron microscopy reveals that they are still characterized by randomly distributed spherical domains of polystyrene, whose diameter increases progressively with the amount of added homopolymer[13]. This means that the polystyrene doesn't form

additional domains but is distributed in those ones present in the
original copolymer.

The same results are obtained with original copolymers containing
a larger amount of polystyrene and therefore exhibiting a more complex
mechanical behaviour. Figure 3 shows the stress strain curves of
K 1102 films cast from THF/MEK 90:10 mixture and characterized by
lamellar morphology; the second deformation follows immediately the
first one. Stress softening and stress hardening effects appear at
low and high values of α respectively. Again also in this case the
modulus of the copolymer can be increased by addition of polystyrene
fractions with a molecular weight similar to that of each polystyrene block of the original material (Figure 4a).

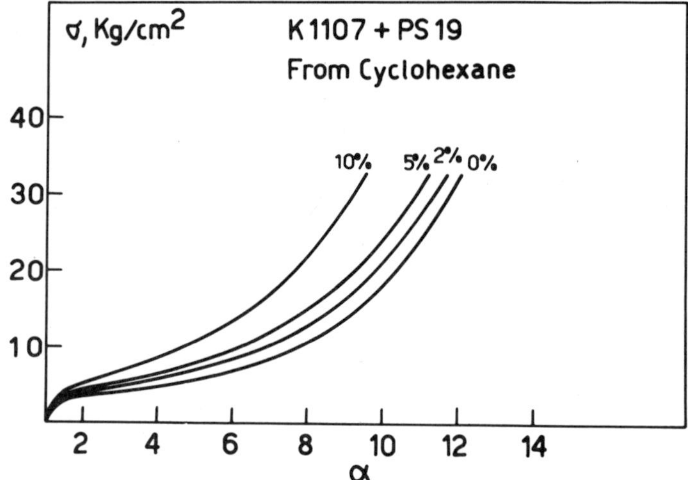

Figure 2.   Stress strain curves of K 1107/PS 19 blends containing
            different amounts of homopolystyrene.

Interpretation of the results: interaction between domains

The experimental results could be explained on the basis of a
model that takes into account the mechanism of the topological
changes which occur during the first deformation[6]. The model is
based on the idea that the contraction of the sample perpendicularly
to the strain direction (Figure 5) causes the polystyrene domains

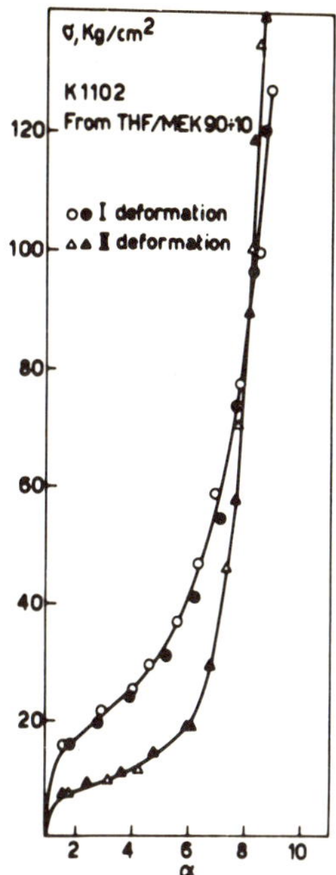

Figure 3.   Mechanical behaviour of K 1102 film cast from THF/MEK
            90:10.   The second tensile cycle is performed immediately
            after the first one.

in the transverse section to approach one another. When the extension ratio α reaches a well defined value, to be correlated with
the composition and the morphology of the sample, the rigid domains
will "impinge" and a further elongation needs a progressively large
amount of mechanical energy since interactions of hard domains are
involved from now on.

To calculate the value of the critical extension ratio α^* at
which the impingement between the nearest domains occurs two fundamental hypothesis have to be made, i.e. at not too high deformations:
i) the rubbery network undergoes an affine deformation
ii)the volume remains constant upon elongation (Poisson ratio equal
   to 0.5).

A furhter semplification of the treatment can be reached if
one assumes that the polystyrene domains has sharp and regular boun-

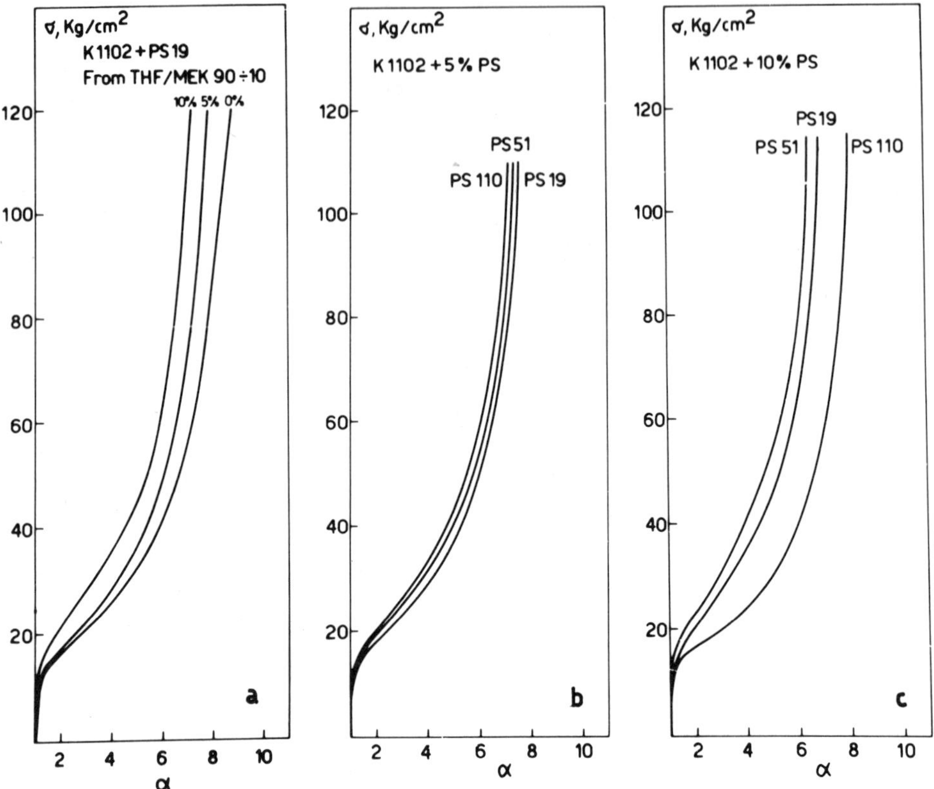

Figure 4.   Stress-strain curves of K 1102/PS blends containing:
a) different amounts of PS 19; b) and c) fixed amount
of polystyrene fractions with different molecular weights.

daries and are homogeneously dispersed in a phantom, perfectly elastic matrix.

Let us consider now spherical domains of even diameter ϕ. When
a stress is applied to the sample along an arbitrary z direction,
on the plane perpendicular to z the glassy domains will begin to

impinge when the distance between the centers of the nearest ones
becomes equal to their diameter. It is easy to demonstrate, on the
basis of the above-mentioned hypothesis, that

$$\alpha^* = (x_o/\phi)^2 \tag{1}$$

where x_o is the initial distance between the centers of adjacent
domains.

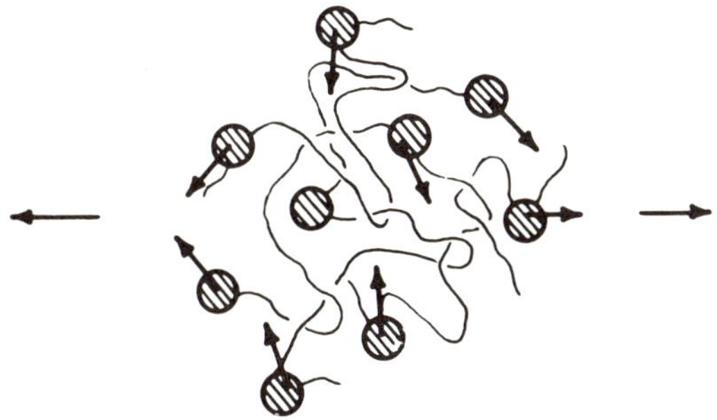

Figure 5. Topological changes during the stretching of a thermoplastic elastomer having spherical polystyrenic domains.

Combining eq. 1 with geometrical considerations it is possible
to estimate α^* for different morphologies[6]. When N spheres are
randomly distributed in the total volume V_T, the average distance
between the centers of two nearest domains will be:

$$\bar{x}_o = \left(\frac{V_T}{N}\right)^{1/3} = \left(\frac{\pi\phi^3}{6}\right)^{1/3} y^{-1/3} \tag{2}$$

where $y = \dfrac{\pi\phi^3}{6}\dfrac{N}{V_T}$ is the volume fraction of the hard phase. It
follows:

$$\alpha^* = 0.65 \, y^{-2/3} \qquad\qquad (3)$$

The application of our semplified model to copolymers and blends with morphology characterized by spherical domains of the hard phase enables us to estimate the α^* value as a function only of the volume fraction of polystyrene. With the aim to verify the reliability of Eq3 a point in the stress-strain curve should be found to be identified with α^*. Since the impingement of the hard domains will cause an increase of the tangent modulus we can identify α^* with the flex point in the deformation curve; this assumption may be not too rigorous but it seems reasonable that the upturn in the curve is somehow related to the critical deformation.

In Table III calculated and experimental values of α^* for the blends of Figure 3 are compared. The agreement seems to be quite satisfactory since we have to consider either that the dimensions of the domains are not even and that their distribution inside the rubbery matrix is not regular. A distribution around the mean value of both ϕ and $\overline{x}_o$ will cause a spreading of α^* and a smooth increase of the tangent modulus thus making more difficult a proper estimate of the experimental critical deformation. Actually Figure 6, in which eq. 3 is plotted in logaritmic form, shows that the experimental value of the slope is 0.44 in good agreement with the foreseen dependence of α^* on volume fraction of the hard component.

Other experimental results[11] support this model which will be discussed in more detail elsewhere[14].

TABLE III

Comparison between calculated and experimental values of α^*

Sample	y	$\alpha^*_{calc.}$	$\alpha^*_{exp.}$
K 1107	0.088	3.29	3.5
K 1107+2% PS 19	0.104	2.94	3.15
K 1107+5% PS 19	0.129	2.55	2.9
K 1107+10% PS 19	0.169	2.13	2.6

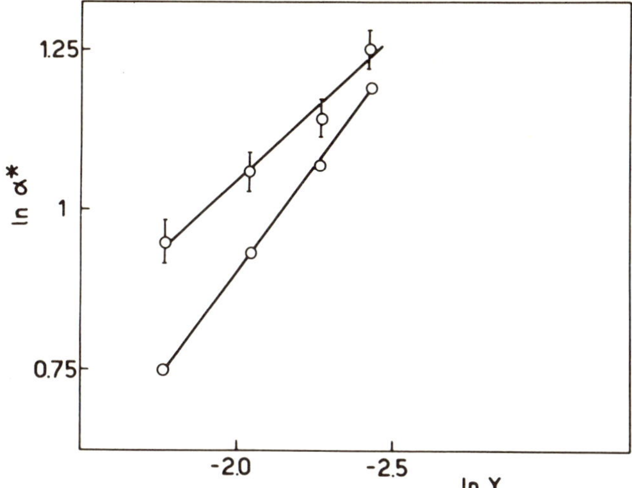

Figure 6. ln α^* vs. ln y plot. O values calculated according to equation (3); Φ experimental results.

<u>Correlation between modulus and molecular weight of added polystyrene.</u>

Figure 7 shows that the elastic modulus of the material can be increased by adding a fixed amount of homopolystyrene with progressively higher molecular weight. Blends of K 1107 with PS 110 and PS 223 show a mechanical behaviour not too different from that with PS 51.

PS 411 modifies rather moderately the mechanical properties of the original copolymer; the modulus becomes smaller than that obtained with the addition of the same amount (2%) of PS 51. This effect seems to be attributed to segregation of homopolymer in additional domains, whose dimensions are much bigger than the original ones and therefore break the homogeneity of the material. Actually for blends with PS 411 the modulus has its maximum value for an addition of only 0.5% of this fraction (Figure 8). Behind this percentage the segregation of the added polystyrene starts to depress the mechanical behaviour of the material.

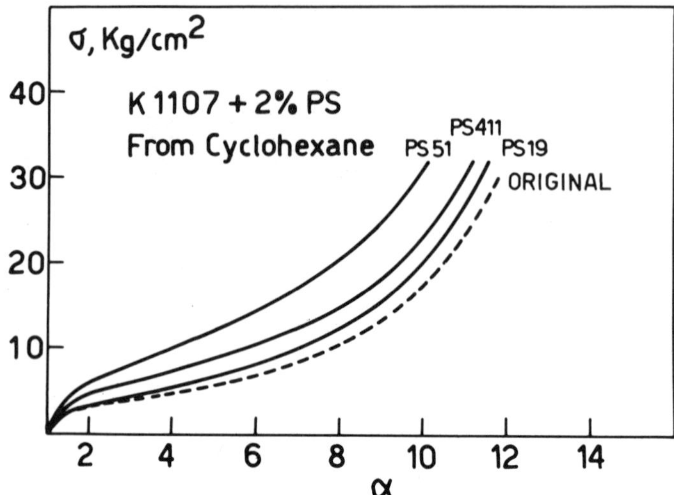

Figure 7.   Stress-strain curves of K 1107/PS blends containing fixed amount of polystyrene fractions with different molecular weights.

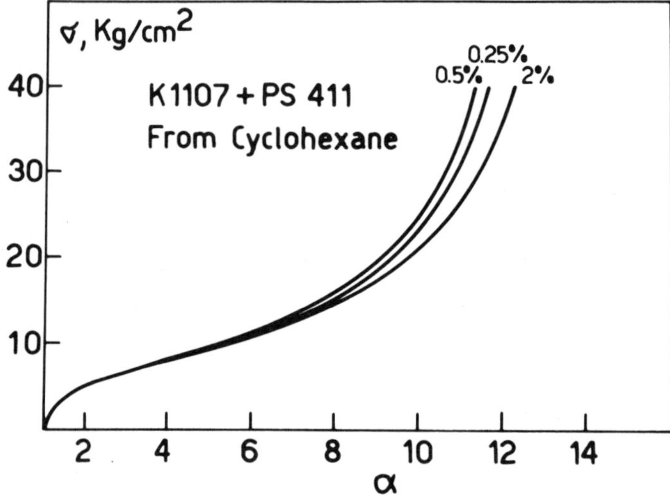

Figure 8.   Stress-strain curves of K 1107/PS 411 blends containing different amounts of homopolystyrene.

This behaviour is general. The maximum amount of polystyrene
which can be dissolved in a copolymer without any segregation decreases rapidly when the molecular weight of the added homopolymer is
much higher than that of the polystyrene blocks of the copolymer.
One can conclude that the improvement of the mechanical properties
of the starting material can be obtained only if the molecular lenght
of the polystyrene macromolecules is not too different from that of
the polystyrene ends of the original three block copolymer.

REFERENCES

1. T. Inoue, T. Soen, T. Hashimoto and H. Kawai, "Block Polymers",
    ed. S. L. Aggarwal, Plenum Press, pg. 73, New York 1970.
2. D. McIntyre and E. Campos Lopez, ibidem, pg. 29.
3. G. Kraus and K. W. Rollmann, J. Polym. Sci., Polym. Phys. Ed.
    14, 1133 (1976).
4. G. Akovali, J. Diamant and M. Shen, J. Macromol. Sci. Phys.
    B13, 117 (1977).
5. M. Niinomi, G. Akovali and M. Shen, J. Macromol. Sci. Phys.
    B13, 133 (1977).
6. G.C. Alfonso and E. Pedemonte, ACS Polymer Preprints, 18, 276
    (1977)
7. U. Bianchi, E. Pedemonte, A. Turturro and M. Tombini, La Chimica
    & l'Industria, 54, 603 (1972).
8. E. Pedemonte, A. Turturro, U. Bianchi and P. Devetta, Polymer,
    14, 145 (1973).
9. E. Pedemonte, G.C. Alfonso, G. Dondero, F. de Candia and
    L. Araimo, Polymer, 18, 191 (1977).
10. G. Dondero, L. Oliviero, P. Devetta, S. Cartasegna and
    E. Pedemonte, La Nuova Chimica, 48, 1 (1972).
11. G. C. Alfonso, E. Pedemonte and G. Siccardi, Atti del IV Conve-
    gno Italiano di Scienza delle Macromolecole, Colleferro (Roma)
    1979, pg. 195.
12. A. Turturro, U. Bianchi, E. Pedemonte and P. Ravetta, La Chimica
    & l'Industria, 54, 782 (1972).
13. E. Pedemonte, G. C. Alfonso and G. Siccardi, Polymer Bulletin,
    to be published.
14. G. C. Alfonso, E. Pedemonte and G. Siccardi, Polymer, to be
    published.

SYNTHESIS OF INTERPENETRATING POLYMER NETWORKS AND

THEIR POSSIBLE USES

P. Penczek

Institute of Industrial Chemistry, Warsaw, Poland

Synthesis of interpenetrating polymer networks (IPN)
from various crosslinkable monomers and oligomers, is discussed. Some special IPNs, e.g. simultaneous interpenetrating networks (SIN), gradient IPNs and IPNs with one
linear component are described. The importance of IPNs,
as toughened elastomers, impact resistant plastics as
well as noise and vibration attenuation materials, is
pointed out.

INTRODUCTION

Interpenetrating polymer networks (IPNs) had been investigated
ans used, in practice, for a long time before this term was introduced by Millar[1] in 1960. The term "IPN" covers a broad range of
polymeric systems, varying in the chemical composition and the structure. The IPNs form a particular class of multicomponent polymeric
systems, consisting of two separate crosslinked polymers with no
covalent bonds between networks. However, binary polymer systems
with only one polymer crosslinked are also included in IPNs as so-
called semi-IPNs.

At present, IPNs are very important from both the purely
scientific and practical points of view. The research of IPN systems
has systematically been conducted by three main teams:

- at the Materials Research Center of Lehigh University, Bethlehem,
  Penn. (USA): L. H. Sperling,
- at the Polymer Institute of the University of Detroit (USA):
  H. L. Frisch, K. C. Frisch and D. Klempner, and
- at the Institute of Chemistry of High Molecular Compounds of the
  Academy of Science of Ukrainian SSR in Kiev (USSR): Yu. S. Lipatov.

The results of the research in this field, have been reviewed
by Sperling[2,3], Klempner[4], Lipatov and Sergeeva[5] a.o. and reported
in the polymer encyclopedia[6].

SYNTHESIS

Subsequently polymerized IPNs

The polymers, which form an IPN, can be synthesized subsequently
or simultaneously. The subsequent synthesis enables polymer chemists
to obtain IPNs from a broad range of starting materials, since the
formation of undesired chemical bonds between the separate polymer
networks is avoided.

The IPNs synthesized by the subsequent polymerization methods
are generally considered as "true" IPNs, if both polymers involved
are crosslinked. In order to obtain such IPN, one polyfunctional
monomer is first polymerized. The obtained crosslinked polymer is
subsequently swollen with another monomer, which is polymerized
inside the existing polymer network. Thus, an IPN consisting of
two networks, having the same chemical composition, can be obtained.
Such systems are called homo-IPN (or HIPN). In this subclass,
crosslinked copoly(styrene-divinylbenzene) systems were investigated[1,7-11]. Both networks were obtained by benzoyl peroxide initiated radical polymerization. It is obvious that covalent bonds
between the separate networks cannot be formed when no frozen free
radicals remain in the primary network. The frozen free radicals
could initiate the grafting of the second monomer, which would
eventually result in the crosslink formation between both networks.
Of course, chain transfer during the second polymerization should
not occur.

More frequently, the IPNs obtained by subsequent polymerization
comes from two chemically different monomers or comonomer systems.

Both networks, the primary one and the secondary one can be obtained
by a radical polymerization. So far, only few such systems have been
investigated. Therefore many interesting possibilities could be
suggested. Sperling and coworkers studied the system consisting of
ethyl acrylate – diethylene glycol dimethacrylate (DEGDM) (primary
or matrix network) and styrene – DEGDM (secondary or penetrating
network)[12]. Both monomer systems were photopolymerized in the presence of benzoyl peroxide. The structure of this system differs
from the IPN obtained in a similar way, from monomers, which have
similar chemical composition: ethyl acrylate – tetraethylene glycol
dimethacrylate (TEGDM) (primary network) and methyl methacrylate –
TEGDM (secondary network)[13-15]. A crosslinked polystyrene – polybutadiene system was also investigated[16].

Recently, subsequently polymerized IPNs from more differing
monomer systems are investigated, being one of those systems a polyurethane. The polyurethane network is first formed and then followed by swelling in an unsaturated monomer and by radical polymerization of the monomer.

Considerable work was done using castor oil for constructing
the matrix network[16-18]. The hydroxyl groups containing oil was
crosslinked with a diisocyanate and then swollen with a styrene –
divinylbenzene mixture with benzoin as an ultraviolet photopolyme-
rization initiator. In another series of experiments, the matrix
network was obtained by crosslinking castor oil with sulfur, whereas
the penetrating network was derived from methyl methacrylate with
TEGDM crosslinker. The investigations on castor oil based IPNs
allowed better understanding of the effect of the chemical composition and the structure of IPNs consisting of a high elastic network
and a glassy network on their mechanical behaviours.

An interesting approach to the synthesis of subsequently polymerized systems was applied by Shilov and Lipatov[20], both networks
were obtained by naphthalene – sodium initiated anionic polymerization. In this way, "living" polymer networks were formed. The primary network was obtained from bis-(triethylene glycol) phthalate
dimethacrylate and triethylene glycol dimethacrylate. The secondary
network was formed from styrene and divinylbenzene. Desired microheterogeneous structure of the IPN was achieved.

Sometimes the borderline between subsequently and simultaneously

polymerized IPNs is not distinct. The matrix network can be obtained
by polymerizing the first monomer in solution in another monomer,
which does not polymerize in this reaction step. Thus, a matrix
network swollen in the second monomer is obtained without carrying
out the swelling operation. Such way should be rather considered
as a variant of the subsequent polymerization method. It seems to
be very promising provided that different polymerisation mechanisms
for both monomers can be chosen. For example, the syntheses described by Byelonovskaya et al[21] should be mentioned. Polycyclotrime-
rization of toluenediisocyanate dissolved in methyl methacrylate,
acrolein, acrylonitrile or propylene sulfide was carried out. A
tridimensional matrix consisting of isocyanurate and benzene rings,
was formed. The monomer contained in the matrix network was polymerized subsequently. In this way, heat resistant polymers have
been obtained.

The products synthesized by Byelonovskaya et al are not "true"
IPNs, because the penetrating polymer is not crosslinked. Similar
"true" IPNs can be, however, easily obtained using crosslinkable
second monomer with the same polymerizing groups as in the monomers
above mentioned. The chemical composition of IPNs obtained by the
polymerization I – swelling – polymerization II process is constant
on a macroscopic scale in the whole polymer sample. The IPNs, which
composition varies throughout the polymer sample, can be obtained
by nonequilibrium swelling of the matrix network followed by rapid
polymerization. Such products are called "gradient IPNs". They
were obtained to be used as noise-damping materials[22].

Simultaneous interpenetrating polymer networks (SINs)

Considerably more research has been done in the field of the
so-called simultaneous interpenetrating polymer networks (SINs).
In this case, the choise of monomers is restricted to compounds,
which do not react together in the polymerization conditions. Thus,
the formation of both networks by radical polymerization is excluded.
The networks are obtained using different polymerization mechanisms.
In most cases, one network is formed by radical polymerization and
another one is obtained by polyaddition or rarely by ionic polymerization. Polyaddition – ionic polymerization SINs were also investigated.

Components, which are polymerized by the radical mechanism, involve mostly a low-viscous monounsaturated monomer and a crosslinking monomer with two or more unsaturated bonds. Typical examples are styrene – divinylbenzene[23], styrene – unsaturated fumarate polyester[24,25], and alkyl acrylate or methacrylate – alkylene diacrylate or dimethacrylate[25-29] systems.

Most networks obtained by polyaddition consist of polyurethanes. A broad range of components was used. Polyesterols or polyetherols (linear or branched polyesters and polyeters with hydroxyl end groups) served as main structural components. In general, prepolymer technique was used: the polyol was prereacted with excess diisocyanate and then was crosslinked with a low-molecular-weight diol or triol in a presence of an organotin catalyst. Unsaturated monomers were added to the prepolymer -crosslinker system. Polyurethane containing SINs were also obtained using castor oil based compositions[23]. If unsaturated components contain hydroxyl or carboxyl group, they should be prereacted with a monoisocyanate before they are joined with the polyurethane, forming components in order to avoid the formation of covalent bonds between both networks. This method was applied for an unsaturated polyester resin to be polymerized simultaneously with an urethane prepolymer[25].

SINs consisting of two urethane-based networks cannot be obtained in a simple way. However, Lipatov et al succeeded in synthesizing such SIN by transforming one of the urethane compositions to be crosslinked into an unsaturated oligomer, which could react via radical polymerization[30]. The unsaturated oligomer was synthesized from ethylene glycol monomethacrylate, toluene diisocyanate (TDI) and polyoxypropylenediol (POPD). The usual polyurethane network was obtained from TDI, POPD and trimethylolpropane. SINs are usually made using the solventless casting technique. In this case, however, crosslinking occurred in films cast from solution. Besides the unsaturated monomer - polyurethane systems, an unsaturated monomer - epoxy resin system was investigated[31]. The epoxy resin was a liquid Epichlorohydrin - Bisphenol A oligomer with high content of pure diglycidyl ether of Bisphenol A. It was crosslinked with phthalate anhydride. As unsaturated monomer, n-butyl acrylate with DEGDM crosslinker was used. The effect of networks formation conditions on the structure and the properties of the system was studied.

In another paper the effect of grafting, that is to say, of a

definite number of crosslinks between the separate networks on the
structure and the mechanical properties of the system was discussed[32]. As a source of the internetwork crosslinks, glycidyl methacrylate was used in that it took part both in the radical polymerization and in the epoxy resin curing.

Recently, SINs obtained by the polyaddition of polyurethane
precursors and the anionic polymerization of a diepoxy compound are
investigated[33,34]. As for diepoxide, pure diglycidyl ether of Bisphenol A was used so that the reaction of isocyanate groups with
hydroxyl groups, which are present in epoxide oligomers, is avoided.
The anionic polymerization of epoxy groups was initiated with a
Mannich-base type tertiary amine. Unusual ideas of SIN preparation
are sometimes described in patent specifications. This can be
exemplified by a SIN consisting of a hydrophilic (meth)acrylic acid
ester or amide N-vinylpyrolidone and a crosslinking agent (network I)
as well as a monomer system, which forms a crosslinked polydimethyl-
siloxane (network II)[35].

During investigations on the composition - structure - properties relationship of SINs, a very interesting problem arose. The
gel point of both systems involved can be reached at the same time
or can be shifted in time. In an extreme case, the formation of the
first network is completed before the second monomer starts to react
(compare the systems investigated by Byelonovskaya et al[21]). If
both systems reach gelation simultaneously, a very fine state of
phase dispersion is obtained. If one monomer system reacts faster,
a larger size of domains is obtained and the mechanical properties
of the SIN improves[2,31]. This regularity concerns SINs with one
glassy and one elastomeric component.

Latex IPNs

From a practical point of view, the IPNs in latex form seem to
be very interesting and have carefully been investigated. In this
class of IPNs, two different groups are distinguished. The first
group can be considered as proper latex IPNs (or LIPN). The preparation of such LIPNs involves emulsion polymerization of monomer I
with some crosslinking agent added. The so-called "seed latex I"
with matrix network particles is formed. To the latex I, monomer II
with appropriate crosslinker is introduced. The polymer particles

in latex I are penetrated by monomer II, which polymerizes and forms
the penetrating network. Thus, the preparation of this group of
LIPNs is similar to the subsequent polymerization method of IPN synthesis and similar monomer crosslinker pairs can be used. First,
various systems consisting of acrylate and methacrylate esters have
been investigated[36]. It seems, however, that latex IPNs similar to
SINs (that is to say, latex IPNs obtained by simultaneous emulsion
polymerization of two different monomer systems) have not yet been
synthesized. The difficulty is obvious: the ways of polymer formation other than radical polymerization, so as ionic polymerization
or diisocyanate polyaddition are generally excluded, in aqueous
medium. On the other hand, latex SINs could be obtained in a non-
aqueous medium. In this case, a broad range of monomer systems and
polymerization mechanism is thinkable. Processes being actually
used for the manufacture of the so-called "high-solids" (concentrated
film-forming emulsions of polyacrylates in benzine) can be applied.

The latex IPNs from the second group are obtained by mixing two
different latexes followed by casting film and crosslinking at an
elevated temperature. For example, compositions consisting of a
polyurethane and polyacrylate can be mentioned[37]. The polyuretha-
neurea was crosslinked with a triol and the polyacrylate was reacted
with sulfur through unsaturated bonds.

The research in the field of latex IPNs afforded a handle to
the synthesis of suspension polymerized IPNs[38]. It should be pointed
out that the film-forming properties of latex IPNs and the molding
possibilities of both latex and suspension IPNs are not satisfying,
if the crosslinking density is too high. In the case of latex IPNs,
only 0.4% crosslinking agent (TEGDM or divinylbenzene) was added to
the unsaturated monomer[39,40].

There have been some IPN - like systems used for a long time
in practice without calling them "IPN's" and without taking them
into account in reviews on IPNs. As an example, nitrile rubber modified phenolics can be mentioned. No evidence has yet been given,
however, such systems belong to true IPNs with no crosslinks between
the vulcanized rubber network and the hardened phenol-formaldehyde
resin. Surely, more examples of this kind can be found.

Semi-IPNs
<u>---------</u>

A very large group of IPNs form the so-called semi-IPNs or
pseudo-IPNs. Semi-IPNs are polymer systems consisting of one crosslinked and one linear or branched polymer with no covalent bonds
between the network and the non-crosslinked polymer. Strictly spea-
king, semi-IPNs should not be considered together with proper IPNs.
They are, however, often discussed in reviews on IPNs and are investigated by the same research teams.

Semi-IPNs have been obtained and investigated in various sub-
classes of IPNs. Numerous SINs, subsequently polymerized IPNs, both
types of latex IPNs a.o. can be obtained using the compositions
above des-ribed, one monomer (e.g. styrene, an acrylate or a methacrylate) containing no crosslinking agent. In this review, only
some semi-IPNs differing substantially from the usual IPN systems,
will be discussed.

In usual semi-IPNs, a bifunctional monomer is polymerized simultaneously with a crosslinkable monomer system or is used as a
penetrating monomer for the previously formed matrix network.

A different group of semi-IPNs form polymer - oligomer systems.
They consist of a linear or branched thermoplastic high-molecular
weight polymer and a polymerizable oligomer. The matrix polymer is
swollen in the oligomer followed by polimerization (mostly radical)
of this last one. Much research in this field has been carried out
by A. A. Berlin at the Institute of Chemical Physics of the Academy
of Science of the USSR in Moscow. The results were reviewed in the
articles on the polymerization of polyreactive oligomers[41,42] and
on the polymer - oligomer systems[43,44].

The polyreactive oligomers used by A. A. Berlin involved numerous linear or branched oligoesteracrylates with more that two methacrylate end groups. Typical examples[45] are triethylene glycol dimethacrylate, bis-glyceryl phthalate tetramethacrylate, bis-pentae-
rythrityl adipate hexamethacrylate or similar oligomers with oligo
(alkylene carbonate), oligourethane or oligodimethylsiloxane backbone.

From practical points of view, very interesting are polyvinyl
chloride - polymerizable oligomer compounds[43,44,46]. Polyvinyl

chloride is first plasticized with the monomer in the usual way (by
milling) followed by extrusion, compression molding etc., and then
the oligomer is polymerized. Of course, the oligomer must not polymerize during processing of the PVC compound at an elevate temperature. Therefore, it must not contain any initiator and the polymerization in the shaped article is started e.g. by irradiation. In
this way, hard PVC articles with elevated heat distortion temperature
can be obtained.

Unsaturated polyester resins may also serve as polymerizable
oligomers in PVC compounds[47,48]. From emulsion PVC and the unsaturated polyester resin (being a solution of maleate – fumarate oligo-
ester in styrene), a plastisol was obtained. Shaped articles can
be made of such plastisols by casting. During casting operation,
gelatinization of the PVC – resin system and copolymerization of unsaturated polyester with styrene occur at the elevated temperature.
The copolymerization is initiated by a peroxide, which was previously
added to the resin.

Using suspension PVC with usaturated polyester resin, the so-
called dry blends were obtained[47,48]. Moldings were made by compression molding, similar processes occurring as in the case of plastisols.

As above mentioned, ideal IPNs with no crosslinks between the
separate networks are very difficult to obtain, if a monomer polymerizes in the presence of a simultaneously formed one or a previously
formed matrix polymer. The same goes, in particular, for the systems
with polyvinyl chloride, which is very susceptible of radical chain
transfer reaction followed by grafting.

Semi-SINs obtained from polyurethane precursors and a radically
polymerizing monomer are also interesting materials. Crosslinked
polyurethane/linear poly(methyl methacrylate) semi-SINs were investigated along with true SINs consisting of a polyurethane and a
methyl methacrylate/trimethylolpropane trimethacrylate composition[49].

A sophisticated semi-IPN was obtained in the field of poromeric
leather-like materials. Polyethylene terephthalate non-woven fabric
was impregnated with an aqueous solution – born unsaturated polyester
resin with strongly hydrophilic properties followed by crosslinking of
the resin on the surface of the fibers. The matrix network of the

crosslinked resin resin was then swollen in a solution of a polyure-
thaneurea in dimethylformamide and the coagulation this polymer
was carried out by means of water. Dimethylformamide was eventually
washed out with water. Synthetic leather substrate with improved
hydrophilic properties was thus obtained[50].

Another complex semi-IPN systems consist of an unsaturated polyester resin and a linear polymer, e.g. a polyhydroxyester with poly
(ethylene adipate) and bisphenol A diglycidyl ether - derived segments[51]. The thermoplastic polymer (so-called low-profile additive)
decreases the shrinkage in polyester resin based molding compounds
as a result of microsegregation, migration of styrene into the thermoplastic phase, and evaporation of styrene on compression molding
accompanied by voids formation[52-55]. It is noteworthy that the chemical structure of the crosslinked polyester resin is affected by
the thermoplastic polymer. This was experimentally shown for the
polyhydroxyester containing system[51].

STRUCTURE AND PROPERTIES

The IPN chemistry is the main subject of this review. Therefore, the physical structure of IPNs is only shortly discussed here.
More information can be found in the reviews above mentioned, especially those by Sperling[2] and Lipatov[5]. An idealized IPN-type structure presumably does not exist. As above mentioned, covalent bonds
between the separate networks are certainly present in most cases,
especially as a result of chain transfer reactions. Moreover, pseudo-
crosslinking between the networks, e.g. through hydrogen bonds, crystalline areas, ionic clusters, or glassy domains commonly present
in an elastomeric matrix are not taken into consideration. The role
of entanglements between the chain segments of both networks may be
also important.

IPNs are two-phase systems with various domain size. The fine
structure and the domain size of an IPN depend on several factor.
Difference between the solubility parameters of both polymers, crosslinking density of the matrix networks, relative polymerization
rates of both components (in the case of SINs), component I - component II ratio and content of graftings or intercrosslinking bonds
(if any) play an important role. For latex IPNs obtained by the
polymerization of monomer II inside the seed latex particles, a
complex morphology is considered. Cellular structures, fine struc-

tures, and a shell-core morphology are assumed[2].

The morphology of IPNs can be altered by introducing an inorganic filler. The results of investigations of filled IPNs led to a conclusion that non filled IPN can be considered as a filled system, the matrix network being a filler for the next formed network[5,56]. The analogy lies, among others, in the effect of the filler surface or of matrix network on the properties of the polymer to be formed. The point is that the structure of the matrix network is changed by the arising penetrating network, whereas the particles of inorganic fillers do not change[57].

Physicomechanical properties of IPNs depend on their chemical composition and morphology. Some characteristic features can be of practical importance.

IPNs consisting of polyurethane and a glassy polymer (crosslinked polyacrylate or epoxy resin) show a composition - tensile strength relationship with one minimum and one maximum[4,5]. The minimum at a high polyurethane content is due to the weakening of the initially strong hydrogen bonds in the polyurethane by the glassy polymer, which acts in this case in the plasticizer-like way. Interpretation of the maximum at high concentration of the glassy polymer is difficult. Segment entaglements or intercrosslinks are taken into consideration.

In IPNs consisting of a glassy and a high-elastic polymer, the systems with elastomer matrix behave as a reinforced elastomer, whereas the systems with glassy polymer prevailing are impact-resistant plastics. In the midrange compositions, the materials are leathery[19]. IPNs consisting of incompatible components with phase separation show two distinct glass transitions. Sometimes occurrence of a third phase connected supposedly with segmental interactions on a molecular level reveals itself in a form of the third glass transition. IPNs consisting of compatible components or containing numerous graftings and intercrosslinks show no phase separation and one broad glass transition only can be found.

The noise and the vibration damping ability over a broad temperature range is a further important feature of some semicompatible latex IPNs[58]. The temperature range reaches from the glass transition of the homopolymer I to the glass transition of the homopolymer II[59].

PRACTICAL USES

Information on suggested and actual uses of IPNs can be found
in patent specifications and rarely in journal articles. Ten patents
are listed in the review by Spearling[3].

The most promising uses of IPNs are noise damping systems in
the form of films, thick multilayer coatings, constrained layers,
and so-called "silent paints"[22,58-60].

An IPN-based adhesive for bonding metals showed increased short-
term and long-term butt-joint strength thanks to low level of internal stresses and rapid internal stress relaxations[25]. Pressure-sensitive adhesives with both elastomeric networks were also suggested[61].
An expecting of practical application give IPNs consisting of network
I with cationic groups and network II with anionic groups. The
fields of application are piezodialysis membranes[62] and ion-exchange
resins[63]. Soft contact lenses are obtained, if both networks are
high-elastic and hydrophilic[35,64]. Swelling in water is an important
feature of such IPNs.

A broad range of applications can result from the improved mechanical properties of IPNs consisting of glassy network I and high-
elastic network II. Tough plastics with elevated tensile strength
and elongation as well as high impact resistance seem to be the most
important group of future applications. IPN additives could serve
as impact modifiers[65]. On the other hand, reinforced elastomers are
worth notice.

In the field of semi-IPNs, the most promising are compositions
consisting of a thermoplastic polymer and a polyreactive oligomer
as discussed above, headed by polyvinyl chloride with polymerizable
plasticizers.

In the next future, growing application of semi-IPNs obtained
from methyl methacrylate and polyurethane precursors may be expected.
The main future use seems to be modified polyurethane foams.

In the field of poromeric synthetic leather, the use of the IPN
principle can lead to better materials, the properties of which would
came nearer to those of natural leather.

Many important uses, which have not revealed so far, may be found in the field of thermally stable polymers.

REFERENCES

1. J. R. Miller, J. Chem. Soc., 1311 (1960).
2. L. H. Sperling, J. Polym. Sci., Macromol. Rev., 12, 141 (1977).
3. L. H. Sperling, J. Polym. Sci., Polym. Symp., 60, 175 (1977).
4. D. Klempner, Angew. Chem., 90, 104 (1978).
5. Yu. S. Lipatov and L. M. Sergeeva, Usp. Khimii, 45, 138 (1978).
6. L. H. Sperling, Encycl. Polym. Sci. Technology, Supplement N.1, J. Wiley & Sons, New York, 1976.
7. K. Shibayama, Kobunshi Kagaku, 19, 219 (1962).
8. K. Shibayama, Kobunshi Kagaku, 20, 221 (1963).
9. K. Shibayama, Zairyo, 12, 362 (1963).
10. K. Shibayama and Y. Suzuki, Kobunshi Kagaku, 23, 24 (1966).
11. K. Shibayama and Y. Suzuki, Rubber Chem. Technol., 40, 476 (1967).
12. L.H. Sperling and D. W. Friedman, J. Polym. Sci.A-2, 7, 425 (1969).
13. L. H. Sperling, D. A. Thomas and H. F. George, Amer. Chem. Soc., Polymer Prepr., 11, 477 (1970).
14. L. H. Sperling, H. F. George, V. Huelck and D. A. Thomas, J. Appl. Polym. Sci., 14, 2815 (1970).
15. V. Huelck, D. A. Thomas and L. H. Sperling, Macromolecules, 5, 340 (1972).
16. A. J. Curtius, M. J. Covitch, D. A. Thomas and L. H. Sperling, Amer. Chem. Soc., Polym. Prepr., 12, 699 (1971).
17. L. H. Sperling, J. A. Manson, G. M. Yenwo, A. Conde and N. Devia, Polym. Prepr., 16, 604 (1975).
18. G. M. Yenwo, J. A. Manson, J. Pulido, L. H. Sperling, A. Conde and N. Devia, J. Appl. Polym. Sci., 21, 1531 (1977).
19. G. M. Yenwo, L. H. Sperling, J. Pulido, J. A. Manson and A. Conde, Polym. Eng. Sci., 17, 251 (1977).
20. V. V. Shilov and T. E. Lipatova, Vysikomol. Soed. A, 20, 62 (1978).
21. G. P. Byelonovskaya, L. S. Andrianova, Zh. D. Chernova, L. A. Korotneva, and B. A. Dolgoplosk, Dokl. Akad. Nauk. SSSR, Ser. Khim., 212, 615 (1973).
22. L. H. Sperling and D. A. Thomas, U.S. Patent 3,833,404 (1974).
23. N. Devia-Manjarres, J. A. Manson, L. H. Sperling and A. Conde, Polym. Eng. Sci., 18, 200 (1978).

24. H. L. Frisch, K. C. Frisch and D. Klempner, Polym. Eng. Sci., 14, 646 (1974).

25. Yu. S. Lipatov, R. A. Veselovskii and Yu. K. Znachkov, Dokl. Akad. Nauk. SSSR, 238, 174 (1978).

26. K. C. Frisch, D. Klempner, S. Migdal and H. L. Frisch, J. Polym. Sci.A-1, 12, 885 (1974).

27. K. C. Frisch, D. Klempner, S. Migdal and H. L. Frisch, J. Appl. Polym. Sci., 19, 1893 (1974).

28. K. C. Frisch, D. Klempner, T. Antczak and H. L. Frisch, J. Appl. Polym. Sci., 18, 683 (1974).

29. K. C. Frisch, D. Klempner, S. Migdal, H. L. Frish and H. Ghiradella, Polym. Eng. Sci., 14, 76 (1974)

30. Yu. S. Lipatov, L. V. Karabanova, T. S. Khramova and L. M. Sergeeva, Vysokolmol. Soed., Ser. A, 20, 46 (1978).

31. R. E. Touhsaent, D. A. Thomas and L. H. Sperling, J. Polym. Sci. Polym. Symp., N. 46, 175 (1974).

32. P. R. Scarito and L. H. Sperling, Polym. Eng. Sci., 19, 297 (1979).

33. H. L. Frisch, J. Cifaratti, R. Palmo, R. Schwartz, R. Foreman, H. Yoon, D. Klempner and K. C. Frisch, "Barrier and Surface Properties of Polyurethane-Epoxy Interpenetrating Polymer Networks", in "Polymer Alloys" (ed. by D. Klempner and K. C. Frisch) p. 97, Plenum Publishing Co., New York, 1977.

34. H. L. Frisch, R. Foreman and R. Schwartz, Polym. Eng. Sci., 19, 294 (1979).

35. Bausch and Lomb Inc., Ger. Offen. 2,518,904; Chem. Abs. 84, 60757.

36. J. A. Grates, D. A. Thomas and E. C. Hickey, L. H. Sperling, J. Appl. Polym. Sci., 19, 1731 (1975).

37. D. Klempner, H. L. Frisch and K. C. Frisch, J. Polym. Sci. A-2, 8, 921 (1970).

38. J. Khanderia and L. H. Sperling, J. Appl. Polym. Sci., 18, 913 (1974).

39. L. H. Sperling, Tai-Woo Chiu, C. P. Hartman and D. A. Thomas, Intern. J. Polym. Mater., 1, 331 (1972).

40. L. H. Sperling, Tai-Woo Chiu and D. A. Thomas, J. Appl. Polym. Sci., 17, 2443 (1973).

41. A. A. Berlin and S. M. Mezhikovskii, Zh. Vses. Khim.O-va, 21, 531 (1976).

42. A. A. Berlin, Vysokomol. Soed.A, 20, 483 (1978).

43. A. A. Berlin, Plaste Kautschuk, 18, 563 (1971).

44. A. A. Berlin, Plaste Kautschuk, 20, 728 (1973).

45. A. A. Berlin, T. Ya. Kefeli and G. V. Korolev, Poliefirakrilaty, Izdatelstvo Nauka, Moscow, 1967.

46. A. A. Berlin, N. N. Zavodchikova and N. N. Yanovskii, Plasticheskiye Massy, 47, 4 (1975).

47. P. Penczek, R. Pyrko and L. Przybora, Polimery, 15, 244 (1970).

48. P. Penczek and W. Tokarski, Polimery, 16, 475 (1971).

49. S. C. Kim, D. Klempner, K. C. Frisch, W. Radigan and H. L. Frisch, Macromolecules, 9, 258 (1976).

50. M. Kosinska, P. Penczek and Z. Klosowska-Wolkowicz, in preparation.

51. W. Krolikowski and M. Pawlak, Faserforschung Textiltechnik - Z.f. Polymerforschung, 29, 39 (1978).

52. K. Demmler and E. Muller, Kunststoffe, 66, 781 (1976).

53. A. Siegmann, M. Parkis, J. Kost and A. T. Di Benedetto, Intern. J. Polym. Mater., 6, 217 (1978).

54. V. A. Pattison, R. R. Hindersinn and W. T. Schwartz, J. Appl. Polym. Sci., 18, 2763 (1974).

55. V. A. Pattison, R. R. Hindersinn and W. T. Schwartz, J. Appl. Polym. Sci., 19, 3045 (1975).

56. Yu. S. Lipatov and L. M. Sergeeva, Vysokomol. Soed.A, 16, 2290 (1974).

57. Yu. S. Lipatov and F. G. Fabulyak, Dokl. Akad. Nauk SSSR, 205, 635 (1972).

58. J. A. Grates, D. A. Thomas, E. C. Hickey and L. H. Sperling, J. Appl. Polym. Sci., 19, 1731 (1977).

59. L. H. Sperling, D. A. Thomas, J. E. Lorenz and E. J. Nagel, J. Appl. Polym. Sci., 19, 2225 (1975).

60. J. A. Grates, Mod. Paint Coat., 65, N. 2, 35 (1975).

61. H. A. Clark, U.S. Patent 3,527,842 (1970).

62. L. H. Sperling, V. A. Forlenza and J. A. Manson, J. Polym. Sci., Polym. Lett. Ed., 13, 713 (1975).

63. G. S. Solt, Brit. Pat. 728,508 (1955).

64. J. J. Falcetta, G. D. Friends and G. C. C. Niu, Ger. Offen. Pat., 2,518,904 (1975).

65. C. F. Ryan and R. J. Grochowski, U.S.Pat. 3,426,101 (1969).

MECHANICAL AND THERMAL PROPERTIES OF POLYCARBONATE -

AROMATIC COPOLYESTER BLENDS

J. Majnusz

Institute of Physical Chemistry and Technology
of Polymers, Silesian Institute of Technology
Gliwice, Poland

Cast blends of polycarbonate and aromatic copolyesters were prepared and examined for their transitional
behaviour by dynamic mechanical testing and differential
scanning calorimetry. Both techniques showed double glass
transitions indicative of incomplete miscibility in the
amorphous phase. After annealing above the melting point
of polycarbonate a single Tg was observed for any ratios
of blend components. Stress-strain relations for unannealed blends have been determined.

INTRODUCTION

Blends of polycarbonates and polyesters have been investigated
since about 6 years. Several patent specifications[1-6], were issued
in 1973-1974, followed by a short reference[7]; more recently, a
series of Paul, Barlow and associates papers[8-11] were published concerning the blends of polycarbonates with five different polyesters.
The investigations on polyestercarbonates have a longer history,
initiated by Goldberg[12-13] and recently followed by other authors[14].

___

Work done at Material Research Laboratory, Chemical Engineering
Department, University of Massachusetts, Amherst,
MA 01003 (USA).

The published papers[8-11] concern the blends of bisphenol A polycarbonates. These polycarbonates form totally miscible blends with polyesters prepared from 1,4-cyclohexanedimethanol and tere-/o-phtalic acids as well as with polypropiolactone. These blends have properties specific for single phase mixtures. The blends of polycarbonates with poly(ethylene or butylene terephtalate) have properties characteristic of homogeneous amorphous phase containing both blend components only when polyesters are in a great excess. Such kinds of polyesters show a tendency towards crystallization and, therefore, their blends with polycarbonate comprise also multiphase mixtures in which polyester crystals are present.

This work deals with cast blends of bisphenol A polycarbonate (PC) and aromatic copolyesters (ACPE). There are two main goals of this work. We looked firstly for methods able to improve the processability of aromatic copolyesters and secondly to determine their structures, knowing the properties of the blends. The properties of the blends were determined by dynamic mechanical measurements and differential scanning calorimetry. In addition, stress-strain relations at an ambient temperature have been given.

Preparation of blends and description of measurements

Commercial bisphenol A polycarbonate (PC), $\overline{M}w$ = 40000, used here was Merlon M40 of the Mobay Chemical Corporation. The aromatic copolyester (ACPE) $\overline{M}n$ = 3700, i.e. a polyterephtalate of 2,2-di(4-hydroxypropane) – (bisphenol A) and 4,4'-bi-naphtol-1 was obtained by low temperature solution polycondensation[15]. It comprises the structural units in a polymer chain as follows:

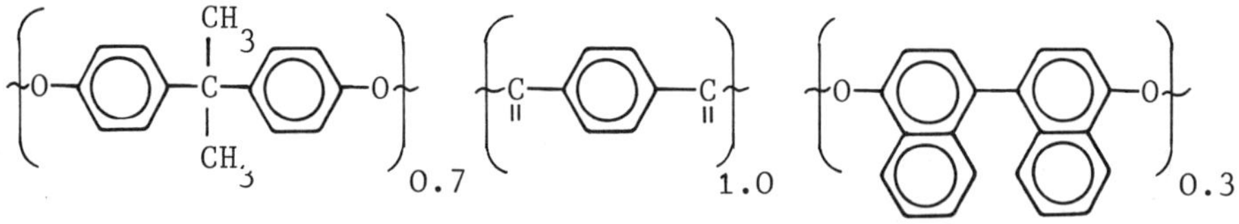

$$\left[\!\!-O-\!\!\left\langle\bigcirc\right\rangle\!\!-\!\!\underset{\underset{CH_3}{|}}{\overset{\overset{CH_3}{|}}{C}}\!\!-\!\!\left\langle\bigcirc\right\rangle\!\!-O-\right]_{0.7} \left[\!\!-\underset{\overset{\|}{O}}{C}-\!\!\left\langle\bigcirc\right\rangle\!\!-\underset{\overset{\|}{O}}{C}-\right]_{1.0} \left[\!\!-O-\text{naphtyl}-\text{naphtyl}-O-\right]_{0.3}$$

| D | T | N |

The NMR spectrum shows that the copolyester has a random block structure. It is soluble in aliphatic chlorinated hydrocarbons, infusible and it decomposes above 650 K. The blends were prepared as films cast from 10% solutions in a mixture of methylene chloride and chloroform. The obtained films were clear but showed an in-

creasing opalescence up to complete opacity as the weight ratio of
components tended to unity. For the PC/ACPE blends of composition
from 2:3 to 2:1 a distinct phase separation was observed. The prepared films were carefully dried under vacuum to remove any solvent
residue. For some measurements the films were annealed at a temperature of 460 K or 550 K and then quenched.

Dynamic mechanical properties were determined by means of the
Vibron Dynamic Viscoelastometer Model DDV II of the Toyo Measuring
Instruments Co., using 100 µm cast film strips as specimens. The
measurements were performed under dry nitrogen atmosphere at a
temperature range of 100 - 550 K using a heating rate of 1 - 2 K
min^{-1} and frequencies of 3.5, 11 and 110 Hz.

The DSC measurements were carried out using the Perkin-Elmer
DSC-2 differential scanning calorimeter. Film discs were used as
specimens and the determinations were usually performed at a
heating rate of 20 K min^{-1}. The glass transition temperature T_g was
taken as the mean value of the temperature range at which a rapid
heat consumption by the specimen was observed during the second
heating.

Stress-strain relations were determined by a Tensilon mechanical tester using as specimens film strips similar to those used
in dynamic measurements.

DISCUSSION

Blend components

Figure 1 presents the changes of viscoelastic properties of
both blend components against temperature at 11 Hz, while Figure 2
shows their thermograms.

From the dynamic mechanical measurements it was found that PC
shows a β-relaxation at 187 K (max. tan δ) and a α-relaxation at
429 K (max. E"). These values agree well with those given in literature[16] considering the different frequencies at which the measurements were carried out. The DSC thermogram of PC shows the T_g
at 421 K, in agreement with the well known data for this polymer.

The ACPE subjected to dynamic mechanical measurements at 11 Hz

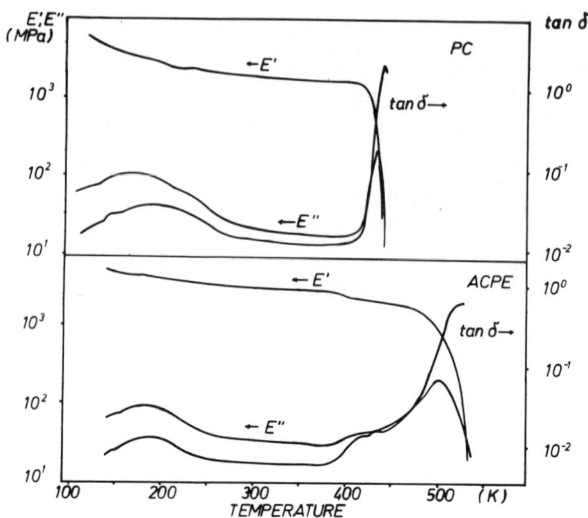

Figure 1.   Dynamic mechanical behaviour at 11 Hz for PC and ACPE films.

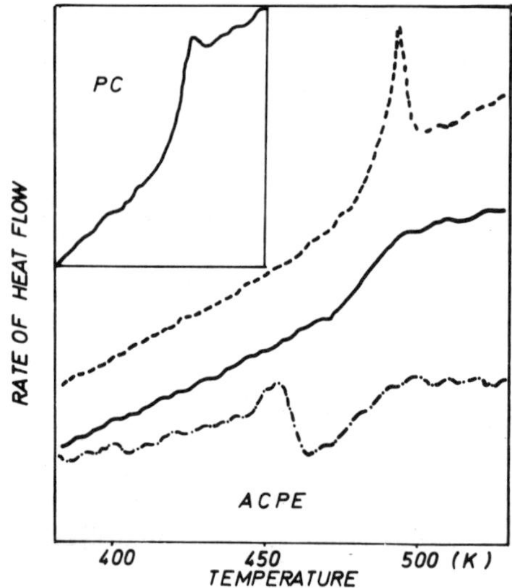

Figure 2.   DSC thermograms for PC and ACPE
— ·— ·—· —· 1st Heat
——— 2nd Heat
— — — — — — 1st Heat after annealing at 460 K for 24 h.

shows the β-relaxation at 201 K (max. tanδ) within a temperature
range close to that of PC, some tanδ variations at 410 - 450 K and
finally the α-relaxation at 503 K (max. E"). In the DSC measurements,
the ACPE shows the glass transition at 486 K during the second
heating. During the first heating within the temperature range of
430 - 460 K, an undetermined transition occurs as shown by a broa-
dened maximum at 455 K. This transition affects the dynamic mechanical properties, causing precisely a variation of tanδ and a small
decrease of the storage modulus E'.

The measurements performed with a large number of aromatic copolyesters[17] characterized by different weight ratios and length
sequences of both D-T and N-T elements in the polymer chains, suggested that they probably form two-phase systems. One phase, containing the D-T elements in excess, has properties close to those of
homopolyester $\sim\!\!\epsilon\,\text{D-T}\,\mathord{\mapsto}_n$, while another one, comprising large and
rigid N elements, did not show significant transitions at tempera-
tures up to 600 K acting as a "stiffening" element. The transition
recorded at T_g concerned the phase consisting of a majority of D-T
elements.

The X-ray examination proved that aromatic copolyesters are
not crystalline, however, when subjected to annealing below T_g, they
formed ordered amorphous regions in the phase formed by a majority
of D-T elements. In turn, these amorphous regions caused, for
example, polarized light depolarization in a polarization microscope.

At the glass transition region the DSC thermogram of annealed
aromatic copolyesters shows a behaviour similar to that of a melting
transition of low enthalpic content as it is for ACPE in Figure 2.

Blends
‾‾‾‾‾‾

The dynamic mechanical measurements were carried out using
films of blends containing 0 - 30 % or 70 - 100 % by weight of PC
which do not show distinct phase separation.

Figure 3 illustrates the changes of storage and loss moduli E',
E" and of tanδ against temperature for blends containing 30 and 70%
PC, respectively.

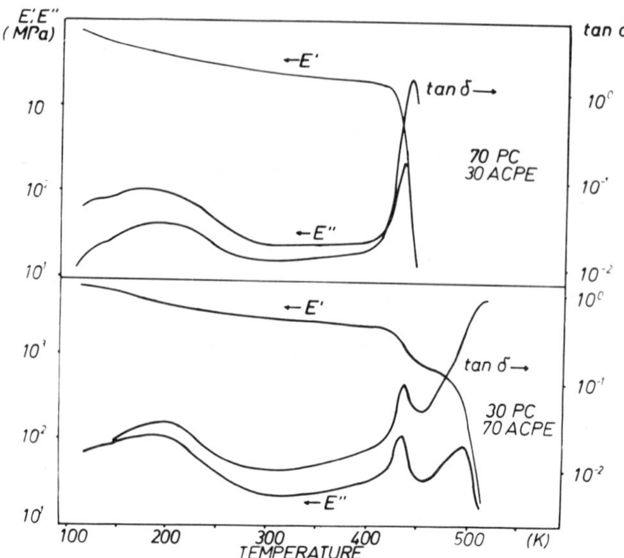

Figure 3.   Dynamic mechanical behaviour at 11 Hz for PC - ACPE
            blends.

In the β-relaxation region there are single peaks at a range
of temperatures characteristic of the β-relaxation of the blend
components. Due to a small difference in peak temperatures of both
components, when recording broad maxima, it may be expected that
the peaks of both components are coincident giving a resultant
maximum.

Figure 4 shows the variation of β-relaxation temperatures of
E" and tanδ against the blend composition. The viscoelastic functions are more interesting in the temperature range at which the
α-relaxation of blend components occurs. In the case of blends
comprising mainly PC only one region, characterized by higher
damping, is recorded, which superimposes the α-relaxation region
for PC. In this region, the decrease of the complex modulus E* is
so significant that it exceeds a measuring range of the apparatus.
Therefore, further transitions cannot be recorded. Since the peak
temperatures of E' or tanδ were only 1 or 2 K higher than those of
PC, this suggests that a mixed phase which should contain a considerable quantity of ACPE is not formed. Thus, the transition of
this phase is determined by the behaviour of PC. On the other hand,

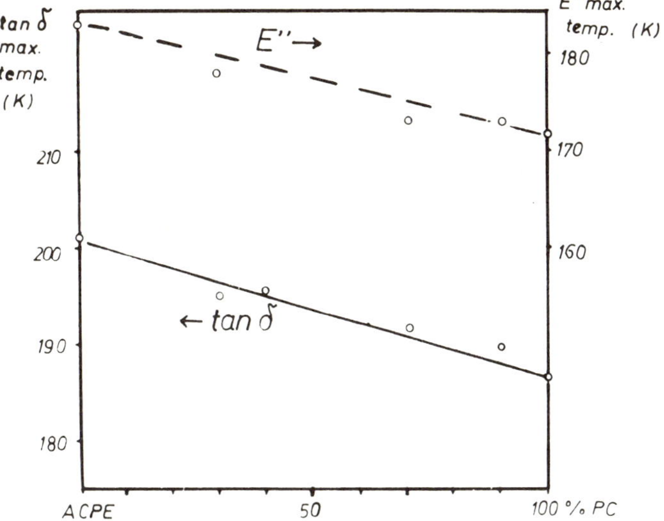

Figure 4.   Effect of blend composition on the β-relaxation temperatures of PC - ACPE blends.

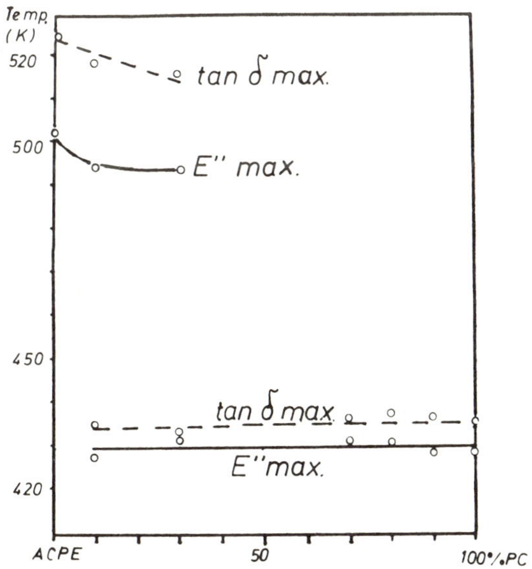

Figure 5.   Effect of blend composition on the α-transition temperatures of PC - ACPE blends.

in blends containing much more ACPE two regions of damping peaks are observed. Double peaks correspond also to two regions of the E' decrease. One region is coincident with a corresponding region for PC, while the second one was found at slightly lower temperatures than that for ACPE.

Figure 5 shows the effect of the blend composition on the temperature at which the maxima of $\tan\delta$ and E'' occur in the region of the α-transitions. The presented results indicate clearly that the prepared blends are characterized by phase heterogenity. To confirm the mechanical properties determined, the blends have been subjected to DSC measurements. During first heatings the endothermic processes in blends are difficult to be analysed due to the superimposing of the T_g of PC and an undetermined transition characteristic of ACPE below its T_g. During the second heatings, performed after a first heating up to 525 K and quenching, it is possible to analyse the recorded transitions and, therefore, further considerations will refer mainly the measurements carried out during a second heating. Similarly to the measurements of dynamic mechanical properties of blends comprising mainly PC only, one glass transition has been recorded within the region which accurately superimposes the T_g region for PC. In the case of blends containing mainly ACPE, i.e. 70%, two transitions were recorded which superimpose the glass transition of both components. In the case of blends containing 90% of ACPE only one transition was found, the mean value of which is slightly lower than that of T_g for ACPE.

Figures 6 and 7 show the DSC thermograms and the relation of the recorded T_g against composition, respectively. These results prove undoubtedly that the cast blends are heterogeneous comprising phases in which a particular component is in excess. Stress-strain measurements carried out on specimens of the same blend compositions as used in dynamic mechanical measurements indicate that the stress values at yield lay between those characteristic for both blend components and are closer to the value of the more resistant ACPE. It is interesting to note that a small addition of ACPE to PC (e.g. blend of 10% ACPE) increases significantly the strength of the material i.e., both stress at yield and complex modulus. In Table I stress-strain data and complex modulus determined dynamically and extrapolated to a frequency of 1 Hz are collected. The obtained results concern only blends which have been formed at an ambient temperature values of 295 - 310 K and dried at temperatures not

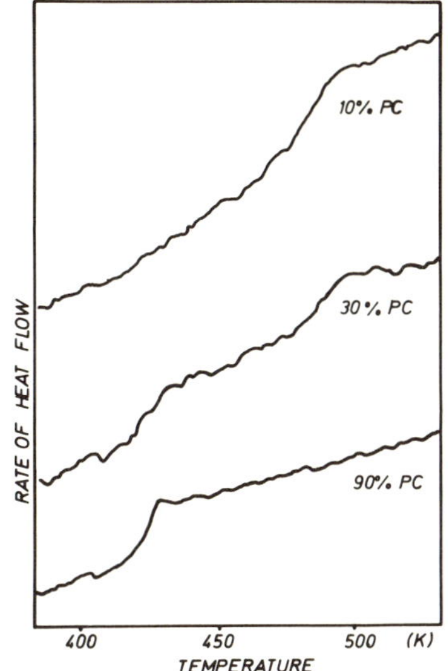

Figure 6.   DSC thermograms for PC – ACPE blends.

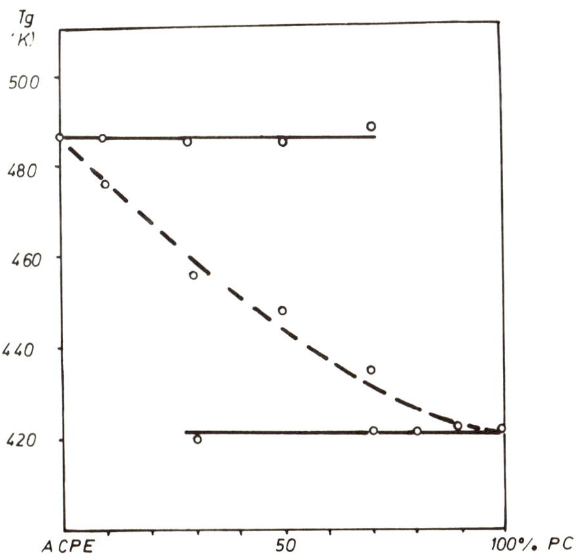

Figure 7.   Glass transition data for PC – ACPE blends DSC results.
___________ unannealed blends
– – – – – blends annealed at 550 K for 15 min.

exceeding 350 K. To evaluate the effects of thermal treatment, the
films were annealed at 460 K (below T_g of ACPE) and at 550 K i.e.,
above T_g of ACPE and simultaneously above melting temperature of
PC. If annealing was performed below T_g of ACPE, distinct phase
separation occurred. This separation is illustrated by the DSC
thermogram of the first heating, see Figure 8.

TABLE I

Mechanical Properties of Polycarbonate – Aromatic Copolyester
Blends and their Components at 295 K.

Blend composition	Complex modulus at 1 Hz	Stress at yield	Strain at yield	Strain at break
	E^*	σ_Y	ε_Y	ε_B
% of PC	MPa	MPa	%	%
100	1800	46.0	6.3	100.0
0	3000	83.5	6.4	18.3
10	2650	73.0	6.3	12.5
30	2560	68.0	5.5	9.4
70	2430	61.0	5.8	10.9
90	2380	64.8	3.5	48.1

A transition temperature decrease for the PC phase and a transition temperature increase for the ACPE phase were obtained. Thus
the annealing caused not only a distinct phase separation but probably also an enrichment of the PC phase with low molecular weight
ACPE fractions. This process yielded in decreasing the T_g of the
PC phase and simultaneously in increasing the ACPE phase transition
temperature and ordering its structure.

The annealing process carried out above the melting temperature
of PC changed completely the properties of the blends. For any

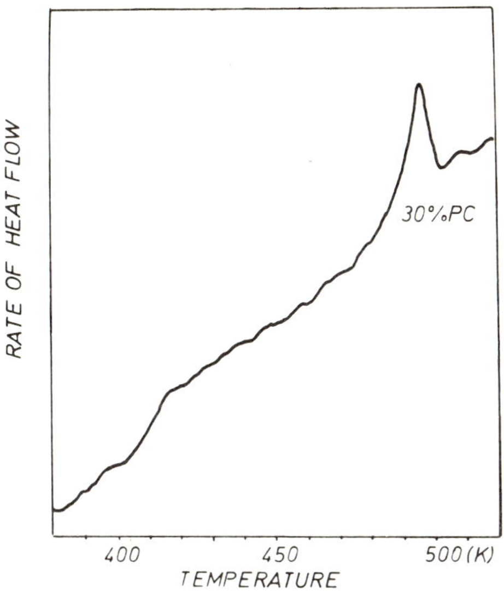

Figure 8.   DSC thermograms for PC 30% - ACPE blend after annealing
at 460 K for 24 h - 1st Heat.

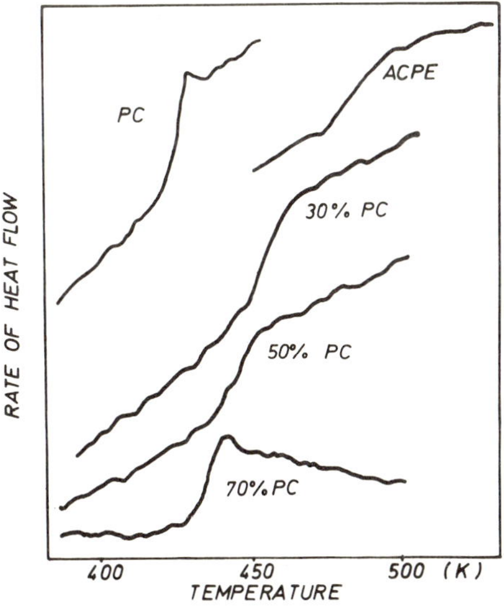

Figure 9.   DSC thermograms for PC - ACPE blends after annealing
at 550 K for 15 min.

ratios of components only one glass transition was recorded within the T_g of both components.

Figure 9 shows the DSC thermograms, while Figure 7, presents the relation of T_g against composition.

The DSC results indicate that one phase with homogeneous mixture properties was formed from two phases of the blend components. The annealing process above the melting temperature of PC causes an increase of film clarity, however a slight opalescence was still present. This can be explained by a probable two-phase ACPE used for blends, one of them comprising D-T elements in excess and a second one with more N elements. Thus, the phase comprising more D-T elements should dissolve itself in polycarbonate while the second ACPE phase comprising immobile rigid N elements did not take part in mixing the phases at 550 K. The obtained blends are to some extent similar to the blends of PC and polyesters showing a tendency towards crystallization. Such blends comprise miscible amorphous phases and crystalline phases causing the blends to be heterogeneous.

CONCLUSIONS

The obtained results have shown that it is possible to prepare the blends of PC and ACPE by casting from solutions. It was proved that a small addition of ACPE to PC in cast films yielded in a significant increase of their mechanical strength. The films were elastic and clear in spite of slight opalescence. The blends consisted of separated component phases if they have not been subjected to thermal treatment. Thermal treatment caused noticeable changes in their properties. If a thermal treatment was conducted below T_g of ACPE, a more distinct phase separation occurred. One amorphous two component phase rich in PC and a second one also amorphous showing some ordering and consisting mainly of structural D-T units of the ACPE have been found. The presence of the phase containing mainly structural N units of the ACPE seemed also to be probable. Thermal treatment carried out above the melting temperature of PC resulted in mixing of two blend component phases containing mobile elements and in formation of one amorphous phase having typical miscible blend properties.

The third probable phase, containing rigid immobile N elements, did not take part in this process as indicated by T_g determined after annealing above the melting temperature of PC. The T_g values found were within those of PC and ACPE. The transition of ACPE was caused by its phase comprising D-T elements in excess. Therefore, it can be said, that this phase partipates in observed mixing processes. The obtained results permitted also to determine the structure of aromatic copolyesters containing binaphtyl units in the chain.

It seems to be interesting to continue the investigations, expecially on the mechanical properties of blends prepared by mixing at temperatures above the melting temperature of PC. The aim of this work will be to prove whether it will be possible to process ACPE blends by conventional methods which are typical for processing thermoplastics.

ACKNOWLEDGEMENTS

The author wishes to express his thanks to the Maria Skodowska Curie Foundation, Warsaw, Poland and the National Science Foundation Washington D.C., U.S.A. for supporting these investigations, and also to Prof. R. W. Lenz and Prof. W. J. MacKnight from the University of Massachusetts, Amherst for their help in performing this work and for discussing the results.

The author is also indebted to Prof. Z. J. Jedliński from the Institute of Polymer Chemistry, Polish Academy of Sciences, Zabrze for discussions as well as to Dr. D. Sęk from the same Institute for preparing the ACPE sample and their characteristics.

REFERENCES

1. A. D. Wambach, Ger. Pat., 2,310,742 (1973).
2. J. V. Koleske, C. J. Withworth Jr. and R. D. Lundberg, U.S. Pat. 3,781,381 (1973).
3. Y. Nakamura, R. Hasegawa and H. Kubota, Ger. Pat. 2,343,609 (1974).
4. M. Matsukane and C. Azo, Jpn. Pat. 54, 160 (1973); Chem. Abstr., 82, 58,832 (1975).
5. J. Yamamoto, M. Ueno and K. Umeda, Jpn. Pat. 80, 162 (1974); Chem. Abstr., 82, 59,321 (1975).

6.   T. Kaneko and M. Takamatsu, Jpn. Pat. 115,146 (1974); Chem. Abstr., 82, 99,299 (1975).

7.   Anon., Chem. Eng. News, 54, 14 (1976).

8.   D. C. Wahrmund, D. R. Paul and J. W. Barlow, J. Appl. Polym. Sci., 22, 2155 (1978).

9.   T. R. Nassar, D. R. Paul and J. W. Barlow, J. Appl. Polym. Sci., 23, 85 (1979).

10. R. N. Mohn, D. R. Paul, J. W. Balow and C. A. Cruz, J. Appl. Polym. Sci., 23, 575 (1979).

11. C. A. Cruz, D. R. Paul and J. W. Barlow, J. Appl. Polym. Sci., 23, 589 (1979).

12. E. P. Goldberg, U.S.Pat. 3,207,814 (1965).

13. E. P. Goldberg, S. F. Strause and H.E. Munro, Polymer Preprints, 5(1), 223 (1964).

14. D. C. Prevorsek, Y. Kesten and B. De Bona, Polymer Preprints, 20(1), 187 (1979).

15. D. Sęk, Europ. Pol. J., 13, 967 (1977).

16. K. H. Illers and H. Breuer, Kolloid Z., 176, 110 (1961).

17. J. Majnusz, R. W. Lenz and M. J. MacKnight, to be published.

PRELIMINARY INVESTIGATION OF POLYCARBONATE-POLYPROPYLENE BLENDS

Z. Dobkowski, Z. Kohman and B. Krajewski

Institute of Industrial Chemistry

01-793 Warszawa, Poland

Properties of polycarbonate (PC)-polypropylene (PP) blends, such as melt flow rate, mechanical and thermal properties, were measured in the whole range of composition. PC-PP blends with low content of PP up to about 10 wt % maintain some valuable properties of PC, e.g. tensile modulus, impact strngth, and Vicat softening temperature. Moreover a single dynamic glass transition temperature Tg_d is observed in the range of low PP content. We suggest that good properties of these blends coincide with good compatibility of components.

INTRODUCTION

There are reports that by blending polycarbonate (PC) with polypropylene (PP), a lower melt viscosity is obtained[1,2] and some properties of blends, such as tensile and impact strength[2,3], are improved. PC-PP blends have been used as extruded insulating films[4], safety helmets[5], or components of wood-like materials[6-9]. PC has also been applied as a modifier of PP[10-12].

The purpose of our preliminary study was to evaluate the influence of composition on the properties of PC-PP blends of domestic components, and to choose a formulation for their eventual application.

EXPERIMENTAL

Materials investigated were Polish bisphenol A PC, PP - J400, and PP - J330, characterized by melt rate (MFR) in the IIRT (USSR) universal plastometer under load 21.2 N (2.16 kgf) at 503 (230), 523 (250) and 553 K (280°C). The MFR values are shown in Figure 1.

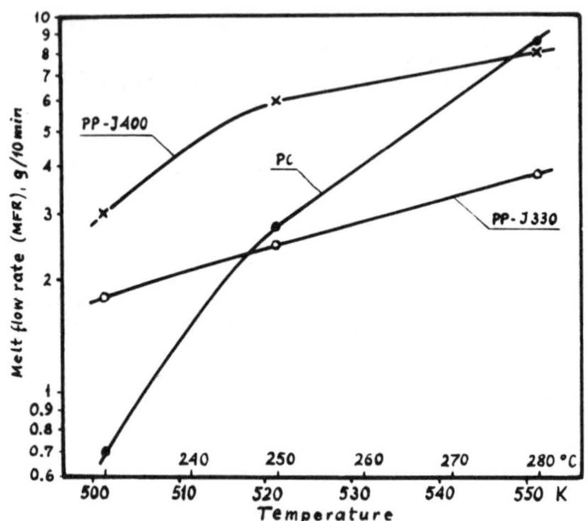

Figure 1. Melt properties of components

The blends were prepared by homogenization of dry pellets, subsequent melting in conventional screw-type extruder at 493 - 523 K (220 - 250°C), and pelletization. Then they were characterized by MFR at 523 K (250°C). The composition of blends varied between 0 and 100 wt % of PP.

The standard test specimens for the measurements of mechanical and thermal properties were formed by injection moulding at 473 - 498 K (200 - 225°C) under injection pressure 100 MPa. The blends were also transformed into 0.5 mm films by compression moulding at 478 - 518 K (205 - 245°C) under pressure up to 20 MPa to form specimens for measurements of dynamic mechanical properties. Instron TM-SM, Charpy impact testing machine PSW 0.4 (GDR), Vicat-HDT

Vaschetti-Grosso and Zwick model 5201 apparatus were used for measurements of their respective properties.

RESULTS AND DISCUSSION

The results of MFR measurements as a function of composition
for blends of PC with PP – J400 and PP – J330 are shown in Figure 2.

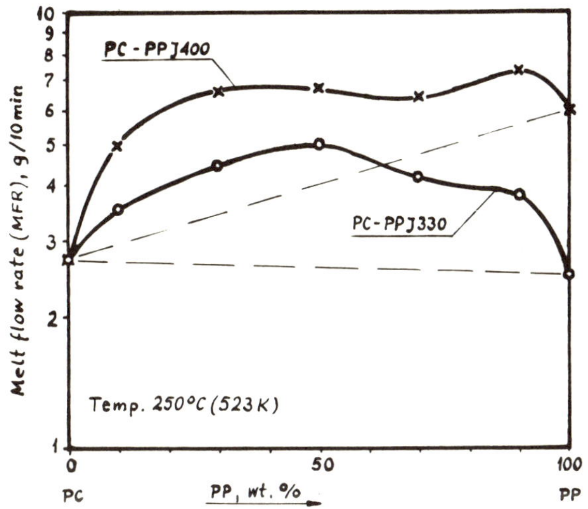

Figure 2.   Melt properties of PC – PP blends.

The MFR of each blend is higher than that of pure component, i.e.
the melt viscosity is always lower. On the curves a distinctive
inflexion at about 70 wt % of PP is observed for both types of
blends. Generally the results agree with published data of Dobrescu
and Cobzaru[1], and Nagumanova et al.[2]. Lower melt viscosity of blends
means that their processability can be improved compared with pure
components.

The dependence of selected properties on composition of PC – PP
blends is presented in Figs. 3 – 6. Mechanical properties, such as
tensile yield, tensile ultimate, and flexural strengths, tensile and
flexural moduli, thermal properties, such as Vicat softening temperature (VST), and flame resistance measured by oxygen index (OI),
are generally higher when the content of PC in the PC – PP blends

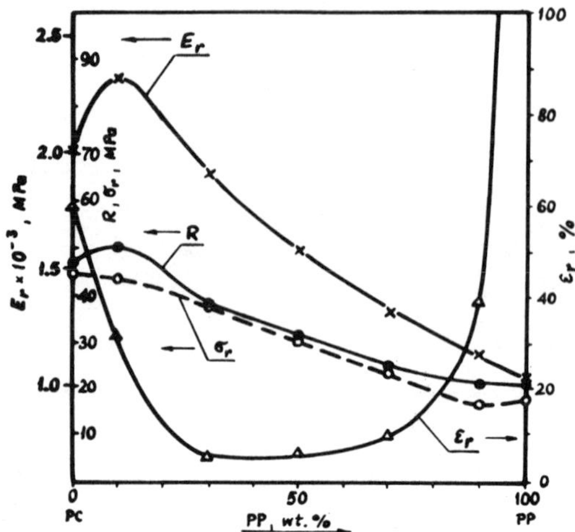

Figure 3.   Tensile properties of PC – PP blends.

—— x —— E_r tensile modulus

—— ● —— R   tensile strength at yield

– –o– – σ_r tensile strength at break

—— Δ —— ε_r elongation at break

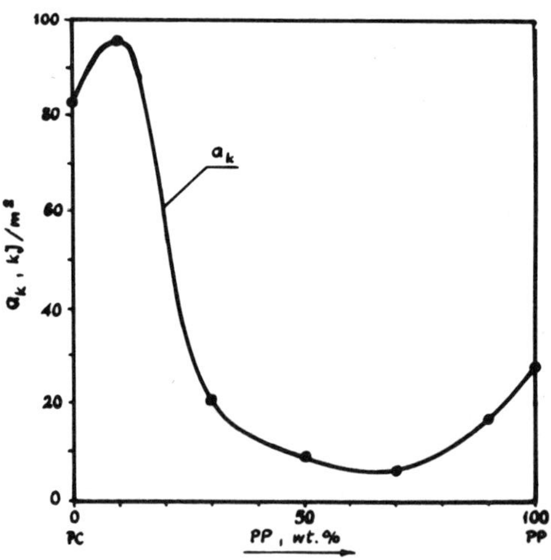

Figure 4.   Charpy impact strength, notched

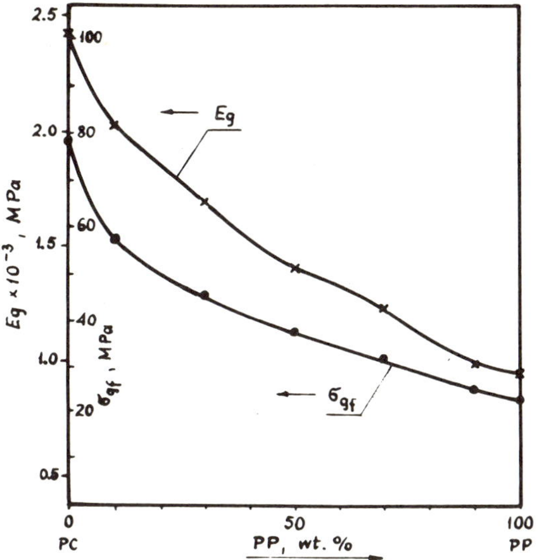

Figure 5. Flexural properties of PC – PP blends

    x   Eg, flexural modulus

    ● σ_{gf}, flexural strength at conventional deflection, f = 6 mm

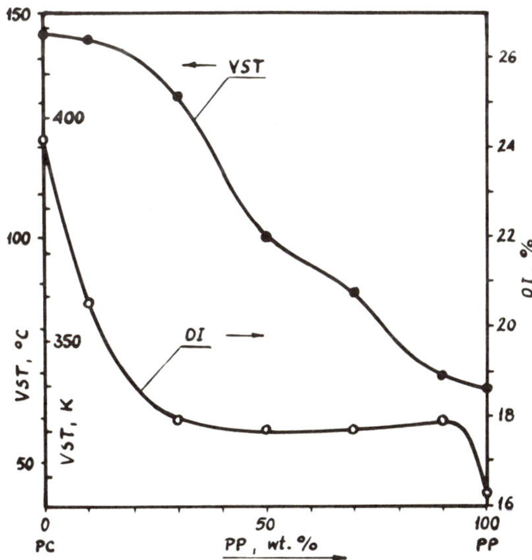

Figure 6. Thermal properties of PC – PP blends

    ●   VST, Vicat softening temperature

    ○   OT, oxygen index

is increased. Distinctive maxima of tensile yield strength R and
tensile modulus E_r (Figure 3), as well as Charpy impact strength
notched a_k (Figure 4) are however observed in the range of small
amounts of PP in the blend, up to about 10 wt % of PP. Such improvement of these properties coincide with the anomalies in density
of PC-PP blends at 298 K (Figure 7). In the case of impact strength
the above mentioned observation is consistent with the data mentioned[9]. Notched impact strength a_k and elongation at break ε_r curves
show a broad minimum in the composition range 30 - 70 wt %.

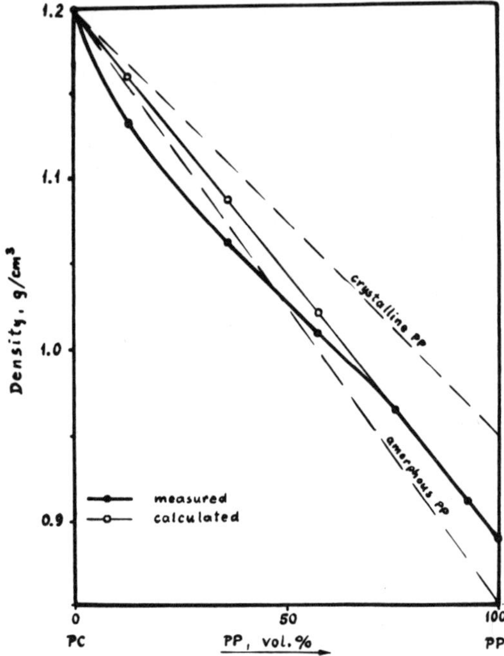

Figure 7.   Density of injection moulded parts of PC - PP blends
            at 25°C (298 K).

Dynamic mechanical properties of PC-PP blends, viz. storage
modulus at shear G' and loss tangent tanδ as a function of temperature
and composition, are shown in Figures 8 - 10. The dynamic glass
transition regions for PC (α_1) at 424 - 428 K, and for PP (α_2) at
281 - 282 K, as well as chain segment relaxation for PP (β_2) at
227 - 228 K, are consistent with those identified by Van Krevelen[13].
The chain segments relaxation for PC (β_1) observed at 198 K, e.g.
in the range of acoustic frequencies[14], is beyond the investigated

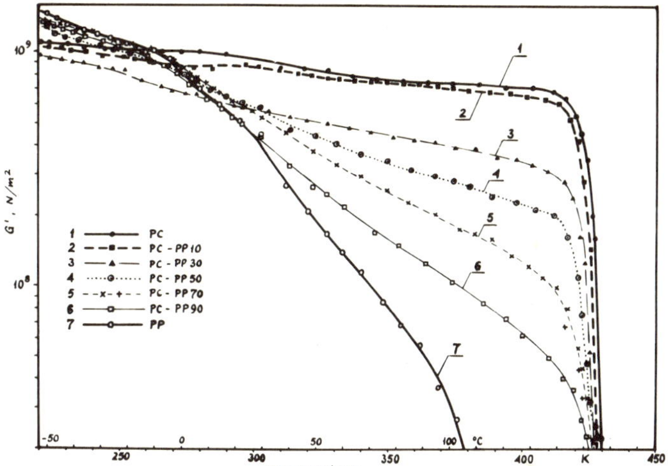

Figure 8. Dynamic mechanical properties of PC-PP blends: shear modulus G'vs temperature.

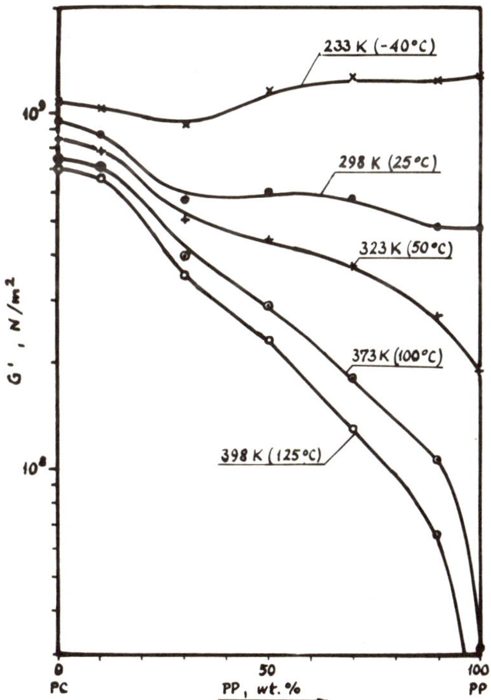

Figure 9. Shear modulus G'vs blend composition.

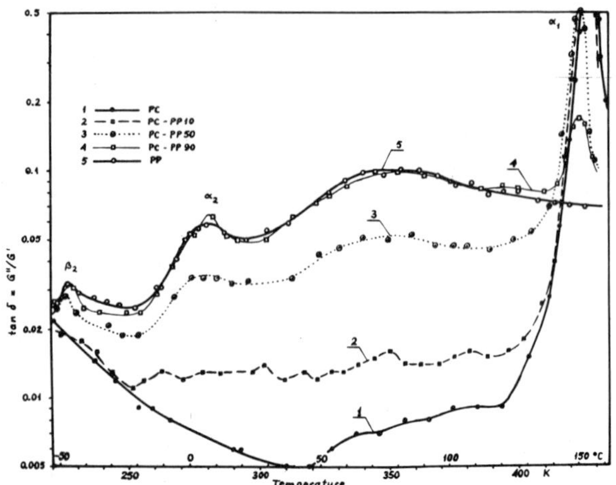

Figure 10.   Temperature dependence of the mechanical loss tangent, tanδ, for PC-PP blends.

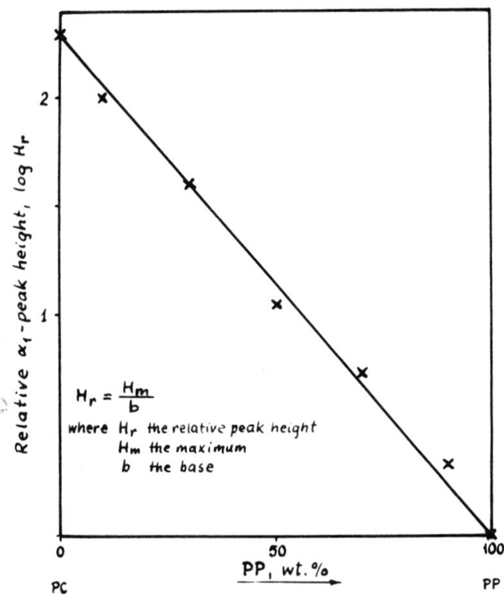

Figure 11.   Relative α_1 peak height vs composition of PC-PP blends.

temperature range. Additional transition peaks attributable to
short-range structural motion occour at a temperature above 282 K
within crystalline regions of PP and below 424 K within amorphous
regions for PC (Figure 10). It has been also been noted that the
relative α_1 - peak heights, log H_r, show linear dependence on the
composition of PC-PP blends (Figure 11). Moreover it has been observed that the blends with high PP content exhibit two glass transition peaks, each nearly corresponding to those of pure components.
Single glass transition peak exists for PC-PP blend with about
10 wt % of PP, cf. Figures 10 and 12. Thus we can adfirm that
PC-PP blends with up to about 10 wt % of PP are macrohomogeneous,
and their components are compatible.

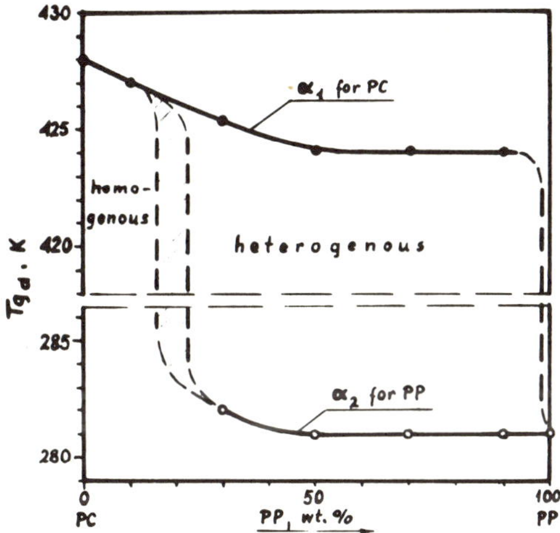

Figure 12. Glass transition regions for PC-PP blends.

CONCLUSIONS

It has been shown that the PC-PP blends with low content of PP
up to about 10 wt % reveal higher tensile modulus and notched impact
strength compared with pure PC, and mantain some valuable properties
of PC, e.g. Vicat softening temperature. Thus they can be applied
as useful engineering thermoplastic material.

An improvement of the above mentioned properties coincides with
the density anomalies and with the existence of a single dynamic
glass transition temperature. Thus it can be suggested that good
properties of blends are connected with the good compatibility of

components. However more detailed studies on PC-PP blend structures
are still needed; the appropriate works are under way.

REFERENCES

1. V. Dobrescu and V. Cobzaru, J. Polym. Sci. Polym. Symp., 64, 27,
    (1978).
2. E. I. Nagumanova, R. G. Timergaleev, E. R. Galimov and
    V. A. Voskresenskii, Izv. Vyssh, Uchebn. Zaved. Khim. Technol.,
    17, 1233 (1974).
3. Jap. Pat. 14 581, (1966).
4. Jap. Pat. 116 537, (1975).
5. U.S. Pat. 3 437 631, (1969).
6. Jap. Pat. 40 775, (1971).
7. Jap. Pat. 40 777, (1971).
8. Jap. Pat. 41 101, (1971).
9. Jap. Pat. 20 419, (1973).
10. L. Alexandru, I. Agachi, V. Ciobanu, D. Costian and T. Rizescu,
    Faserforsch. Textiltech., 16, 433 (1965).
11. Brit. Pat. 982 753 (1965).
12. Can. Pat. 958 204, (1974).
13. D. W. Van Krevelen, "Properties of Polymers", Elsevier Publishing
    Co., Amsterdam, 1972.
14. Z. Dobkowski, unpublished report, 1977.

MULTICOMPONENT POLYESTER SYSTEMS WITH MESOGENIC UNITS

E. Chiellini[*], R.W. Lenz[**] and C. Ober[**]

[*]Istituto di Chimica Organica Industriale

Università di Pisa, Pisa, Italy

Copolyesters of either linear hydrocarbon glycols
or glycolethers with terephthalic acid and p-hydroxybenzoic acid have been prepared by melt-phase and catalyst-free ester interchange process starting from mixture of preformed glycol terephthalate polymers and
p-acetoxybenzoic acid. The reaction leads to systems
whose homogeneity, solubility and tendency to give mesophases are dependent upon the chemical composition of
the reaction mixture and the structure of the preformed
polyester. Either physical mixtures of two polymers or
a random copolymer may be obtained depending upon the
reactivity of the polyester and the reaction conditions
used.

INTRODUCTION

Polymeric multicomponent systems are commonly understood as
blends of two or more polymers with suitable additives which may be
partially compatible. There are systems, however, which may be
considered from a macrochemical view-point to be "monocomponent",
in that they contain only one kind of polymeric material, but are

---

[**]Materials Research Laboratory, Chemical Engineering Department,
University of Massachussetts, Ma., 01003 USA.

capable of having, under definite thermodynamic conditions, the
physical properties of mechanical mixtures of at least two components.
As an example of this possibility, we wish to report some preliminary
results obtained in the preparation of polyesters containing mesogenic groups which are able to form ordered mesophases. Such systems
can be taken, in fact, as useful and simple models of heterogeneous
systems in which the chemical interactions between the various components are not completely ruled out. As a consequence, by considering
here the way in which such products are obtained, an interesting
example is reported of the possible chemical interactions which can
arise during thermal treatment of multicomponent systems based on
linear polyesters.

Very recently a great deal of interest has been devoted to the
preparation and physical characterization of condensation or step-
growth polymers with thermotropic "liquid-crystal" properties associated with the presence of mesogenic aromatic moieties in the main
chain[1-3]. Some of these systems have been obtained by insertion of
a bifunctional aromatic monomer into poly(ethylene terephthalate).
For these polymers, very little systematic information has been reported on the possible influence of structural, stereochemical and
segment flexibility of the chain backbone on the thermotropic properties. In connection with our interest in the preparation of hemo-
compatible polymeric materials to be tested for fabrication as arti-
ficial kidney devices[4,5], we undertook the investigation of the possible routes leading to the insertion of stiff aromatic groups, such
as p-oxybenzoate units, into soft blocks of polyterephthalates based
on glycols having different hydrophilic character (Structure A and
B).

$$\text{HO}\left[\overset{\text{C}}{\underset{\text{O}}{\|}}-\!\!\bigcirc\!\!-\overset{\text{C}}{\underset{\text{O}}{\|}}-\text{O}(\text{CH}_2)_n-\text{O}\right]_y\!\!\text{H} \quad \text{and} \quad \text{HO}\left[\overset{\text{C}}{\underset{\text{O}}{\|}}-\!\!\bigcirc\!\!-\overset{\text{C}}{\underset{\text{O}}{\|}}-\text{O}(\text{CH}_2\text{CH}_2\text{O})_n\right]_y\!\!\text{H}$$

<u>A</u> n = 2,3,4,5,6   <u>B</u> n = 2 and 4

EXPERIMENTAL

The A and B type polymers were obtained by "melt-phase" transesterification reaction[6] between dimethyl terephtalate and an excess of the glycol in the presence of either a titanium alcholate
or a calcium acetate/antimonium oxide catalyst under vacuum at 270°C.
In either case the reaction was carried out in a 100 ml resin kettle

equipped with thermometer, mechanical stirrer, distillation take-off,
and nitrogen inlet. In a typical experiment 19.8 g (0.102 mole) of
recrystallized dimethyl terephthalate (DMT), 31.9 mg (0.207 mmole)
of calcium acetate (0.10% solution in water) and 7.5 mg (0.026 mmole)
of antimony trioxide (0.05% solution in water) were introduced into
the reactor. To the mixture, after purification from moisture by
evacuation and a subsequent dry-nitrogen purge, was added 0.105 mole
of the diol, and the reactants were subjected to the following sequential heating conditions: 200°C (3 hr), 220°C (0.5 hr) and 270°C
(0.2 hr). Alcoholysis began at 200°C as soon as the contents of the
flask were molten. The evolved methanol was trapped at liquid nitrogen temperature and measured to evaluate the progress of the reaction.
Finally at 270°C vacuum was applied to reach 0.2-0.5 mm Hg pressure
over a period of 0.2-0.3 hr, and these conditions were then mantained
for an additional 5 hrs. The reaction contents, after cooling under
a nitrogen sweep, were removed from the kettle and purified by dissolution in trifluoroacetic acid or chloroform and reprecipitation
in a large excess of methanol. When either titanium tetra-isopro-
poxide or butoxide was used as catalyst [molar ratio DMT/Ti = (1-1.2)
$\cdot 10^3$], it was added to the preheated (170°C) reaction mixture and
shorter reaction times were odopted (2 hr both at 200°C and 270°C).
In all cases white powders were isolated (yield > 90% with respect
to DMT).

By using the catalytic procedure a third series (C) of copoly-
meric terephthalates based on ethylene glycol and different homolo-
gous diols or diethylene glycol was prepared.

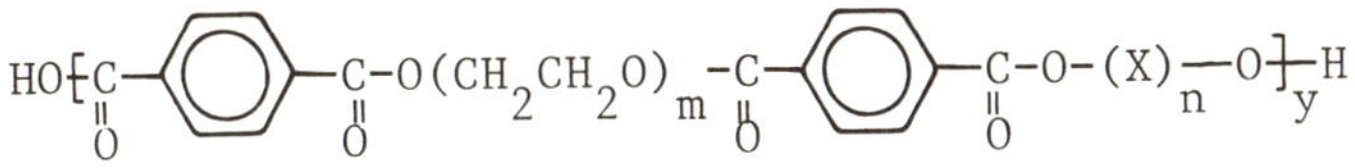

$$HO\{C-C_6H_4-C-O(CH_2CH_2O)_m-C-C_6H_4-C-O-(X)_n-O\}_y H$$

C $(X)_n = -(CH_2)_3$, $(CH_2)_4$, $(CH_2)_6$ and m = 1

    $(X)_n = (CH_2)_2$, and m = 2

Details of experimental procedure, polymer melting points, and intrinsic viscosity data for the A and B samples are listed in Table 1.
Good agreement was found between the properties in Table 1 and those

TABLE 1

Products of the melt reaction of poly(terephthalate)s of different glycols with p-acetoxybenzoic acid under catalyst-free conditions [a].

Run	(Methylene)$_n$ glycol Type	Initial poly (methylene)$_n$ terephthalate				Reaction product				
		T_m [b]	$[\eta]$ [c]	Structural unit		Yield	Soluble fraction [d]			
							% of the total	Oxybenzoate units [e]	T_m [b]	$[\eta]$ [c]
		(°C)	(dl/g)	mmol	%-mole	(%)		(%-mole)	(°C)	(dl/g)
1	HO(CH$_2$)$_2$OH	275	0.39	62.5	50	69.2	74.0	37.5	237[f]	0.28
2	HO(CH$_2$)$_3$OH	234	0.56	45.4	50	73.8	77.5	15.9	214	0.50
3	HO(CH$_2$)$_4$OH	227	0.34	45.4	50	100	57.1	<1	227	0.32
4	HO(CH$_2$)$_6$OH	157	0.55	40.3	50	67.4	75.0	6.5	139	0.51
5	(HOCH$_2$CH$_2$)$_2$O	89	0.96	29.6	50	78.3	73.4	<1	89	0.95

[a] Reactions carried out at 260-280°C under vacuum for 4 hrs. [b] Melting point from DSC analysis as indicated by the point of return of the endotherm to the base line. [c] In tetrachloroethane/phenol solution (40/60 by weight) at 25°C. [d] In TFA or chlorinated solvents. [e] Determined by NMR in TFA. [f] Of the unclear melt still existing at 350°C.

TABLE 2

Products of the melt reaction of ethylene terephthalate copolymers with p-acetoxybenzoic acid under free-catalyst conditions [a].

		Initial copolyester [b]					Reaction product with p-acetoxybenzoic acid					
							Yield	Soluble fraction [e]				
Run	(Methylene)$_n$ glycol Type	Structural unit amount	%-mole	Endotherm T_m-range	ΔH_f	$[\eta]$ [d]	% of the total	Oxybenzoate units [f]	Endotherm T_m range	ΔH_f	$[\eta]$ [d]
		(mmol)		(°C)	(cal/g)	(dl/g)	(%)	(%-mole)	(°C)	(cal/g)	(dl/g)
1	HO(CH$_2$)$_3$OH	10.0	50.0	137–182	8.2	0.42	66.4 73.2	12.5	137–175	5.8	0.35
2		8.8	70.0				58.7 83.3	7.7	137–177	6.6	0.38
3	HO(CH$_2$)$_4$OH	28.8	48.0	165–187	8.6	0.29	73.0 100	41.5	137–189	4.8	0.16
4		28.8	70.9				85.0 100	21.9	127–187	7.5	0.30
5	HO(CH$_2$)$_6$OH	22.7	50.0	100–122	5.6	0.76	60.0 77.8	31.0	94–143	6.0	0.27
6		22.7	70.1				72.6 100	20.0	98–139	8.0	0.35
7	(HOCH$_2$CH$_2$)$_2$O	28.0	50.2	65–83	0.9	0.61	78.9 76.9	34.2	58–77	0.3	0.19
8		24.3	70.1				59.0 100	14.5	67–90	1.8	0.25

[a] Reactions run at 260–280°C under vacuum for 4 hrs. [b] Poly{[ethylene glycol-co-(methylene)$_n$glycol (50/50)] terephthalate}

[c] From initial departure from the baseline to the return to the baseline of the melting endotherm in the DSC endotherm.

[d] In tetrachloroethane/phenol solution (40/60 by weight) at 25°C. [e] In boiling TFA. [f] Determined by NMR in TFA solution.

reported for these homopolymers[6]. In Table 2 are data for the C
series of copolymers.

Both A and B polymers and the C series of copolymers were
reacted in catalyst-free conditions at 270-280°C with an equimolar
amount of p-acetoxybenzoic acid. In the C series reactions were
also carried out on mixture containing nearly 30% of the latter
component. In all cases the reactions were carried out under magnetic stirring in a 100 ml flat-bottomed cylindrical flask equipped
with two ground glass fittings assembled at the top of the apparatus
for the nitrogen inlet and condensate take-off, which was connected
to a receiver cooled by liquid nitrogen and ready for vacuum application.

In a typical experiment 12.0 g (62.5 mmole) of poly(ethylene
glycol terephthalate)(20-40 mesh powder) and 11.3 g (62.5 mmole) of
p-acetoxybenzoic acid were intimately mixed and introduced into the
reactor purged three times with dry nitrogen after heating at 110°C
at reduced pressure (0.1-0.05 mm Hg). The flask was then carefully
placed into a fused-salt bath preheated at 275°C and kept at that
temperature under a nitrogen blanket with magnetic stirring until
a low melt viscosity was obtained (0.5 hr). A vacuum of 0.1-0.05
mm Hg was applied, and the reaction mixture was mantained at that
temperature under continuous stirring for about 4 hr.

The high melt viscosity samples which resulted in all cases
were opaque, light tan and characterized by strong adhesion to glass.
They were removed from the reactor by heating overnight with a three
to four fold volume of trifluoroacetic acid (TFA) at reflux temperature. Two fractions of the polymeric products were obtained in
most cases, one soluble and the other insoluble in TFA. Each was
isolated and purified by either reprecipitation or by suspension
respectively, in a large excess of methanol. For the specific
example above the total yield was 69.2% and the soluble portion
amounted to 74.0% of the total.

The chemical composition of the soluble fraction was determined by NMR in TFA solution on the basis of the relative intensities
of the signal of the aromatic protons in the 3 position of the 4-
oxybenzoyl moiety which were sufficiently separated from the signals
of all other aromatic protons.

RESULTS

For polymers A and B the expected ester interchange with insertion of measurable amounts of p-hydroxybenzoic acid units into the
polyester took place only for poly(ethylene terephthalate). For all
other polymers the starting copolymer was either recovered almost
completely unconverted or the product contained only a very small
amount of oxybenzoate units in it. In these cases the p-acetoxybenzoic acid was almost totally converted into the corresponding insoluble and infusible homopolymer, so a physical mixture of the two
polymers was obtained. These results seem to indicate that the
transesterification reaction occurred by a concerted mechanism and
was strongly affected by the polarization of C-O bond in the polymer
involved in the ester interchange.

In contrast to these results, the oxybenzoylation reaction occurred readly in the C series of statistical copolymers of terephthalic acid with ethylene glycol and different poly(methylene) or poly
(ethyleneoxy) glycols on reaction with p-acetoxybenzoic acid. In
most cases, with reaction mixtures containing up to 50 mole percent
of oxybenzoate units, the reaction product was completely soluble
in either trifluoroacetic acid or in halogenated solvents as indicated in Table 2. The products contained mole fractions of oxybenzoate units which were somewhat lower than those in the reaction
mixtures. For copolyesters containing approximately 50 mole percent
of oxybenzoate units there existed a typical indication of liquid-
crystal behaviour as shown by the Schlieren textures detected under
the polarizing microscope on the unclear melt. The properties of
these liquid crystalline polymers will be discussed in more detail
in a subsequent publication.

The oxybenzoylation of the starting ethylene terephthalate
copolymers leads in all cases to marked changes in the profiles and
positions of the DSC melting endotherms of the initial polymers.
A considerable broadening of the endotherm of the starting copolyester, as shown by the data in Table 2, was accompanied by the occurrence of two or more melting transitions, which were reproducible.
In Figure 1 is shown a typical set of endotherms for polymer C5,
before and after the oxybenzoylation. It is seen that after oxybenzoylation, the sample developed several endotherms after first heating and cooling cycle. When this sample, polymer C5 of Table 2,

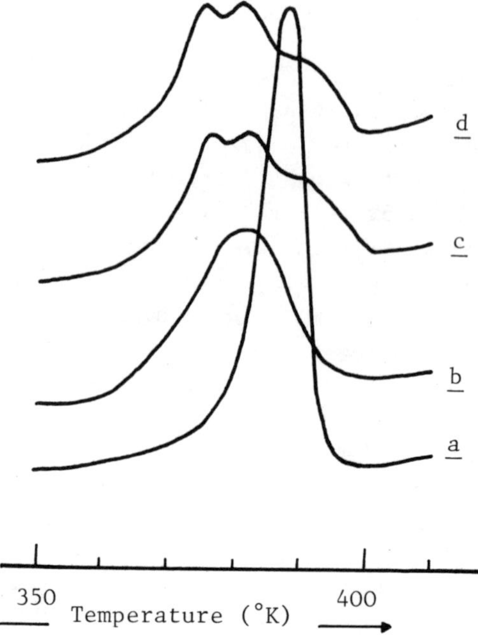

Figure 1. DSC curves for: <u>a</u> Polymer B4 (Table 1): poly [(ethylene
                              glycol-co -hexamethylene glycol)tereph-
                              thalate] (50/50).
                          <u>b</u> Polymer C5 (Table 2) with 31 mole %
                              oxybenzoate units: first heating of
                              annealed sample.
                          <u>c</u> Polymer C5: second heating cycle
                          <u>d</u> Polymer C5: third heating cycle

was examined on the hot stage of a polarizing microscope it was
observed a form of highly birefringent melt at a temperature in the
range of the highest temperature peaks of curves c and d in Figure 1.
The melt retained its birefringency up to approximately 80°C above
the melting point and is apparently thermotropic or liquid crystalline. Investigations on the liquid crystalline properties of these
terpolymers are continuing and will be described in detail in a future publication.

REFERENCES

1. W. J. Jackson Jr., and H. F. Kuhfuss, J. Polymer Sci., Pol. Chem. Ed., 14, 2043 (1976).

2. F. E. McFarlane, V. A. Nicely and T. G. Davis, "Contemporary Topics in Polymer Science", E. M. Pearce and J. R. Schaefgen Eds., Plenum Press, New York, USA (1976).

3. R. W. Lenz and K. Feichtinger, ACS Polymer Prep., 20, 114 (1979).

4. E. Chiellini, F. Galli, F. Ciardelli, R. Palla and F. Carmassi, Inf. Chim., 176, 221 (1978).

5. R. Bianchi, A. Bionda, F. Carmassi, R. Palla, C. Donadio, G. Galli, E. Chiellini, G. Molea and G. Mariani, J. Dialysis (in press).

6. J. G. Smith, C. J. Kibler and B. J. Sublett, J. Polymer Sci.A1, 4, 1851 (1966).

TRANSPORT PROPERTIES AND MORPHOLOGY OF POLYMERIC BLENDS

C. Carfagna, A. Apicella, E. Drioli, H.B. Hopfenberg[*],
E. Martuscelli[**] and L. Nicolais

Istituto di Principi di Ingegneria Chimica, Università
di Napoli, Italy

Some transport properties of the seemingly compatible glassy polyblends of atactic Polystyrene (APS) with
both Poly(2,6-dimethyl-1,4-phenylene oxide) (PPO) and
isotactic Polystyrene (IPS), obtained by casting solutions of mutual solvents, have been studied. In particular, morphology changes associated to sample preparation and physical treatments (annealing and swelling in
n-hexane) have been related to their transport properties.

It has been shown as annealing may improve the solvent resistence of the blend and as the effect can be
exalted by varying its composition.

Thermal analysis by means of Differential Scanning
Calorimetry (DSC) has been used to characterize the
blends regarding crystallinity, glass transition and
presence of residual casting solvent.

---

[*]Present address - Chemical Engineering Dept., N.C. State University, 20650 Raleigh, N.C., U.S.A.

[**]Present address - Laboratorio di Ricerche su Tecnologia dei Polimeri e Reologia, Arco Felice, Naples, Italy.

INTRODUCTION

Whereas transport of gasses and vapours in a polymer is generally described by ordinary Fick's laws at temperatures above its glass transition[1], the transport of penetrants in polymeric organic glasses is frequently considered anomalous[2]. Dual Sorption Mode, time dependent boundary condition, polymer relaxation and crazing are examples of phenomena related to the diffusion of low molecular weight substances in glassy polymers. The knowledge of the morphological modifications associated with the presence of a penetrant, or, more in general to changes induced by physical aging is of particular interest in their applications. Moreover, the formulation of polymer blends appears to be an interesting approach for changing and improving the properties of the individual components. For example, the water vapor permeability of blends of nylon and polyethylene is lower than permeability of either homopolymers[3]. Similarly, blends of cellulose triacetate and cellulose acetate exibit salt rejections, in eccess of 99.9% significantly higher than either homopolymers[3].

Detailed and meaningful characterization of polymer blends may be, in principle, more reliable if the components are molecularly dissolved in a precursor casting solution than if one relies upon the arbitrarily imposed mixedness by direct thermal forming. The possibility of domain phase separation during drying of the casting solution of the mixed polymers must be considered. Polymer blends are often considered compatible if the resulting mixture meets the use criteria of the intended application. Arbitrarily chosen criteria are then based on observations as that of a single glass transition temperature or that of an optically transparent material. Domains rich in one polymer specie dispersed within a matrix rich in the second component, may not be distinguishable by these techniques (i.e. a single glass transition temperature, monotonically varying in blend composition, has been noted for the PPO--APS system, while the presence of domains rich in both homopolymers has been determined using different techniques[4]). Recently the Authors studied[5,6] the thermal and n-hexane sorption behaviours in PPO-APS and APS-IPS films cast from suitable mutual solvents. In this paper the transport properties of these glassy compatible blends are emphasized and related to morphological properties of the respective homopolymers and blends detected by Differential Scanning Calorimetry (DSC).

Thermal analysis by means of D.S.C. is, in fact, a useful tool to investigate the morphological modifications, that such two component systems may undergo when subjected to particular treatments. The "as cast" may contain residual trapped solvent and solvent induced crystallinity developed during the casting and the drying procedures used to form the films[5,7].

The knowledge of physical aging is then crucial to understanding and improving the barrier and mechanical properties of such multicomponent materials.

THERMAL ANALYSIS AND TRANSPORT PROPERTIES OF SOLVENT CAST BLENDS

Differential Scanning Calorimetry thermograms obtained using a DSC-2 Perkin Elmer calorimeter are presented in Figure 1 for "as cast" films in the first (1a) and the second (1b) scans of the PPO-APS blends and for the two homopolymers. While the second cycle scan reveals only a single transition for all blend composition,

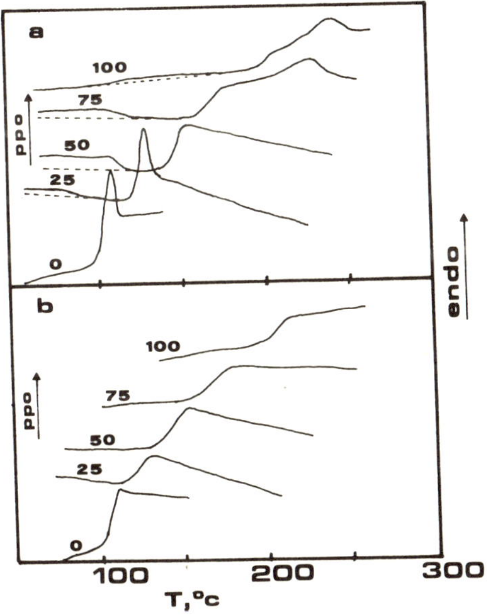

Figure 1. First (a) and second (b) D.S.C. scans for "as cast" IPO-APS blends of different composition and homopolymers. Heat rate 20°/min.

first cycle scans are complicated by additional endothermic processes both above and below the glass transition region. The scans
for the pure PPO, and the PPO rich blend, reveal distinct endotherms
at temperature significantly higher than the respective glass transitions. These high temperature thermal events are related to the
melting of crystalline PPO (8), presumably "solvent induced"[9] during
the casting and drying procedures. The fast quenching which was
imposed to the sample between the two scans, indeed, erases this
endotherm indicating that the crystallinity was not developed in
this different condition and reflecting the difficulty of PPO to
undergo thermal crystallization[8,10]. Low temperature endotherms
are broadly distributed over the temperature range 80° - 110°C for
all "as cast" samples in the first scan and are related to diffu-
sion-controlled liberation of trapped casting solvent. In Figure 2,
first and second scans are compared for an "as cast" APS sample.
Solvent liberation from the still glassy system (T<Tg) is clearly
evident in the first scan at temperatures around the boiling point
of the casting solvent (trichloro-ethylene).

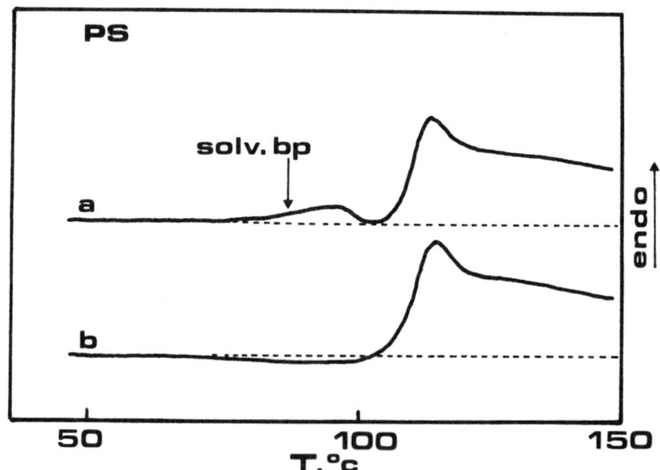

Figure 2.   First (a) and second (b) D.S.C. scans for "as cast"
            APS. Casting solvent boiling point is indicated by
            the arrow. Heating rate 40°/min.

The glass transition temperatures calculated for both the homopolymers and the intermediate components are reported in Figure 3.

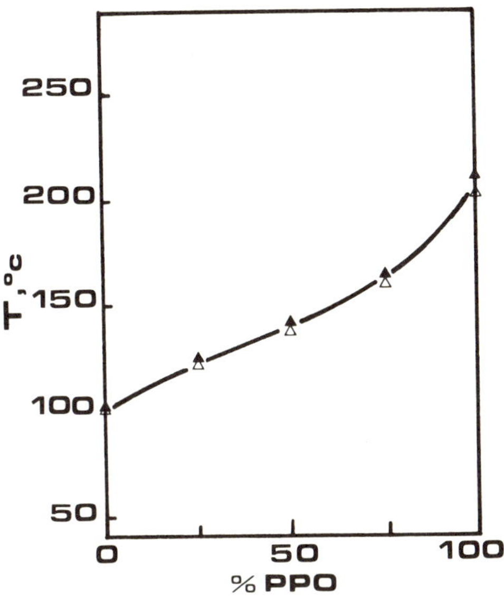

Figure 3. Glass transition temperatures calculated in the first (open triangles) and second (full triangles) scans vs. blend composition for the PPO-APS system.

Despite the distinct character of thermograms in Figure 1, the values of calculated Tg seem independent of scanning history. Previously trapped solvent is essentially eliminated before the Tg is reached during the first cycle scan and the Tg calculated from the first cycle is, therefore, not significantly affected by plasticization related to the residual trichloroethylene. There is, however, a slight systematic increase in the measured Tg, calculated from the second cycle scans (full triangles), compared with Tg' calculated from the original thermograms (open triangles). This increase is more apparent as the PPO content is increased. These results are consistent with the suggestion of more rapid liberation of trichloroethylene from PS rich samples.

Moreover, elimination of crystallinity on the first cycle will contribute additional PPO to the seemingly compatible, single phase amorphous glass. The subtle increases in measured Tg's are, therefore, consistent both with a systematic liberation of casting sol-

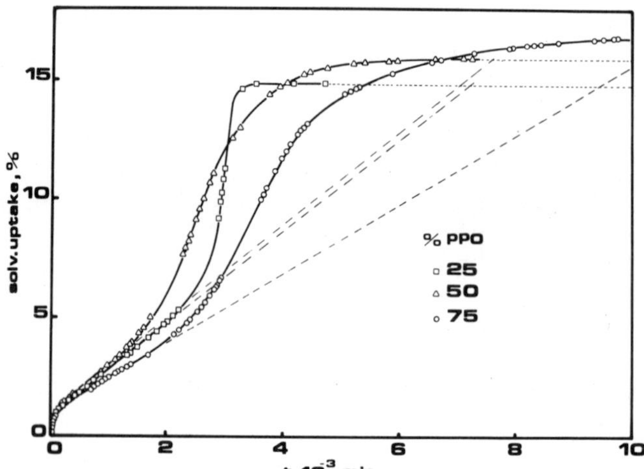

Figure 4. Vapor sorption kinetics for n-hexane in three different blend composition films of the PPO-APS system. T = 40°C, vapor activity = 0.97.

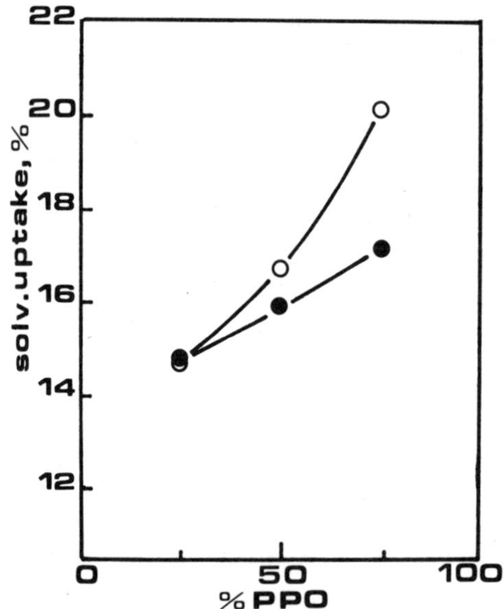

Figure 5. Vapor sorption equilibria for n-hexane vs. blend compositions for the PPO-APS system at T = 40°C and a = 0.97. Annealed above T_g 48h (●) and previously swollen and vacuum dried (o) samples.

vent and with the increased concentration of PPO in the blended phase after melting of PPO crystallinity. Vapor sorptions in thin films
have been studied using a quartz spring microbalance. Vapor sorption kinetics describing the penetration of n-hexane at T = 40°C
and a = .97, in films of the three intermediate blend compositions,
annealed at 20°C above respective glass transition, are shown in
Figure 4. After the thermal annealing, a systematic density increase has been measured, using a density gradient column filled with
appropriate NaCl solutions, and is reported in Table 1 for the three
intermediate blend compositions.

TABLE 1

Blend composition % PPO	"As cast" $\varrho = g/cm^3$	Annealed 48 hrs 20° above respec- tive T_g. $\varrho = g/cm^3$	% Increase
25	1.0564	1.0581	0.16
50	1.0621	1.0660	0.37
75	1.0664	1.0700	0.34

Moreover n-hexane apparent equilibrium solubilities are presented
in Figure 5 for the annealed samples and for samples n-hexane pres-
wollen and dried under vacuum. Annealing has a more pronounced effect on the apparent equilibrium as the PPO content is increased.
Presumably, the lower glass transition temperature, associated to
PS rich samples, permits significant consolidation during identical
casting and drying procedures. Sorption curves (Figure 4) are characterized by an essentially constant rate of penetrant uptake,
which is associated with limiting "Case II" sorption, where a constant rate advancing swollen front is noted[11], followed by an acceleration preceding equilibration (Super Case II). An initial
Fickian sorption behaviour is also identified.

The early and final anomalies are to be related to the sample
dimensions used in this study. For very small samples (i.e. microspheres) diffusion in the glassy polymer and relaxation of the penetrate material have been successfully separated[12], then overimpo-
sed coupled behaviours are expected in samples of small dimensions

such as those used in our case (20μ).

The fundamental process involved in Case II transport has been explained by Sarti[13] in terms of resistence of the polymer to mechanical deformation. Since the main role played by the solvent is to exert an osmotic stress on the polymeric matrix[13], the relaxation controlled sorption kinetics may be expressed by the rate of the advancing swollen fronts. The rate of penetration calculated from the linear portion of the sorption curves and equilibrium solubilities are reported for the three compositions studied in Table 2.

TABLE 2

Blend composition % PPO	Front Velocity: v, mm/min	n-hexane equilibrium concentration in the advancing fronts: C, % dry wt.
25	1.4×10^{-6}	14.8
50	1.0×10^{-6}	16.4
75	8.5×10^{-7}	17.2

Although higher osomotic stresses are suspected for the higher penetrant equilibrium solubilities, the rate of the advancing swollen fronts is found progressively lower as the PPO composition is increased. Such higher solvent resistence of PPO rich samples may be explained in terms of its high mechanical strength[4]. In addition, since sorption has been associated to relaxation occurring at the interface with the still unpenetrated glassy core, relaxation is then expected to be much more evident in samples showing lower glass transition temperatures. At the temperature of the test (T = 40°C) the PS rich sample is, in fact, 80°C below its glass transition, while the PPO rich sample is 120°C far from its respective transition. Sorption kinetics of n-hexane and equilibria have been studied[7] for "as cast" IPS-APS blends. Maxima in the plots of weight changes versus time, were apparent for all compositions. These maxima were explained as a consequence of the liberation of previously trapped residual casting solvent consequent to palsticizing by n-hexane penetration. Neutro activation analysis for chlorine[7] confirmed that the solvent (o - chlorotoluene) content was reduced dramatically after n-hexane exposure.

Both the rate of sorption and the apparent sorption equilibrium
were greatly reduced as the isotactic polystyrene content was increased. Limiting relaxation controlled Case II sorption kinetics were
observed in the polyblends once thermal annealed or "solvent" annealed[7]. Although the form of the sorption curves was quite similar
for the two physically different annealing, the sorption rate and
amount of sorbed penetrant were both greater for the thermally annealed films.

Both atactic and isotactic polystyrene are susceptible to pene-
trant-induced morphological changes.

Hopfenberg[14] reported that low molecular weight n-alkane are
crazing agents for APS and IPS, while Overbergh[15] reported solvent
induced crystallization in isotactic polystyrene.

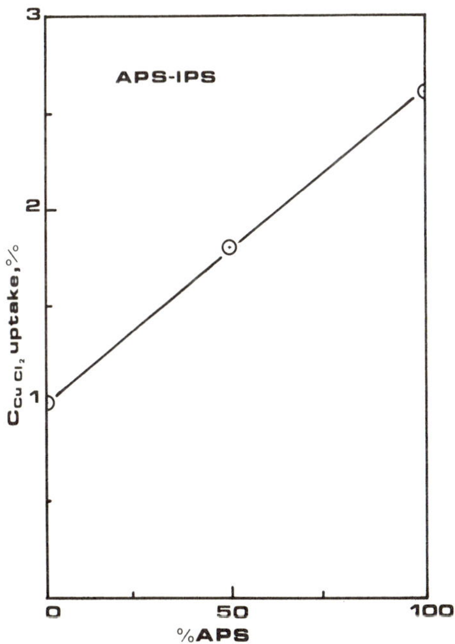

Figure 6. CuCl$_2$ solubilities in IPS, APS and an intermediate blend
           from sorptions in saturated water-salt solutions.
           T = 40° C.

Crystallization of the isotactic component in the blend densifies the overall matrix while crazing of the atactic polymer results in voiding and, therefore, in a significant reduction of both the mechanical and the more interesting barrier properties. In fact, a more compact structure will depress the gas or solute solubilities decreasing the driving force for Fickian diffusion. Figure 6 shows the increase of $CuCl_2$ solubility for APS content increase of IPS-APS blends, as expected for the less crystallizable material.

The effect of solvent on the morphological changes in APS-IPS blends were studied by immerging them in liquid n-hexane for variable times. Crystallinity development induced by solvent immersion was suspected since the immersed and dried samples developed a white opalescence.

Liquid nitrogen fractured films of untreated and n-hexane immersed samples were viewed by scanning electron photomicrographs under varying magnification.

Scanning electron photomicrographs of the fracture surfaces are presented in Figure 7 for the untreated and treated homopolymers as well as for treated intermediate blends. The opalescence appears to be related to solvent craze induced development of non-intercommunicating microvoids in the atactic rich samples. There is no evidence of void development in the isotactic sample nor in the isotactic rich specimen. The same results may be observed by means of a differential scanning calorimetry technique.

Consistent with early publication of Overbergh et al.[16] on n-hexane induced crystallinity in isotactic polystyrene, there is no significant difference for the effect of crystallization temperature (Figure 8) on ΔH_f between the treated and untreated samples. Nevertheless a slight systematic ΔH_f decreases with IPS composition increases are observed for previously swollen blends and the IPS homopolymer.

A possible explanation may be provided by assuming the formation of crystals with varying degrees of structural defects[16] or in whose lattice solvent molecules are also built in[17] during swelling procedures. The subsequent additional crystallinity induced by annealing below its melting point (cold crystallization) may be affected in different measure depending on IPS content which is the cry-

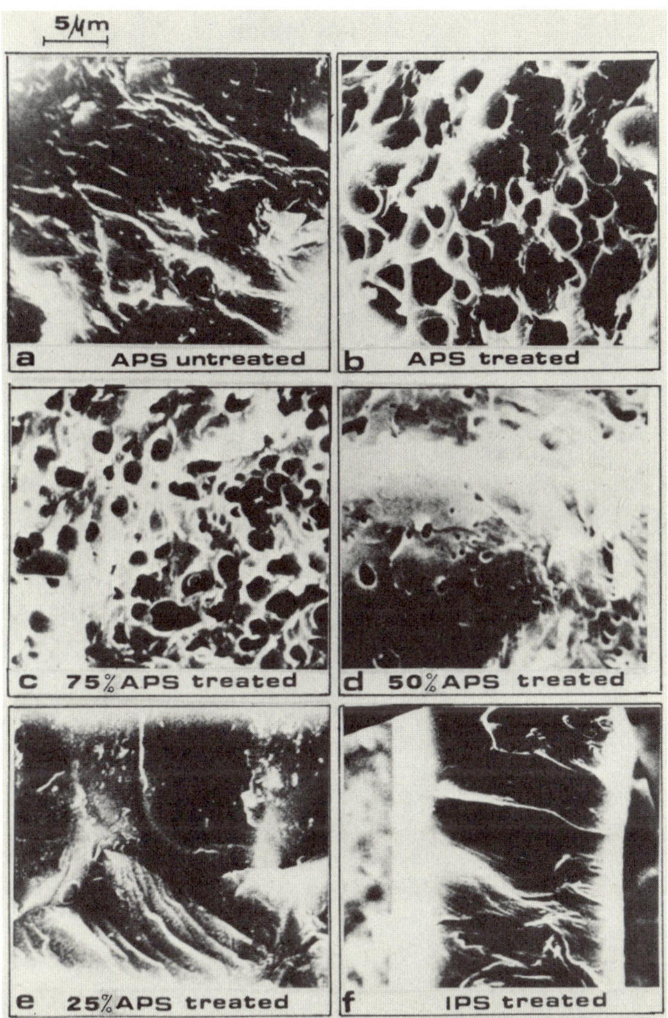

Figure 7.   Scanning electron photomicrographs of IPS, APS and intermediate blends films. APS untreated (a), APS (b),
75% APS (c), 50% APS (d), 25% APS (e) and IPS (f) soaked
20 minutes in liquid n-hexane at T = 40°C.

stallizable component.

Density increases and whitening of isotactic samples exposed
to n-hexane have also been observed[7] and related to possible solvent
induced crystallization.

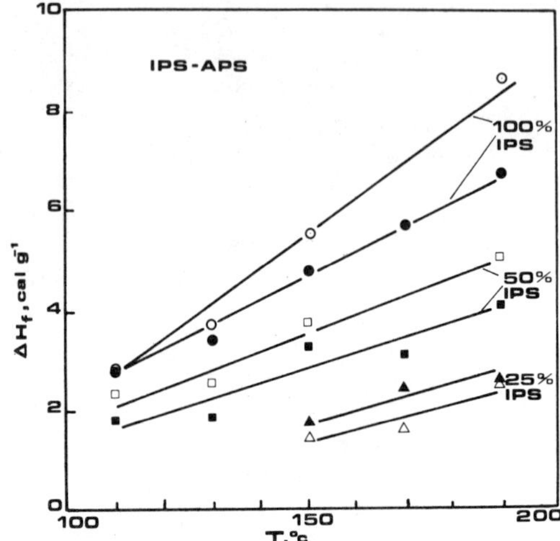

Figure 8. Effect of crystallization temperature on enthalpy of
fusion of IPS and IPS-APS blends treated in liquid
n-hexane at T = 40°C (full symbols) and untreated
(open symbols).

In conclusion, these morphological studies are focal to the
continuing experimentation with these intriguing systems. It has
been shown that isotactic polymer retards crazing of the atactic
component, whereas atactic polymer interfers with the crystallization of the isotactic rich formulations. Compaction by annealing
or densification by thermal or "solvent" induced crystallinity may
improve both the environmental attack resistance and the barrier
properties of such polyblends.

REFERENCES

1. J. Crank, "The Mathematics of Diffusion", Oxford University
    Press (1975).
2. J. Crank and G. S. Park, "Diffusion in Polymer" Academic Press,
    New York (1968).

3.   D. R. Paul and N. Seymour, "Polymer Blends" Academic Press, New York (1978).

4.   J. Stoelting, F. E. Karasz and W. L. Macnigt, Pol. Eng. and Sci., $\underline{10}$, 3, 133 (1970).

5.   A. Apicella, E. Drioli, H. B. Hopfenberg, E. Martuscelli and L. Nicolais, Pol. Eng. and Sci., $\underline{18}$, 1006 (1978).

6.   E. Martuscelli, G. Demma, E. Drioli, L. Nicolais, S. Spina, H. B. Hopfenberg and V. Stannett, Polymer, $\underline{20}$, 571 (1979).

7.   D. L. Faulkner, H. B. Hopfenberg, V. T. Stannet, Polymer, $\underline{18}$, 1130 (1977).

8.   W. Wrasidlo, Adv. Pol. Sci. (B) (1974).

9.   E. Turska, H. Janeczek, Polymer, $\underline{19}$, 81 (1978).

10. F. E. Karasz, J. D. Mangara, Polymer Preprints, $\underline{12}$, 317 (1971)

11. C. H. M. Jacques, H. B. Hopfenberg, V. T. Stannet, Pol. Eng. and Sci., $\underline{13}$, 81 (1973).

12. A. R. Berens, H. B. Hopfenberg, Polymer $\underline{19}$, 489 (1978).

13. G. C. Sarti, Polymer, $\underline{20}$, 827 (1979).

14. H. B. Hopfenberg, "Membrane Science and Technology" Plenum Press, New York (1970).

15. N. Overbergh, H. Berghmans, G. Snets, Polymer, $\underline{16}$, 703 (1975).

16. P. Magree, J. Boon, Proc. IUPAC Int. Symp., 835 (1970).

17. A. D. Gray, K. J. Casey, J. Pol. Sci. (B), $\underline{2}$, 381 (1964).

THE PHASE STRUCTURE OF POLY(N-VINYLCARBAZOLE)-POLYCARBONATE
BLENDS AS STUDIED BY THERMALLY STIMULATED CURRENTS AND
THERMO-OPTICAL ANALYSIS

Jacek Ulanski and Marian Kryszewski

Institute of Polymers, Technical University

90-539 Łódź, Poland

The phase structure of poly(N-vinylcarbazole) - polycarbonate blends for applications in electrophotography
was investigated. The results of Thermally Stimulated
Currents measurements are interpreted in terms of polymers miscibility. It is deduced that PVK - PC mixtures
are not homogeneous, however the blends did not show inhomogeneity in polarizing microscope. This conclusion
is confirmed by Thermo-optical Analysis which indicated
the appearance of two glass transitions in the mixtures.
On the other hand, the measurements of the Lower Critical Solution Temperature show that the phase separation
occurs in the samples in the temperature range of 210-
-240°C depending on PVK content.

The presented results suggest, that PVK and PC are
partially miscible at lower temperatures, but the system
is thermodynamically unstable and at higher temperatures
the phase separation occurs.

INTRODUCTION

Thermally Stimulated Currents technique was elaborated as a
method of investigation of charge carrier trapping phenomena, by
Nicholas and Woods[1]. In our investigations of electrical proper-

ties of poly(N-vinylcarbazole) (PVK) and poly(bisphenol A carbonate) (PC) blends we have obtained TSC spectra which allow us to draw
some conclusions concerning phase structure of the mixtures. The
PVK - PC blends can be used in electrophotography because they
combine good photoelectric properties of PVK and have improved mechanical properties due to PC. The homogeneity of the blend, which
is always the main problem in polymeric mixture, is of particular
importance in our case because it determines mechanical (flexibility) and optical (transparency) as well as electrical properties.
The inhomogeneity can decrease the mobility of charge carriers and
introduces additional trapping centres at interfaces (Maxwell-Wagner effect). We have analysed our TSC spectra in order to obtain
some informations about these phenomena. To confirm the obtained
results we have carried out Thermo-optical Analysis (TOA)[2], and we
have determined the temperatures at which phase separation occurred.

EXPERIMENTAL

The 10 μm thick films were obtained by casting from solution
of appropriate amounts of PVK and PC in chlorobenzene. Then the
blends were carefully dried at 120°C in vacuum for 24 hours in order to remove residual solvent. The obtained films were flexible
and transparent, and did not show neither crystallinity nor inhomogeneities in polarizing microscope. The content of PVK in the
mixtures was changed from 10 to 80 wt. %. The samples for TSC measurements had semitransparent gold electrodes evaporated in vacuum
and a guard electrode (sandwich-type samples). The TSC measurements were carried out using the standard equipment described elsewhere[3]. As it was shown in the same work, TSC spectra are strongly
dependent on experimental conditions.

In the present paper the TSC spectra were obtained using two
different experimental procedures. The first one consisted in polarizing the sample with simultaneous illumination (electric field
E_p = 2 x 10^7 V/m, visible light) at liquid nitrogen temperature.
Then the short circuited sample (E_d = 0) was heated in the dark
with the constant heating rate, 6 deg/min, and the TSC spectrum was
recorded. Under these experimental conditions the TSC maximum characteristic for PC was observed[3].

The second set of TSC spectra was obtained after illumination

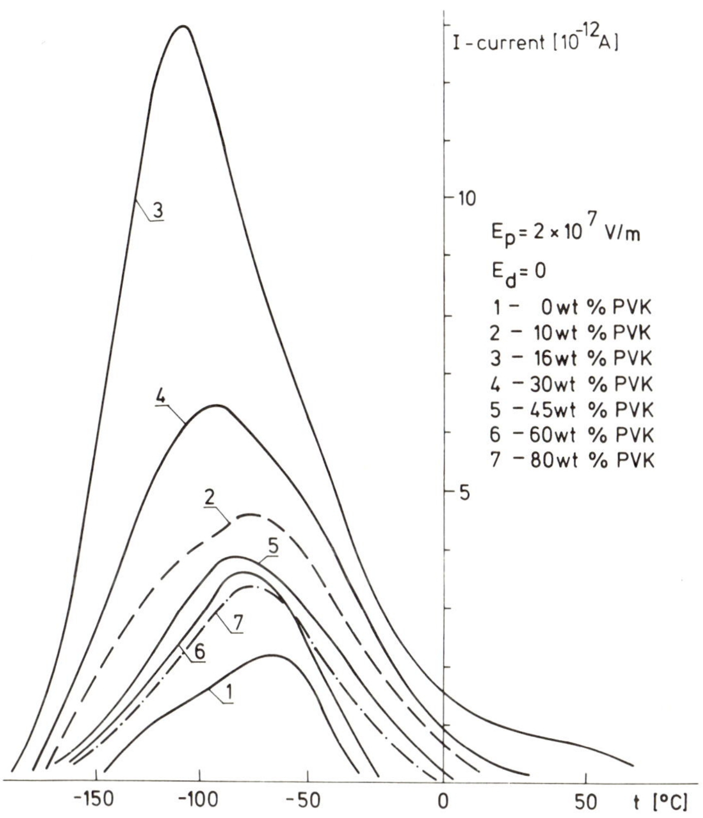

Figure 1.   TSC spectra for PC–PVK blends of different composition.
The samples were polarized and illuminated at −196°C
and then heated in the dark in short circuit.

of the sample with visible light (without polarizing field, $E_p = 0$) at -196°C. Then the sample was heated in the dark with the drain field $E_d = 2 \times 10^6$ V/m. These experimental conditions are proper for observation of a TSC maximum in PVK,[4].

For TOA measurements a polarizing microscope was used, which was equipped with a heated sample support and a Se-photometer connected with a x-y recorder (y axis). A thermocouple for recording the temperature of the sample was connected with this recorder (x-axis). The TOA experiments were performed in the following way: the sample was locally oriented by scratching with a needle and then placed in microscope between crossed polaroids. Light was transmissed through the system only through the birefringent scratch regions. When the temperature was increased so that the polymer segmental motions became sufficient to relax the stress-strain-induced birefringence in the scratches, the transmitted light intensity rapidly decreased. The intersections of the tangents to the obtained curves at its inflection points determine transition temperatures.

The same experimental set with additional side illumination was used to evaluate the temperatures at which the phase separation occurred (cloud points)[5].

RESULTS AND DISCUSSION

<u>Thermally Stimulated Currents</u>

TSC spectra for the blends obtained without drain voltage, after polarization and illumination of the sample at -196°C, are presented in Figure 1. For comparison, the TSC maximum obtained for pure PC in the same experimental conditions is also shown[3]. Pure PVK did not exhibit any maximum in the TSC spectrum, when the experiment has been performed in the same way. The origin of the peaks shown in Figure 1 can be unambiguously related to charge trapping on polycarbonate chains. In the range of low PVK concentration (up to 16%) in the blend, the magnitude of the peak attributed to PC increases with the increase of PVK content. It may arise from the fact that PVK introduces charge carrier photogeneration centres to the bulk of the sample. As a consequence, an amount of free charge carriers generated during illumination, which

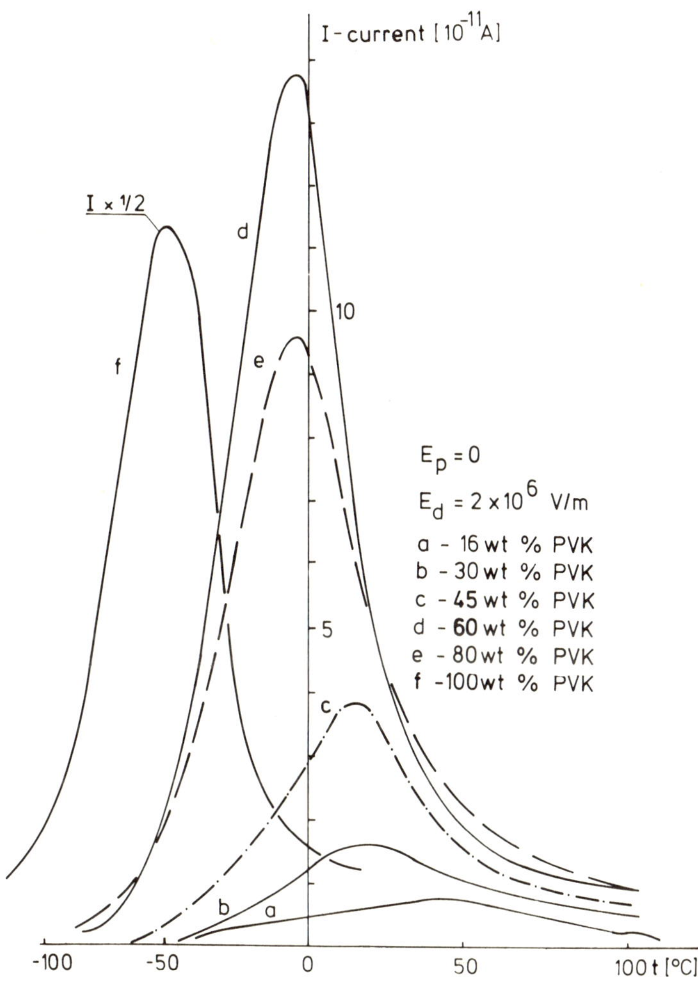

Figure 2. TSC spectra for PC–PVK blends of different composition.
The samples were illuminated at −196°C (without polarization) and then heated in the dark with drain voltage.

can be trapped in the sample, is bigger than in pure PC. For higher contents of PVK, as the concentration of considered trapping centers decreases simultaneously with diminishing PC content, the magnitude of the peak decreases.

The TSC spectra obtained after illumination of the sample at -196°C (without polarizing field) and with drain voltage while heating, are shown in Figure 2. The spectrum characteristic for pure PVK, obtained under the same experimental conditions by Zieliński and Samoć[4] is also shown. One can see that maximum intensity increases with increasing of PVK content in the blend. Taking into account the dependence of the obtained spectra on experimental conditions (e.g. dependence on drain field intensity), one can say, that this peak is connected with the escape of the carriers from the traps localized on PVK chains. Some shift of the peak position towards lower temperatures for higher content of PVK may be due to increase of charge carrier mobility. The results of partial heating experiments show, that the energetic depths of the traps in the samples with high PVK concentration are nearly the same as compared to those calculated for the samples containing the excess of PC[6]. From the presented results one can draw conclusion that in TSC experiments for PVK - PC mixtures we have observed two maxima: first one characteristic for pure PC and the second one characteristic for pure PVK. All observed changes in position and intensity of the peaks can be simply related to changes of PVK (or PC content in the samples). These findings can be hardly explained in terms of molecular mixing of both polymers. Differences in intermolecular PVK - PC interactions in the mixture as compared with PC - PC or PVK - PVK interactions should induce the changes of trapping phenomena, first of all activation energies should be significantly changed. As this is not the case one has to conclude that both components of the mixture exist in separated phases. On the other hand any new maximum which should be expected in inhomogeneous system due to Maxwell-Wagner[7], did not appear in the discussed TSC spectra.

Thermo-optical Analysis

The results of TOA experiments are presented in Figure 3. The obtained curves prove the existence of two phase transitions at the temperatures slightly shifted as compared with the glass

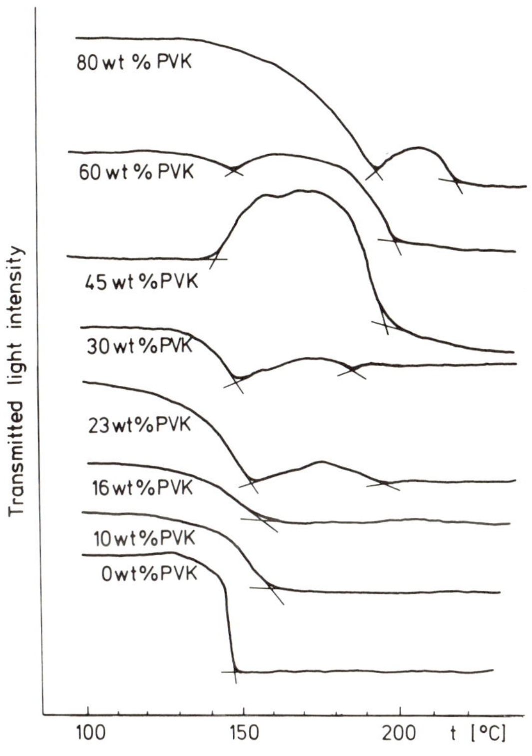

Figure 3.   TOA curves for PC-PVK blends of different composition.
The curves are shifted along vertical axis to make the
picture clear.

transition temperatures of pure PC (149°C) and PVK (225°C). For
the extreme compositions one can observe only one inflection point
of the curve, corresponding to the excess component. This can arise
either from too low sensitivity of TOA method, or from the fact that
the polymers are miscible to some extent.

Using simple visual technique we have also examined PVK - PC
blends for Lower Critical Solution Temperature (LCST) behaviour by
determination of cloud points[5]. We have observed a distinct phase
separation during heating in the temperature range 210-240°C, depending on composition. Such behaviour indicates, that PC and PVK
are, at least partially, miscible at lower temperatures, but the
mixtures are thermodynamically unstable[8]. It can arise from the
fact, that interactions between the polymers that stabilize the
mixture at low temperatures fall off at higher temperatures, as
suggested by some authors[8,9].

CONCLUSIONS

It is shown that PVK and PC are partially miscible at low temperatures, but phase separation can occur at high temperatures.

The results of TOA and TSC experiments indicate, that some inhomogeneities, probably of very small size, exist in the mixture.
These inhomogeneities have negligible influence on the transparency
of the investigated blends and their presence in blends does not
introduce additional trapping centers for generated carriers. Thus
the system PVK - PC preserves the properties required for electro-
photographic materials.

The PVK - PC blends exhibit LCST behaviour, which suggests
the existence of favorable interactions between the components.
However, these interactions are too small to modify the structure
of charge traps in both polymers.

The TSC technique, usually used for investigations of charge
trapping phenomena, is shown to provide also some useful information on the structure of investigated blends.

REFERENCES

1. K.H. Nicholas and J. Woods, Brit. J. Appl. Phys., $\underline{15}$, 783 (1964)
2. A.R. Shultz and B.M. Beach, Macromolecules, $\underline{7}$, 902 (1974)
3. J. Jeszka, J. Ulański and M. Kryszewski, J. Eletrostatics, in press
4. M. Zieliński and M. Samoć, J. Phys. D., $\underline{10}$, L 105 (1977)
5. R.E. Bernstein, C.A. Cruz, D.R. Paul and J.W. Barlow, Macromolecules, $\underline{10}$, 681 (1977)
6. J. Ulański, W. Rybka and M. Kryszewski, to be published
7. J. Van Turnhout "Thermally Stimulated Discharge of Polymer Electrets" Elsevier Pub., Amsterdam (1974)
8. D. Patterson and A. Robard, Macromolecules, $\underline{11}$, 690 (1978)
9. E. Nolley, D.R. Paul and J.W. Barlow, J. Appl. Polym. Sci. $\underline{23}$, 623 (1979).

STRESS RELAXATION OF GLASS BEAD COMPOSITES

C. Migliaresi, G. Guerra, L. Nicodemo
and L. Nicolais

Istituto di Principi di Ingegneria Chimica
Univeristy of Naples, Naples, Italy

The stress relaxation of epoxy and polystyrene filled with various percentages of glass microspheres was
investigated over a wide range of temperatures. The effect of the second phase in thermosetting matrices is to
increase the glass transition temperature of the system.
In the region of linear viscoelastic response, the presence of the filler in all cases has no effect on the
shift factors used for constructing the master curves
according to the time temperature superposition principle.

All the data were shifted to a single master curve
at a reference temperature and filler content by using
independent shift factors function of either temperature
or filler content.

INTRODUCTION

The mechanical properties of polymer composites filled with a
particulate filler or with short fibers are greatly dependent on
the adhesion of phases. Insufficient interfacial adhesion is usually reflected in umpaired ultimate properties[1,2] and in a high loss
factor, tanδ, due to additional friction mechanism at the interface[3,4]. On the other hand, strong interfacial adhesion, which is ne-

cessary for the composite to possess good mechanical properties, may account for a depression of molecular mobility and for a change in the packing of matrix macromolecules adjacent to the interface. The existence of a surface layer with modified physical properties which is formed at the boundary polymer-solid support is borne out by the optical and mechanical anisotropy of thin cast films[5,6]. In composites epoxy resin-glass beads a portion of the matrix is effectively immobilized and made relatively impermeable to liquid water; the thickness of the ordered layer is 0.4 μm for beads 40 μm in diameter[7]. A strong interfacial interaction is believed to underlie the increase (ordinarily less than 20°C) in the glass transition temperature T_g of the matrix with increasing of filler content[8,9,10]. The increase in T_g of a series of polymers filled with silica[11] has turned out to be directly proportional to the polymer-filler interaction energy.

Ferry reports that[12], for a filled crosslinked polymer in its transition region, the general shape of the relaxation distribution curve is not affected by the filler, but that the modulus is increased and the time dependence is shifted to lower frequencies. A slight rise in the transition temperature is also reported. In contrast, Schwarzl reports that[13], for a series of concentrations of salts in a rubbery matrix, there is no effect of filler on the glass transition temperature. Nielsen[14], in discussing the effect of filler on the creep and dynamic properties of filled polyethylene, reports a modulus effect similar to that reported by Ferry[12]. He also suggests a possible change in the matrix properties due to filler-matrix interactions. This corresponds to the idea of Rigbi[15] who proposed a broadening of the relaxation spectrum due to the filler interactions. Analogous results have been obtained by Migliaresi et al[10] and Kolarik et al[8] who show an increase of T_g in the glass bead/epoxy and polymeric filler/polyurethane rubber systems as consequence of filler presence.

In this paper the viscoelastic response of glass beads imbedded both in thermoplastic and thermosetting matrices is investigated by means of stress relaxation experiments. In fact since the use of polymeric composite systems as structural material is becoming more and more important it is of interest to predict the long term mechanical behaviour of such materials by performing short time experiments. The characterization of linear, time dependent mechanical responses of polymers has its origin in the studies of

Voigt[16], Boltzmann[17], and Maxwell[18]. Over the past years the theory of linear viscoelasticity has been experimentally verified as a satisfactory model for the response of polymers in the rubbery state, but small attention has been devoted to the properties of composites in the viscoelastic region.

In this work, stress relaxation experiments have been performed on polystyrene and epoxy resin filled with glass beads. The data have been superimposed using a shifting procedure which takes into account the change in T_g in thermosetting composites and both the temperature and the filler effect.

EXPERIMENTAL

The composite materials were prepared using Vacu-Blast glass microspheres with particle sizes in the range of 32 to 40 microns. The glass fillers were first cleaned in solvents to eliminate dirt and other unwanted organics on the surfaces.

The resins used were Shell's epoxy resin Epon 828 cured with triethylenetriamine and Montedison's Edistir FA, a general purpose polystyrene.

Test specimens of the epoxy composites were cast directly in the form of dumbells with an overall length of 5 cm and a gage cross-section of 0.90 cm width by 0.20 cm thickness. The Epon 828 monomer was mixed with the appropriate weigth of glass microspheres, and the crosslinking agent triethylenetriamine in the proportion 14 grams per 100 grams of Epon 828 was added. The mixture was reacted at $20°C$ for 24 hours and then postcured for 6 days at $100°C$. Under these conditions the glass transition temperature of the unfilled epoxy resin, measured in a Perkin-Elmer DSC-1 differential scanning calorimeter at a heating rate of $40°C/min$, was $125°C$. It was also found that the glass transition temperature of the postcured resin did not change upon short term exposure up to $210°C$. Thus, it was assumed that exposure to testing temperatures did not cause a change in the degree of crosslinking of the materials. In addition to the unfilled resin, composites containing different percentage by volume of microspheres were prepared. The compositions of the prepared specimens were verified by measuring the densities of randomly chosen samples using a Beckman Model 930 pycnometer.

The polystyrene composites were prepared by mixing injection
molding granules of the polystyrene with the appropriate volume of
glass microspheres in a two roll mill mantained at 160-170°C. Both
the time and the temperature of the mixing operation were held constant and a rough-surfaced flat sheet was produced. This material
was then placed in a compression mold at 200°C and a pressure of
3.5×10^6 N/m^2 for 15 min to produce smooth flat sheets 13.5 cm
long by 5.7 cm wide by 0.27 cm thick. Test specimens of unfilled
polystyrene and composites containing 10, 20 and 30 per cent by volume of glass microspheres were then cut from these sheets.

Stress relaxation experiments were performed on an Instron model 1112 equiped with a thermostatically controlled oven. The imposed strain was constant in all experiments and was equal to 0.0015.

RESULTS AND DISCUSSION

In Figure 1 stress relaxation moduli relative to unfilled epoxy resin obtained at different temperatures ranging from T = 80°C
to T = 140°C, are reported. The data for composites are not shown
but have a similar behaviour. The results are typical of polymers
and polymer composites of this kind, producing a glassy plateau at
low temperatures, a strong temperature dependence in the transition
zone and a rubbery plateau at elevated temperatures.

These data, for each filler content, have been shifted hori-
zontally using a time-temperature superposition procedure, to obtain
the master curve shown in Figure 2 at a reference temperature of
80°C. The effect of the filler is to increase the modulus and to
shift the apparent glass transition temperature to higher values
(see Table 1). This result is in agreement with previously reported
data[10]. It has been proposed[4,10] that the glass microspheres could
induce, in the thermosetting resins, an effect equivalent to a higher degree of chemical crosslinking. It is probable than in the
vicinity of the heterogeneous glass surfaces an increased degree
of crosslinking occurs, thus reducing the mobility of the polymer
chains in the cured material. This is in consistent with other
findings reported in literature for systems showing a possibility
of chemical reaction[19] or a physical modification of the matrix[20].

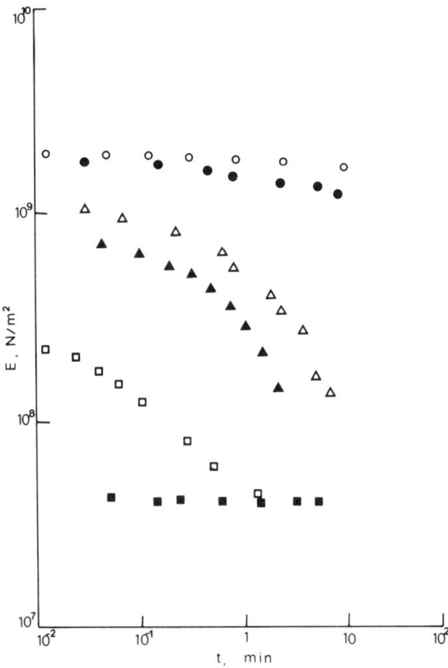

Figure 1. Stress relaxation moduli, E, versus time, t, for unfilled epoxy resin at T = 80°C (o), T = 100°C (●), T = 120°C (Δ), T = 130°C (▲), T = 135°C (□) and T = 140°C (■).

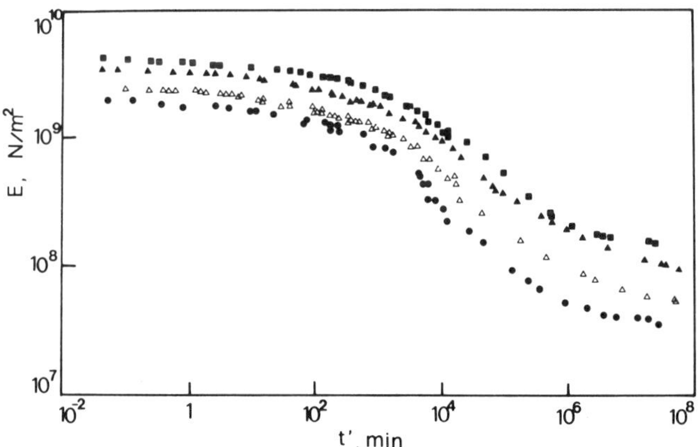

Figure 2. Stress relaxation modulus, E, versus reduced time, t', for Epoxy/glass bead composites at φ = 0 (●), φ = 0.1 (Δ), φ = 0.2 (▲) and φ = 0.3 (■).

TABLE 1

Glass transition temperatures for PHEMA/Glass bead and
Epoxy/Glass bead composites at different filler content

ϕ	PHEMA/Glass beads* T_g (°C)	Epoxy/Glass beads T_g (°C)
0	106	125
0.1	109	130
0.2	112	135
0.3	113	135

* see Reference 22

The shift factors used for the superposition procedure are shown
in Figure 3 in an Arrhenius plot. In the low temperature region,
where the material is in the glassy state (i.e. 80-125°C), the plot
is linear and an activation energy of 19 K cal/mole can be calculated from the slope. The shift factors are independent of the filler
content. This implies that, in the linearly viscoelastic region,
the relaxation behaviour of the polymeric composites is only dependent on the matrix properties. All the data were obtained at an
imposed strain of 0.0015, which is within the limits of the linear
response of the materials studied[21]. Viscoelastic data reported
previously on PHEMA/glass bead composites[22], obtained in a non-linear
region of viscoelasticity (about 0.14 strain), exhibited a strong
dependence of the shift factors on filler content. The limiting values of the tangent moduli in the glassy and rubbery states can be
correlated with filler volume fraction using the Kerner equation[23]
for the glassy state and the Eilers equation 24 for the rubbery state. The Kerner equation may be written as[23]:

$$E_r = \frac{E_c}{E_p} = \frac{1 + AB \, (\phi_f)_{eff}}{1 - B \, (\phi_f)_{eff}} \qquad (1)$$

where E_c is the elastic modulus of the composite, E_p is the elastic
modulus of the polymer and A, B and $(\phi_f)_{eff}$ are given by:

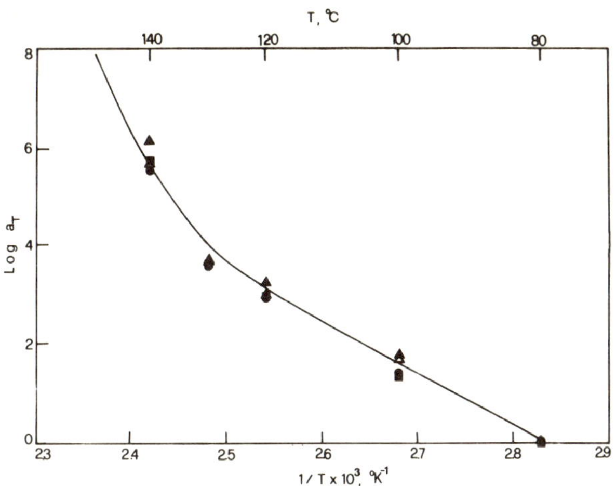

Figure 3. Shift factor a_T versus temperature for Epoxy/glass bead composites at $\phi=0$ (●), $\phi=0.1$ (Δ), $\phi=0.2$ (▲) and $\phi=0.3$ (■).

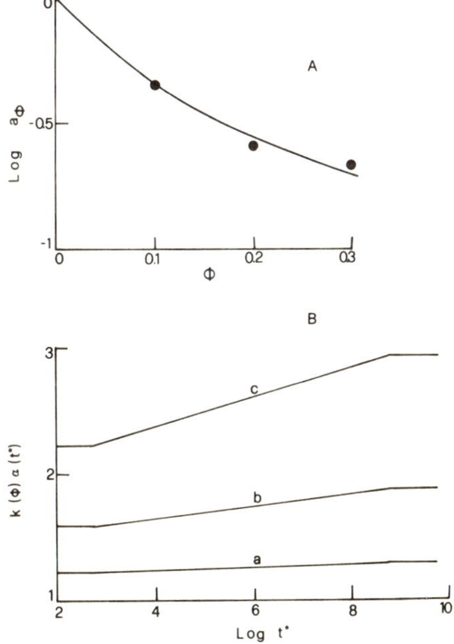

Figure 4. a) Horizontal shift factor a_ϕ for Epoxy/glass bead composites at different filler content
   b) Vertical shift factor, according to eq. 6, for Epoxy/ /glass bead composites at $\phi=0.1$ (a), $\phi=0.2$ (b), $\phi=0.3$ (c) relative to the unfilled epoxy.

$$A = \frac{7 - 5\nu_p}{8 - 10\nu_p} \tag{2}$$

$$B = \frac{(E_f/E_p) - 1}{(E_f/E_p) + A} \tag{3}$$

$$(\phi_f)_{eff} = \left[1 + \frac{(1 - \phi_m)}{\phi_m^2}\,\phi_f\right]\phi_f \tag{4}$$

where ν_p is the Poisson ratio of the polymer, E_f is the elastic modulus of the filler, ϕ_m is the maximum packing fraction for the filler and ϕ_f is the actual volume fraction of the filler.

The Eilers equation for the rubbery state may be written as (24)

$$E_r = (1 + \frac{1.25\,\phi f}{1 - \phi_f/\phi_m})^2 \tag{5}$$

Values of $\nu_p = 0.3$, $E_f/E_p = 20$ and $\phi_m = 0.637$ have been used[4].

The superposition of the stress relaxation data for different filler concentrations will be now described. The horizontal shift along the time scale, a_ϕ, was carried out by translating the midpoints of the transition zones of the composites stress relaxation moduli curves to the same time as that of the unfilled material. The values of a_ϕ are reported in Figure 4a. The particulate fillers, a part from causing the shift due to the increase in glass transition, increases the relaxation moduli in both the glassy and rubbery state. The reinforcing effect can be taken into account by using the Kerner equation in the glassy state, the Eilers equation in the rubbery state and a linear combination of the two in the transition zone (Figure 4b). Combining the two contributions, one can generate a single master curve for the relaxation modulus of the particulate filled system, $E_c(t^*)$. The analytical form is given by:

$$E_c(t^*) - K(\phi)\,\alpha(t^*)\,E_m(t^*) \tag{6}$$

where $k(\phi)$ is the Kerner equation, $\alpha(t^*)$ is the proportionality factor used to account for the change in the matrix from the glassy to the rubbery state (i.e. $k(\phi)\alpha(t^*)$ corresponds to the Eilers equation in the rubbery state), $E_m(t^*)$ is the relaxation modulus of the matrix at the time t^* given by $t^*=t'\,a_\phi$.

In Figure 5 all the relaxation data reduced according to equation 6 are shown at a reference temperature of $80°C$ and relative to the unfilled epoxy. The stress relaxation moduli, $E(t)$, of polystyrene/glass bead composites have been measured in the range $60-80°C$. The imposed strain was 0.0015, which is within the limit of linear viscoelasticity[21]. The values of relaxation moduli for polystyrene are reported versus time at three different test temperatures in Figure 6. Similar results have been obtained with the composite stystems. These curves were shifted according to the time-temperature superposition principle and the results are shown

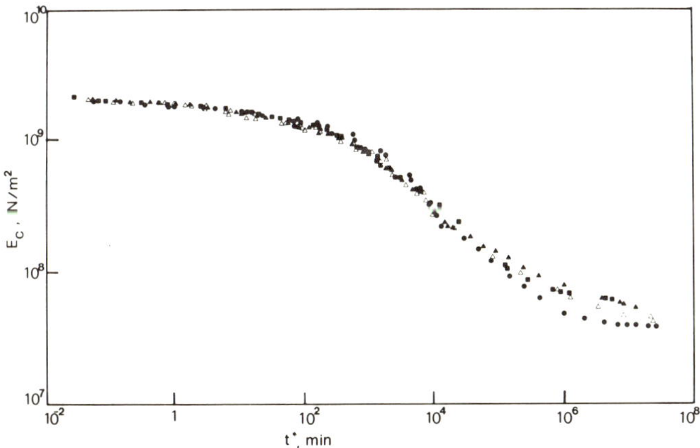

Figure 5. Reduced relaxation curves at a reference temperature
           of $80°C$ for Epoxy/glass bead composites at $\phi=0.1$ (Δ),
           $\phi=0.2$ (▲), $\phi=0.3$ (■) relative to the unfilled epoxy
           (●).

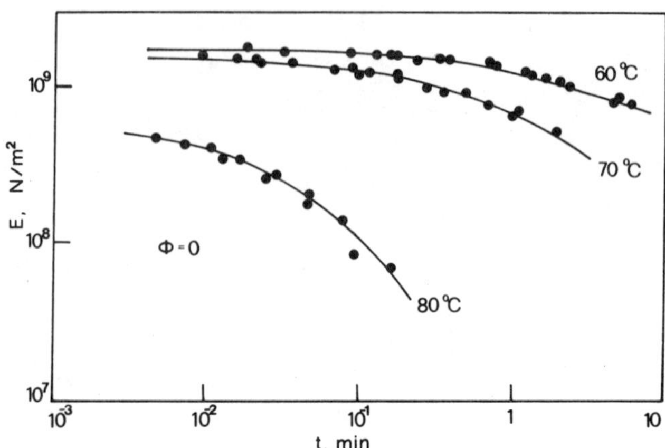

Figure 6. Stress relaxation moduli, E, versus time t, for unfilled
polystyrene at different temperatures

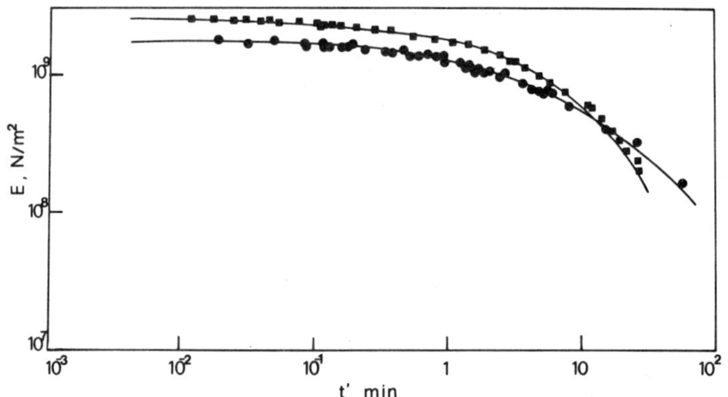

Figure 7. Stress relaxation modulus, E, versus reduced time, t',
for unfilled polystyrene (●) and polystyrene/glass
bead composites at φ=0.3 (■)

in Figure 7 for unfilled polystyrene and composites containing 30 per cent of glass beads at a reference temperature of 60°C. The shift factors used for all materials studied are reported in Figure 8 indicating once more no effect of the filler on the magnitude of the shift factors. Moreover for the system the presence of the beads does not modify the T_g of the polymer. The reduced relaxation curves have been also shifted vertically to take into account the mechanical effect of the filler on the moduli. The value of $K_{(\phi)}$ are reported in Table 2. The resulting master curves are shown in Figure 9 as reduced relaxation moduli versus reduced time. It can be observed that while all the data are superimposed at low times, there is a significant difference in relaxation moduli between unfilled and filled polymers in the transition region. This behaviour is similar to that previously reported for the shear viscosity of glass bead suspensions in non-Newtonian liquids. One possible cause fot the stronger dependence of the relaxation modulus (and also perhaps the viscosity in the other case) is the fact that the experiments were run at constant total deformation. Since the presence of rigid fillers increases the local stress level, the filler effect is tantamount to running the experiments at increasing higher stress level. The faster rate of decline of the modulus at the higher stress level is plausible and is consistent with the apparent decrease in heat deflection temperature (HDT) with increasing load.

TABLE 2

Vertical shift factor $k(\phi)$, used for constructing the Figure 9, for polystyrene/glass bead composites at different filler content.

ϕ	$k(\phi)$
0.1	1.08
0.2	1.3
0.3	1.61

In conclusion it has been shown that while in the glassy region the mechanical properties of composites for both thermoplastic or thermosetting matrices are very similar, in the transition

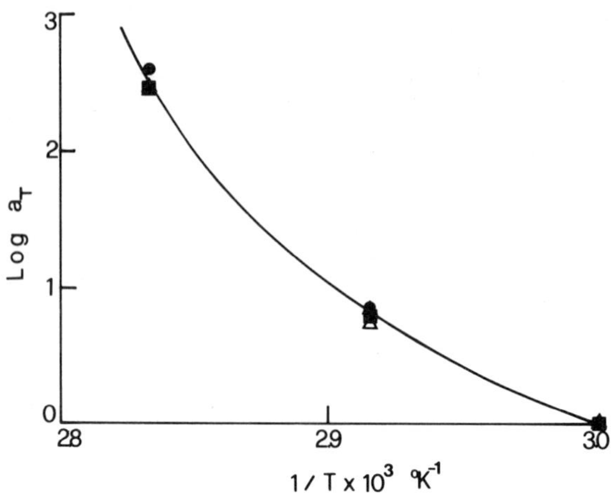

Figure 8. Shift factor a_T versus temperature for polystyrene/glass
bead composites at ϕ=0 (●), ϕ=0.1 (Δ), ϕ=0.2 (▲) and
ϕ=0.3 (■).

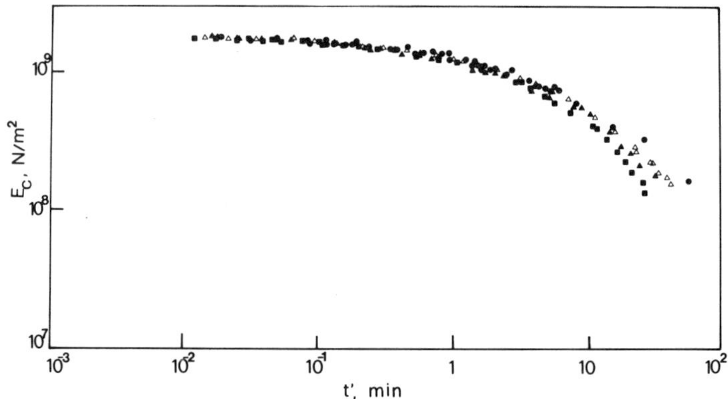

Figure 9. Reduced relaxation curves at a reference temperature
of 60°C for polystyrene/glass bead composites at
ϕ=0 (●), ϕ=0.1 (Δ), ϕ=0.2 (▲) and ϕ=0.3 (■).

region the mechanical behaviour is completely different. In fact
while the addition of glass microspheres to thermoplastic polymers
has no effect on the glass transition temperature of the system[25],
the presence of a second phase enhances the T_g of composites with
thermosetting matrices. Moreover the filler does not affect the
values of the shift factors used in the time temperature superposition procedure of all the system studied. The elastic properties
of the thermosetting composites were correlated in the glassy state
by using the Kerner equation and in the rubbery state by using the
Eilers equation.

REFERENCES

1. T. L. Smith, Trans. Soc. Rheol., 3, 13 (1959).
2. L. Nicolais, E. Drioli and R. F. Landel, Polymer, 14, 21 (1973)
3. J. L. Kardos, W. L. McDonnel and J. Raisoni, J. Macromol. Sci.
      Phys., B6, 397 (1972).
4. L.E. Nielsen "Mechanical Properties of Polymers and Composites"
      Marcel Dekker, New York (1974).
5. Yu. M. Malinskii, Uspekhi Khim., 39, 1511 (1977).
6. A. N. Cherkasov, J. Kolarik and J. Janacek, Int. J. Polym.
      Mater., 5, 295 (1977).
7. J. Kolarik, S. Hudecek, F. Lednicky and L. Nicolais, J. Appl.
      Polym. Sci., 23, 1553 (1979).
8. J. A. Manson and E. H. Chiu, J. Polym. Sci. C, 41, 95 (1973)
9. D. H. Droste and A. T. Di Benedetto, J. Appl. Polym. Sci., 13,
      2149 (1969).
10. C. Migliaresi, L. Nicolais, L. Nicodemo and A. T. Di Benedetto,
      Polym. Eng. Sci., in press.
11.   Yim, R.S. Chahal and L.E. St. Pierre, J. Colloid. Interface
      Sci., 43, 583 (1973).
12. J. D. Ferry, "Viscoelastic Properties of Polymers", J. Wiley
      & Sons, New York (1970).
13. F. R. Schwarzl, H.W. Bren, C. Nederveen, G. Schippert, L. Struik
      and C. W. Van der Wal, Rheol. Acta, 5, 270 (1956).
14. L. E. Nielsen, Trans. Soc. Rheol., 13, 141 (1969).
15. Z. Rigbi, Trans. Soc. Rheol., 9, 379 (1965).
16. W. Voigt, Abhandl Gottingen Wiss, 36 (1889).
17. L. Boltzman, Pogg. Ann. Physik, 7, 624 (1976).
18. J. Maxwell, Phylos. Trans. Royal Soc. of London, 157, 52 (1857)
19. C. W. van der Val, H. W. Brec and F. R. Schwarzl, J. Appl. Po-
      lym. Sci., 9, 2143 (1965).

20. E. Cuddihy and J. Moacanich, J. Polym. Sci., Part C, $\underline{14}$, 313 (1966).
21. J. V. Gauchel, DSc Thesis, Washington University, St. Louis, Mo., USA (1972).
22. D. Acierno, L. Nicolais, V. Vojta and J. Janacek, J. Polym. Sci., Polymer Physics ed., 66, 703 (1975).
23. E. H. Kerner, Proc. Phys. Soc., $\underline{69B}$, 808 (1956).
24. H. Eilers, Kolloid Z. Z. Polym., $\underline{97}$, 313 (1941).
25. L. Nicolais, R. F. Landel and P. G. Orsini, Polym. J., $\underline{7}$, 259 (1975).

STUDIES ON THE PROPERTIES AND STRUCTURE OF ADHESIVE FILMS
CONTAINING ELASTOMER/RESIN BLENDS

J. Kozakiewicz, B. Kujawa-Penczek and P. Penczek

Institute of Industrial Chemistry
01-793 Warszawa, Poland

Novel interpretation is given for the maximum of
the adhesive bond initial strength observed at particular elastomer/resin ratio for contact adhesives containing elastomer/resin blends.

It is suggested that this phenomenon is due to the
solvent retention rather than to the film structure which
appeared to be homogeneous. The results of solvent retention and film tack measurements obtained for polychloroprene solvent-based resin-modified adhesives are presented to prove this explanation.

INTRODUCTION

The properties and the internal structure of adhesive films
containing elastomer/resin blends have been studied carefully by several authors since the results of these investigations can be of
great industrial interest. It was found that the tack of most pressure-sensitive solvent-based adhesives containing elastomer/resin
blends showed maximum at particular resin content[1-5]. On the basis
of electronmicroscopic data and mechanical properties of adhesive
films this phenomenon was explained as being due to the phase separation occuring inside a broad interval of the resin concentrations
in elastomer or the elastomer concentrations in resin[4,6]. This ex-

planation was questioned by other authors[7-10] who suggested rather an optimum wettability of adhesive film as responsible for maximum tack.

Despite pressure-sensitive adhesive, only little work has been done so far on solvent-based contact adhesives also containing elastomer/resin blends. For polyurethane adhesives it was previously shown that initial strength of the adhesive bonds reached maximum value at particular elastomer/resin ratio[11].

We attempted to investigate this more closely and found that the same phenomenon could be observed as well for other contact adhesives obtained from both crystallizing and non-crystallizing rubbers. The results we got for epichlorohydrin rubber adhesives have already been published[12] and those for polychloroprene rubber adhesives are presented in this paper.

EXPERIMENTAL

To compound adhesives, 18% solutions of resin (butylphenolic resin Viaphen PA-101 of melting point 50-60°C, Reichhold-Albert Chemie or coumarone resin of m.p. = 108-110°C) and polychloroprene (Neoprene AD-20, Bayer) in suitable solvents; ethyl acetate/gasoline/toluene = 1/1/1 or cyclohexane/acetone/toluene = 5/3/2 were mixed in various proportions. Polychloroprene solutions contained ca. 1% ZnO and 1% MgO.

Tensile "butt" strength of adhesive bonds was determined at room temperature using "Instron" dynamometer. Aluminium specimens applied for this test were etched in $K_2Cr_2O_7/H_2SO_4$ solution according to the typical procedure.

For solvent retention measurements, small amounts (approximately 0.4 g) of adhesive were put onto glass shields and weight changes with time were noted.

Tack of adhesive films was determined by "rolling ball" method. The ball (4 g glass ball of 14 mm in diameter) was rolled on the adhesive film applied to the glass and the distance it stopped was the measure of the film tack.

To investigate the structure of adhesive films, SEM technique
was used. Adhesives were cast onto this glass shields. Samples
were mounted on specimen studs and vacuum coated with a thin layer
of copper.

RESULTS AND DISCUSSION

In Figure 1, the initial strength of Al-Al bonds obtained using
butylphenolic resin-modified polychloroprene rubber adhesive containing various solvent mixtures is plotted against resin content in
the adhesive film expressed in phr (parts per hundred parts of rubber). It can be noticed from Figure 1 that the maximum of initial
bond strength occurs for both adhesives though various solvent compositions were used for their compounding.

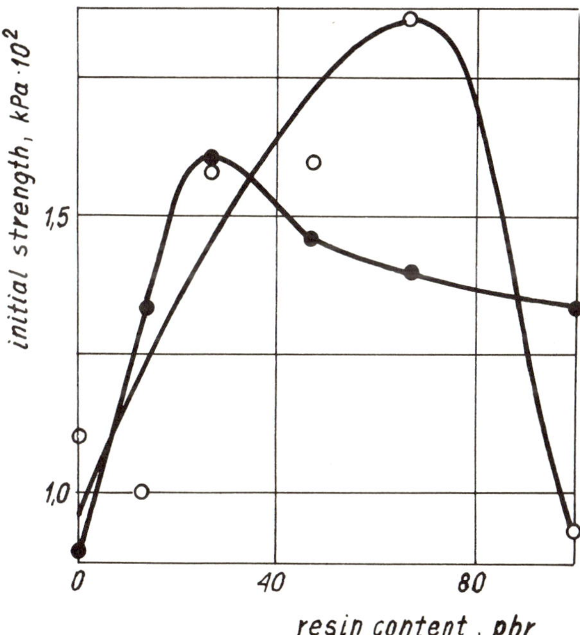

Figure 1. Initial butt strength of Al-Al bonds vs butylphenolic
           resin content in adhesive films.
           ● – solvent composition EA/Gas/T = 1/1/1
           o – solvent composition CH/Ac/T = 5/3/2

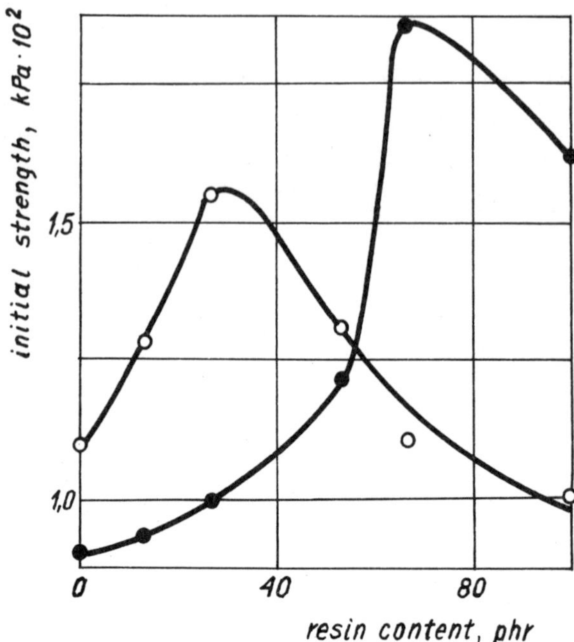

Figure 2.   Initial butt strength of Al-Al bonds vs coumarone resin
             content in adhesive films.
             ● – solvent composition EA/Gas/T = 1/1/1
             o – solvent composition CH/Ac/T = 5/3/2

The same effect was observed for polychloroprene adhesives
containing coumarone resin (see Figure 2) and for resin-modified
polyepichlorohydrin adhesives[12].

Since electronmicroscopic studies of adhesive films did not
show their structure to be heterogeneous, it was concluded that the
interpretation of the initial strength maximum should not be based
on phase-separation phenomenon as suggested before.

Previously, some data have been published for pressure-sensitive
adhesives[13] suggesting that both tack and peel strength are strongly
affected by little amount of solvents left in adhesive films after
drying. For solvent-based contact adhesives investigated in the
present paper, the initial strength of the bonds was measured shortly
after assembling and only 5-30 min open time was used to evaporate

solvents from adhesive films. The retention of solvents in the
adhesive films was then very high and we believed it could have been
the factor responsible for the occurrence of initial strenght maximum. Therefore, further experiments were carried out to clarify
this problem.

Indeed, interesting results were obtained when solvent retention in the adhesive films measured after 20 min open time was plotted against resin content. For both epichlorohydrin[12] and polychloroprene rubber, (see Figure 3 and 4) solvent retention showed maximum and a good correlation between initial bond strength and solvent
retention was observed.

The correlation data are presented in Table 1 and exemplified
graphically in Figure 5.

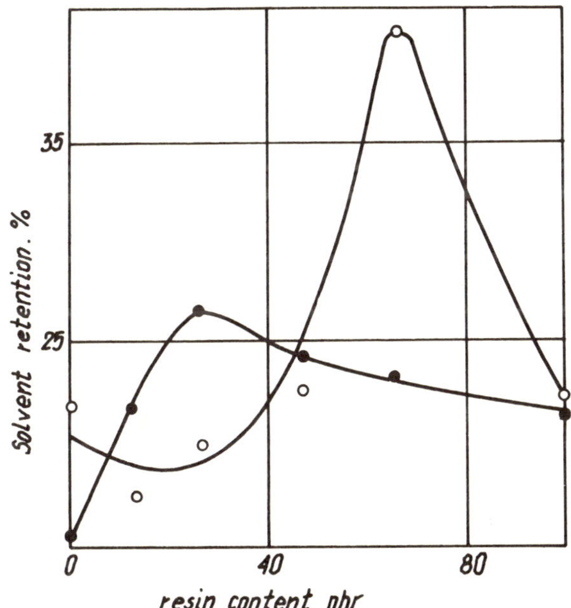

Figure 3. Solvent retention vs butylphenolic resin content in
           adhesive films.
           ● – solvent composition EA/Gas/T = 1/1/1
           o – solvent composition CH/Ac/T = 5/3/2

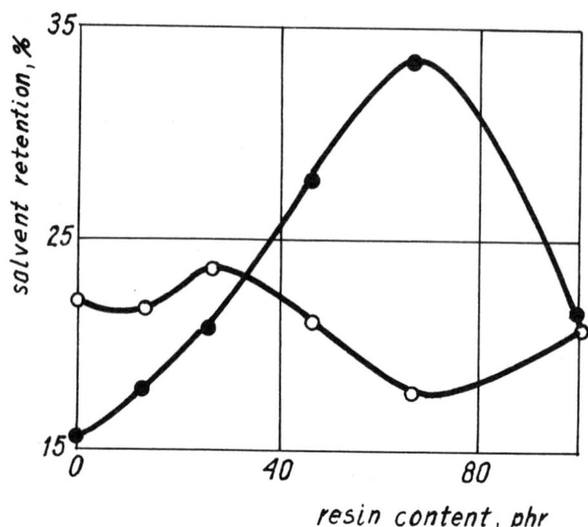

Figure 4.   Solvent retention vs coumarone resin content in adhesive
films.
● – solvent composition EA/Gas/T = 1/1/1
o – solvent composition CH/Ac/T = 5/3/2

TABLE   1

Correlation data for initial bond strength (kPa) and solvent retention (%).

Resin	Solvent composition	Correlation coefficient ν	Parameters in correlation equation $y = ax + b$	
			a	b
buthylphenolic	EA/Gas/T=1/1/1	0,991	6,26	−5,48
buthylphenolic	CH/Ac/T =5/3/2	0,658	3,17	59,23
coumarone	EA/Gas/T=1/1/1	0,871	5,08	7,38
coumarone	CH/Ac/T =5/3/2	0,473	4,22	32,23

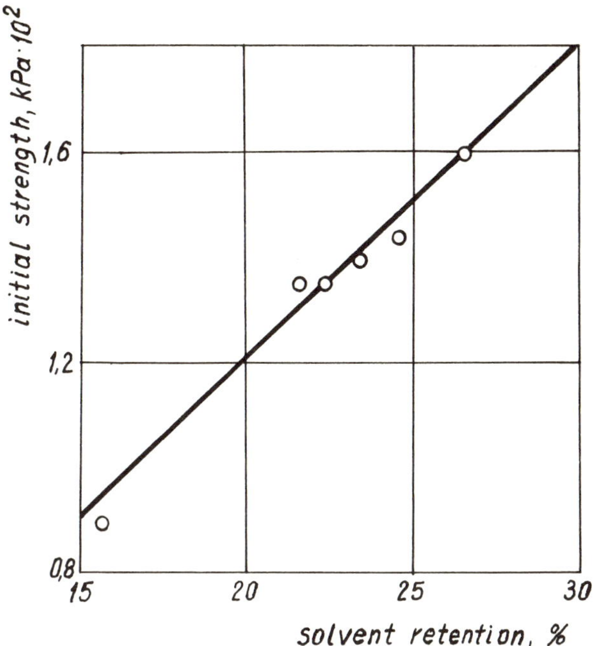

Figure 5.   Example of correlation between initial strength and sol-
vent retention
Resin-butylphenolic resin Viaphen PA 101
Solvent composition - EA/Gas/T = 1/1/1

The better correlation for adhesive containing ethyl acetate/
gasoline/toluene mixture than for those containing cyclohexane/ace-
tone/toluene mixture is difficult to obtain for a clear explanation,
if it is assumed that both solvent mixtures have approximately the
same solubility and hydrogen bonding parameters and lie very close
on the solubility diagram of polychloroprene. The reason can be
different crystallization rates of polychloroprene in both mixtures
but this suggestion needs further studies before we can prove it.
At present we are continuing investigations on the initial strength
solvent retention relationship in polyurethane adhesives of high
crystallization rate and soon expect to get an answer to this question.

Despite the good correlation observed for initial strength and
solvent retention, there seems to be no effect of retention on the
tack of adhesive films as measured by rolling-ball method. The tack

did not reveal the maximum over the investigated range of resin content in the adhesive film (see Figure 6). This can be easily understood supposing that only thin dry layer on the surface of adhesive film is disturbed while tack is measured[14]. Despite initial strength of the bond, tack cannot then be affected by solvents which are present under this thin layer. The lack of tack maximum is also a further evidence for the homogeneous structure of adhesive films.

All the data presented above proved that maximum of initial bond strength should presumably be explained in terms of viscoelastic or mechanical properties of adhesive film rather than of its internal structure.

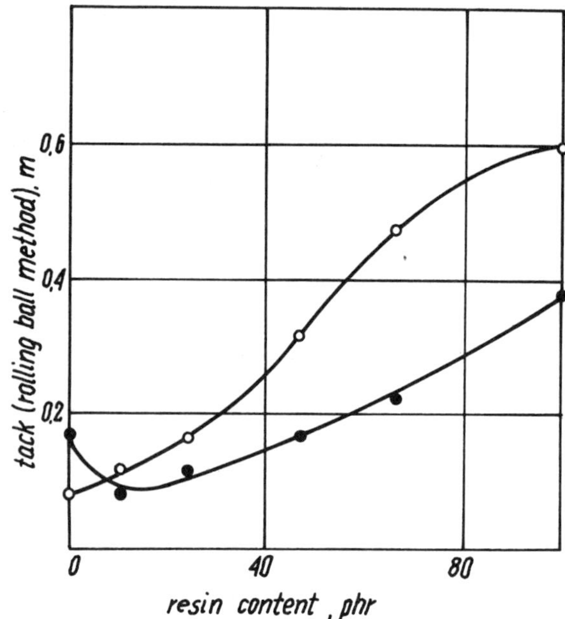

Figure 6.   Tack vs butylphenolic resin content in adhesive films
● – solvent composition EA/Gas/T = 1/1/1
o – solvent composition CH/Ac/T = 5/3/2

CONCLUSIONS

The following interpretation can be given for the observed strong effect of elastomer/resin ratio on the initial bond strength in solvent-based contact adhesives. After applying the adhesive,

the concentration of solvents in the adhesive film (i.e. solvent
retention) decreases very quickly, affecting at the same time an increase in the mechanical strength (i.e. cohesion) of the adhesive
film and a decrease in its wettability. Since it was shown above
that solvent retention depends not only on the adhesive open time,
but also on the resin content, cohesion and wettability of the adhesive film are considerably affected by elastomer/resin ratio in opposite directions. It is believed that this process must result
eventually in the occurrence of initial bond strength maximum.

The schematic representation of these conclusions is presented
in Figure 7. The direct effect of elastomer/resin ratio, which is
not as important to the initial bond strength as the complex effect
through solvent retention, is marked with a broken line in Figure 7.

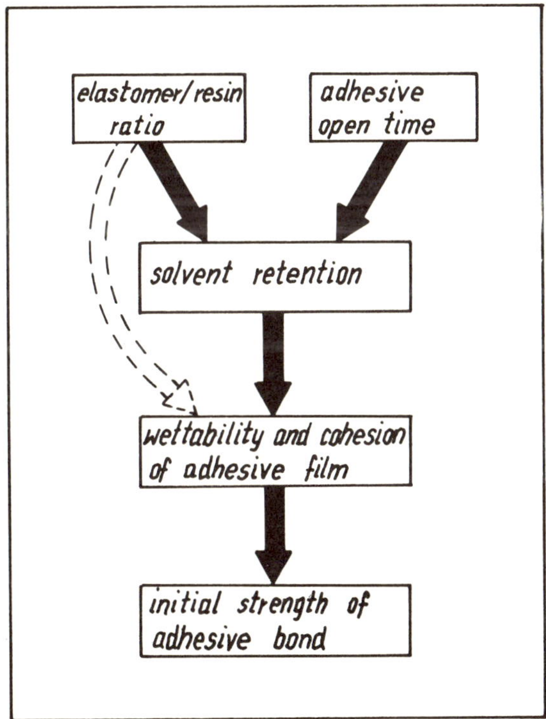

Figure 7. Effect of various factors on initial strength of adhe-
           sive bonds.

REFERENCES

1. F.H. Wetzel, Rubb. Age., 82, 291 (1957).
2. F.H. Wetzel, B.B. Alexander, Adh. Age, 7, 1, 28 (1964).
3. C.W. Hock, A.N. Abbott, Rubb. Age, 82, 471 (1957).
4. C.W. Hock, J. Polym. Sci. C, 3, 139 (1963).
5. C.W. Hock, Adh.Age, 7, 3, 21 (1964)
6. K. Kamagata, H. Kosaka, K. Hino, M. Toyama, J. Appl. Polym. Sci., 15, 483 (1971).
7. C.A. Dalquist in "Adhesion Fundamentals and Practice", Ministry of Technology, McLaren, London 1969.
8. S.S. Wojucki "Autohesion and Adhesion of High Polymers", Interscience Publ., New York, 1963.
9. K. Fukuzawa, T. Kosaka, Prepr. of 6th Symp. on Adhesion and Adhesives, Osaka 1968.
10. K. Fukuzawa, T. Kosaka, Prepr. of 7th Symp. on Adhesion and Adhesives, Tokyo 1969.
11. B. Kujawa-Penczek, Polimery, Warsaw, 13, 215 (1968).
12. J. Kozakiewicz, P. Penczek, Adh. Age, 20, 7, 29 (1977).
13. M. Toyama, T. Ito, M. Suzuki, H. Moriguchi, Prepr. of 9th on Adhesion and Adhesives, Tokyo 1971.
14. K. Kamagata, T. Saito, M. Toyama, J. Adhesion, 2, 279 (1970).

POLYMERIC MODIFIER FOR FILLED POLYPROPYLENE

A. Galęski and R. Kaliński

Centre of Molecular and Macromolecular Studies

Polish Academy of Sciences, 90-362 Łódź, Poland

The present development of the modification of polyolefins by filling has been reported. Based on the literature data and the authors results, it has been concluded
that the further improvement of the mechanical properties
of filled polyolefins cannot be achieved by further increase of the adhesion of filler to polymer. The apparent
increase of the volume of filling and stiff joint filler-
polymer, both due to the increase of the adhesion, limit
the requested changes of the mechanical properties.

The addition of a liquid interfacial layer has been
proposed in order to reduce these effects. The behaviour
of a system containing: filler particle, a liquid interfacial layer and the polymer matrix under tensile and
impact tests has been analysed showing that the system
should exhibit high ultimate elongation and high impact
strength.

Polypropylene filled with chalk has been taken as
the exemplary system and oligomer of ethylene oxide as
the liquid modifier. The sample containing modified
chalk shows: the independence of elastic modulus of
chalk content, high ultimate elongation, high impact
strength and decreasing tensile strength with increasing chalk content. It has been concluded that the
superior function of the liquid modifier in the system

431

of polyolefin and filler is to inhibit the crack initiation and its further propagation.

INTRODUCTION

For many years the selection of polymer fillers was mainly
based on the experimental trials. But, such a way does not lead to
the real progress in solving the problems of polymer modification
by means of the fillers. Some physical properties of polymers filled
with glass fibers and carbon whiskers showed that the main cause of
imperfections of such system is a weak adhesion between filler and
polymer[1-6].

In order to improve the adhesion it is necessary to know all
the interactions at the interphase. All types of interactions
between filler and polymer can be briefly divided into four groups
depending on the type of filler and its interaction with the matrix[7].

1. The simple physical inclusions of filler particles or their
    aglomerates in nonpolar polymer matrix cause the weakening of
    the system acting as a diluent.
2. Physical inclusions of filler particles with a simultaneous
    wetting of the particle surface by polymer cause some stiffe-
    ning of the system i.e. increases Young modulus and decreases
    ultimate elongation.
3. Strong physical adhesion of polymer to the surface of the filler
    particles causes a considerable reinforcement of the system.
4. Chemical bonding of the matrix to the filler particles (e.g. in
    a system of natural rubber and carbon black causes a tremendous
    change in the mechanical properties).

When selecting a filler for a given polymer the type 3 or 4
of interaction is obviously preferred since the strongly bonded
systems have relatively good mechanical properties. Basically there
are four ways to improve the adhesion between polymer and filler:
- an increase of the total adhesion forces can be achieved by the
  reduction in size of filler particles and/or the development of
  their surfaces (increasing roughness or even porosity).
- an increase of the adhesion forces is caused by an increase in
  wetting of the filler particles by the polymer matrix. (Type 2
  or 3 interaction). The modification of the polymer surface, being
  in contact with the filler, is practically difficult, however,
  for polyolefins several methods have been elaborated:

chemical oxidation, treatment with a flame plasma, electrical
discharge[8-13] or activation of polymer in bulk by grafting polymer
molecules with some polar functional groups (carboxyl-, chlorosul-
phonyl-, chlorophosphate- or halogens-groups)[14-17]
- an increase in the adhesion could be achieved by such a modification of the filler surface that in the process of mixing at elevated temperature causes chemical changes of the neighbouring
  macromolecules making possible thus the strong adhesion of the
  physical nature (type 3) or chemical bonding of the filler particles to the matrix,
- an increase of the adhesion is achieved by the introdution of the
  third component which exhibits a good adhesion to the filler and
  polymer. If each particle of the filler is surrounded by a thin
  layer of the third component, it acts as a glue.

In the case of polar polymers the adhesion to the most mineral
fillers is satisfactory. Other polymers, including polyolefins do
not exhibit a strong adhesion to the mineral fillers. The coupling
(glue joint) falls apart after few hours.

Many various types of fillers are used for the filling of the
plastic materials[18-20]. In the case of polyolefins the commonly
used fillers are: talc[21-25], asbestos[26-29], kaolin[30-33], metal
powders[34,35], metal oxides[37,38], mica[29,39,40], aluminium hydroxide[41],
carbon black[42,43], silica[44], glass in the form of spheres or
fibers[22,29,45-48], cellulose[49], straw with wood chips, wood sawdust
and gypsum[50], lignin[51], wood flour[52], peanut shells[53]. Usually one
gets an increase of Young modulus, or an increase of hardness but
a decrease of tensile strength and an increase of brittleness of a
composite. Such properties result from the adhesion character at
the interface. The interaction at the interface can be changed in
many ways giving sometimes good results. In the case of polyolefins
the covering of the filler particle with various materials is
usually applied.

Organo-Titanate layers

Organo-Titanate layers are used for most fillers to polyolefins[54-56]. If applied, in the amount of 0.5 - 3% they cause an
increase of the adhesion between polyolefin and filler, followed
by an increase of the elastic modulus and a decrease of impact
strength and ultimate elongation.

Silane layers

Interesting materials to cover the mineral fillers for polymers are organosilanes[57-62]. The silanes used have two functional groups R' and OR. Usually R' is a reactive organic group amine, vinyl, epoxide, methacrylate bonded to silicon atom via a short aliphatic chain; OR group is hydrolysing alcoholate group. Organosilanes are bonded to the surface of the filler through OR groups while R' groups react with the polymer matrix. OR and R' groups are selectively chosen for a particular filler and polymer.

Organic layers

Stearic acid and stearates of Ca, Na, Ba and their compositions are the most popular modifiers. These compounds function as coupling agents of organic and mineral phase. They also show a lubricating property in processing.

Inorganic layers

A simple way of surface modification of the filler particles is the application of complexed aluminium or magnesium silicates[63,64]. They give some acidity to the system. Some of the fillers, being neutral at a normal temperature, show reactivity at processing temperature. Usually these are mineral magnesium silicates and clays[65-69]. A particular role is played by the metal oxides which are reduced during mixing process with polymers, causing a partial oxidation of polyolefins. Thus the polymer is able to create the adhesive bonds with the filler particles. Such specific properties exhibit: copper oxide[34], aluminium oxide[37,70] and chromium trioxide[44].

Polymeric layers

The most interesting way for the modification of the filler surface is to cover it with another polymer which exhibits good adhesion to the filler and to the matrix. The second polymer can be introduced as a monomer and polymerized according to the radical mechanism, initiated with conventional catalysts or according to cationic mechanism initiated by activated mineral surface. The

following monomers were used: piridyne, divinyl benzene, acrylic
acid, acrylates.

Polymeric layers, or perhaps oligomeric with molecular weight
500-800, are usually added up to 3% by weight of the filler. The
thickness of the layer is from 20 to 30 Å[57,71]. Polyolefins filled
with fillers modified in such a way have high impact strength. The
polymeric layer can be introduced in a separate technological process or, in the case of polyolefins, mixing them with a filler[72,73].
Some attempts have been made to use block copolymers containing
blocks with good adhesion to the filler and blocks with good adhesion
to polymer[74-76].

In order to improve the mechanical properties of filled polyolefins several other modifications were applied[77-81]. In all cases
the improvement of adhesion between filler and polymer is achieved.
It leads to the increase of elastic modulus, the decrease of ultimate elongation and the small decrease of tensile strength. In the
case of brittle layers surrounding filler particles, a decrease of
impact strength is also observed. Only for polymeric modifiers an
increase of impact strength is achieved.

Too strong adhesion between filler and polymer makes the composites hard but brittle. It was shown[82] that the crack initiated
in a tough phase, assuming a good adhesion to the polymer, can pass
through the interphase and further propagate undisturbed. However,
if the adhesion is weaker, the crack tip is oriented along the interphase and usually the fracture ends passing few particles. That
fracture mechanism was described by Gordon[83]. He showed that the
interfacial adhesion should not exceed one fifth of the matrix
strength. Then the cracks are directed along many interfacial layers
and the work necessary to fracture the sample becomes high. The
phenomena of cracking in the system with a very good adhesion is
illustrated by the following[84]: on the surface of a see sand there
is a thin layer of magnesium silicate, $MgSiO_3$, which was formed
during a long term interaction with see water rich in magnesium ions
Mg^{++}. This layer is responsible for the strong adhesion to polyethylene since it shows some acidity. The good adhesion manifests
itself because the crack initiated during the fracture experiment
(at a liquid nitrogen temperature) is propagated across a sea sand
particle. The surface of fractured sample is shown in Figure 1.
The surface is not well developed and is smooth. The tested samples
show low values of the impact strength. A layer of polyethylene

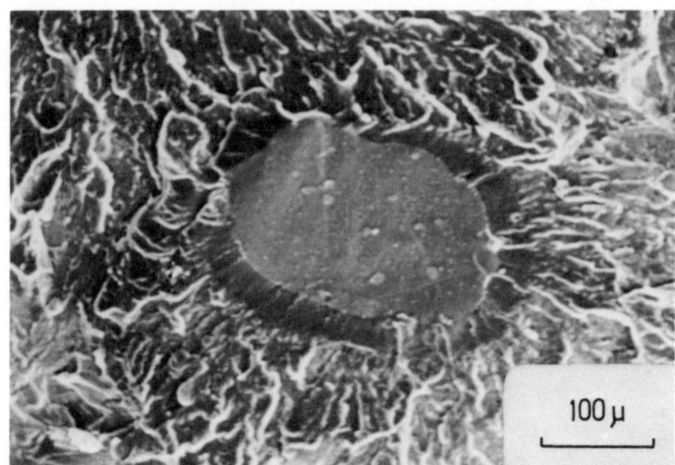

Figure 1. SEM of the surface of freeze fractured polyethylene
filled with sea sand .

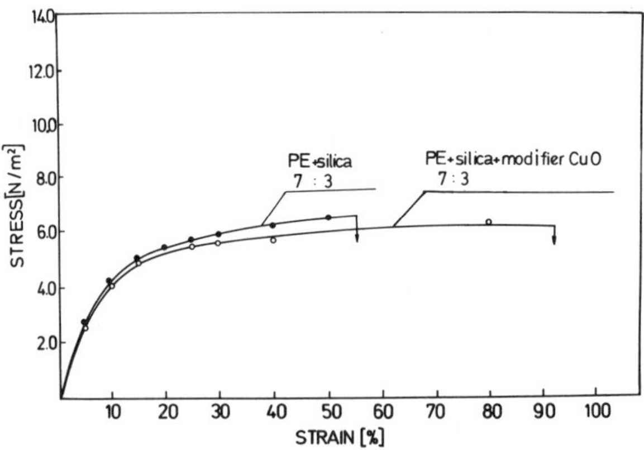

Figure 2. Stress-strain curves of samples of polyethylene filled
with silica and covered with copper oxide (CuO).

with changed supermolecular structure is seen around the sand particle.

We have developed similar good adhesion in a system of polyethylene of low density and silica covered with a thin layer of copper
oxide[85]. The mechanical properties of the system were improved, in
comparison with the composite prepared without the cover of copper
oxide (see Figure 2), the changes, however, are not very high
although the adhesion between polymer and matrix was increased dramatically.

Significant but not very large, profitable changes were found
for some other types of modification. Composites containing 50%
or more of the filler do not have properties characteristic for
thermoplastic materials but rather have an agglutinate-like type of
structure and hence the possible application in the building industry.
The cause of the improvement less than the one expected can be analysed based on Figure 3. Let us assume that there is a good adhesion between the filler particle and the matrix. When adhesion is
good, polymer changes locally its mechanical properties since the
polymer chains are bounded to a tough filler particle. That polymer
layer surrounding the filler particle gets the mechanical properties
rather closer to filler than to polymer. The stronger is the adhesion the thicker is the layer. Hence, increasing the adhesion
between the filler and the polymer one increases the apparent volume
of filling. The existence of that layer is also the cause of local
changes of the supermolecular structure of the polymer around the
filler particle. Moreover, the adhesive bond is now between two
stiff solid states due to the stiffening of the interfacial layer.
If the extentional force is applied to the system (see Figure 3)
the stress is concentrated in the polar areas of the filler particle.
The small deformation is enough to produce very high stresses in
those areas, shortly leading to breaking of the adhesive joint and
finally to the fracture of a sample. Summarizing one can conclude
that an increase of the adhesion between the filler and polymer
beside an increase of the tenacity of the material causes:
1. an increase of an apparent volume of filling,
2. a stiff inelastic ahesive joint between filler and polymer
   which is easily broken in stress concentration areas.
These factors limit the improvement of mechanical properties of the
composites by means of the increase of adhesion. The above analysis
showed that the further improvement of the mechanical properties is
possible by eliminating the two factors but not by further increase

of adhesion between the filler and polymer. It could only be done
by changing the properties of the interfacial layer maintaining the
good adhesion.

As modifier we propose a layer of a well wetting liquid both for
the filler and the polymer. A liquid layer of a sufficient thickness
could play a role of a soft elastic interfacial layer. The schematic situation is presented in Figure 4.

Let us consider the case of the extensional stresses. Similarly
as above mentioned, there are stress concentrations in polar areas;
however, even for a relatively high deformation, the adhesive contact polymer-liquid-filler is not lost. A liquid modifier moves
from equatorial to polar regions maintaining the contact. Since
the modifier is a liquid, a low molecular weight substance, the adhesive forces cannot reach far into the liquid. Hence, there would be
a thin interfacial layer and no large increase of the apparent volume
of filling would be observed. The adhesive joint polymer-liquid-
filler would now have viscoelastic properties. The crack initiated
on the surface of the sample, after reaching the liquid layer, will
be stopped and the energy of crack propagation absorbed. Only at a
very high crack propagation rate the liquid can crack. So the impact
strength will be improved. Liquid phase of a modifier should be
particularly effective in a reconstitution of the adhesive bonds
after the application of excessive forces. A substance chosen as
a modifier should have an high boiling (or degradation) temperature,
a low melting temperature, a low vapour pressure, a good adhesion
to the polymer and the filler both in expoitation and processing
conditions.

EXPERIMENTAL

We have chosen oligomer of ethylene oxide (OEO) of molecular
weight 300, boiling temperature about 260°C and low melting temperature (much lowered due to overcooling). The modifier does not
evaporate or degradate easily at processing temperature (about 200°C)
and remains liquid at exploitation temperatures. Oligomer of ethylene oxide shows a good adhesion to low and high density polyethylene and polypropylene. The technical process of covering of filler
particles is easy since the OEO is water soluble.

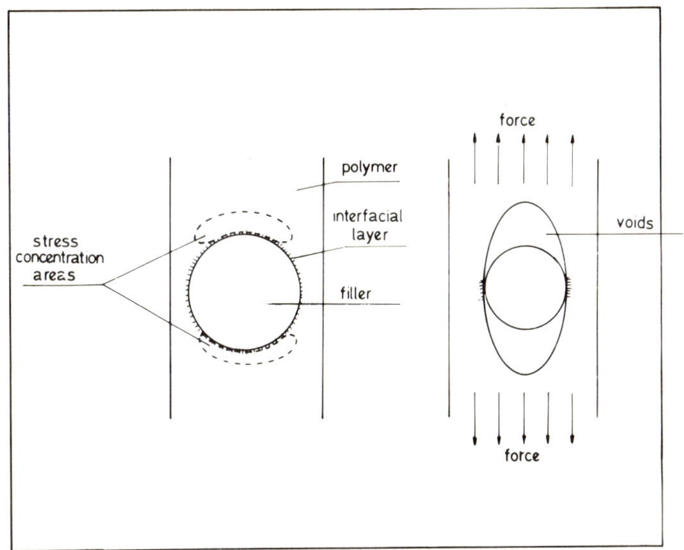

Figure 3. Scheme of the filler particle embedded in polymer matrix. Good adhesion is assumed. Stretching forces applied.

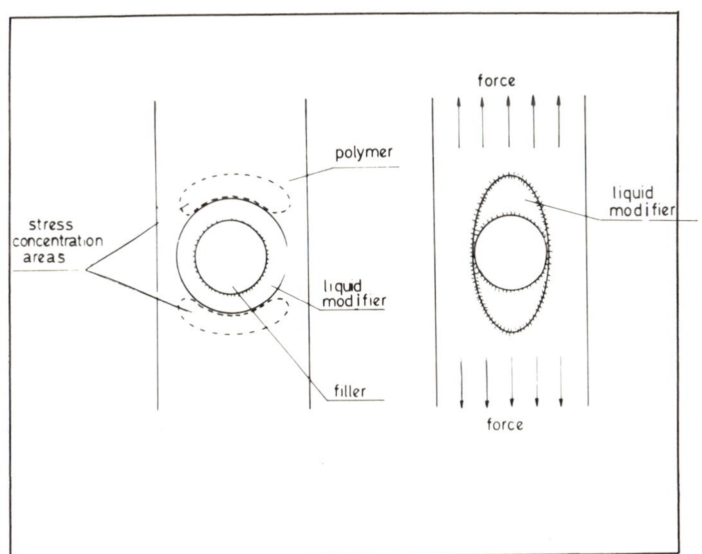

Figure 4. Scheme of the filler particle covered with liquid modifier embedded in polymer matrix. The behaviour of the system under tension is shown.

Isotactic polypropylene (PP) Malen J-400 (Polish product), low density polyethylene Politen (Polish Product), precipitated chalk and talc were used in further studies. Chalk and talc were characterized by the determination of the particle size distribution.

Modification of the filler was conducted in a water suspension by addition from 0 to 10% of oligomer per weight to the filler. Water was removed by evaporation and the filler was used for filling polypropylene by means of a screw extruder fitted with a static mixer on the orifice. The samples for tensile and impact experiments were prepared by melt pressing from granules of the composites. A series of samples using an unmodified chalk were prepared for comparison purpose.

RESULTS

Figures 5a, b, c, and d show the mechanical parameters of composites low density polyethylene (70%) and talc (30% by weight) depending on the modifier content. It follows from the graphs that the samples containing of 0,5-1,0% of modifier have the best properties. Particularly large changes are observed for the ultimate elongation and the impact strength for notched samples according to Charpy method. Knowing the particle size distribution of talc one can show that 0,5% to 1% weight of the modifier corresponds to the uniform layer around each particle 450-900 Å thick. Similar tests were performed for other compositions. In all cases the best mechanical properties were achieved using 0,5-1% (by weight of the filler) of the modifier. The ultimate elongation and impact strength for polypropylene filled with chalk (the ratio of polypropylene to chalk 6:4) modified with various amounts of oligomer of ethylene oxide and for unfilled polypropylene are presented in Table 1.

The best results are obtained for compositions containing 5 - 10% of OEO which gives 500 - 1000 Å thick layer around each particle of chalk.

A series of compositions of polypropylene and chalk containing 5 and 10% of oligomer of ethylene oxide with various degrees of filling were prepared. All modified samples exhibit a stress strain behaviour typical for plastics. The samples undergo necking and flow process during deformation. The stress-strain behaviour for 5:5 composites of polypropylene and modified chalk is illustrated in

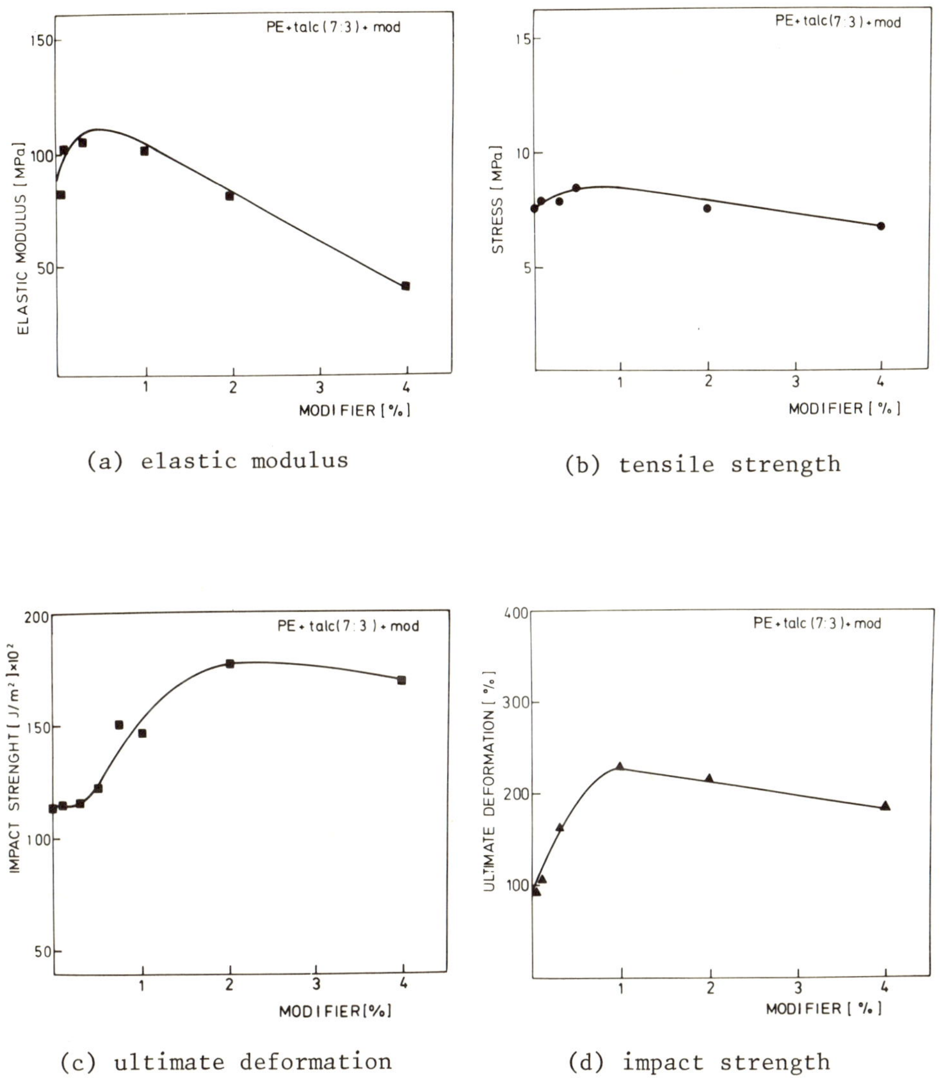

(a) elastic modulus (b) tensile strength

(c) ultimate deformation (d) impact strength

Figure 5. Mechanical parameters for low density polyethylene filled with 30% by weight of modified talc drawn against content of oligomer of ethylene oxide.

Figure 6. The stress-strain curves for unfilled polypropylene and filled with unmodified chalk are drawn for comparison.

TABLE 1

Ultimate elongation and impact strength of filled polypropylene samples.

Composition 6 : 4	Ultimate elongation (%)	Impact strength acc. Charpy $(J/m^2) \times 10^2$
PP	850	55,7
PP + chalk	360	57,0
PP + chalk + 1% OEO	380	68,0
PP + chalk + 5% OEO	430	72,0
PP + chalk + 10% OEO	470	72,5

Some mechanical parameters for modified composition of polypropylene and chalk are presented in Figures 7a, b, c and d as functions of the amount of chalk in the composition. The elastic modulus is shown in Figure 7a.

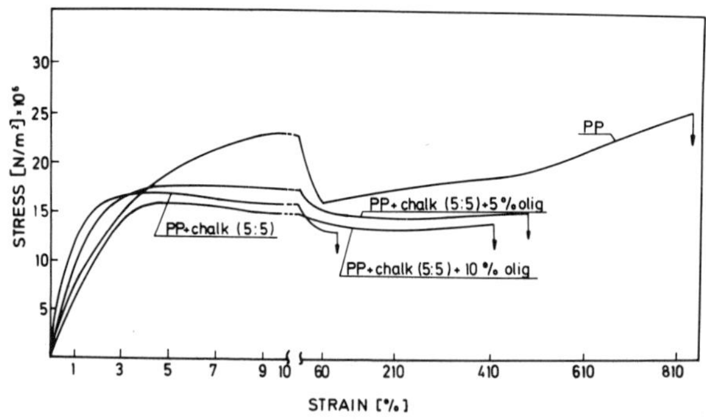

Figure 6. Stress-strain curves for polypropylene filled with 50% of chalk by weight modified with various amount of oligomer of ethylene oxide.

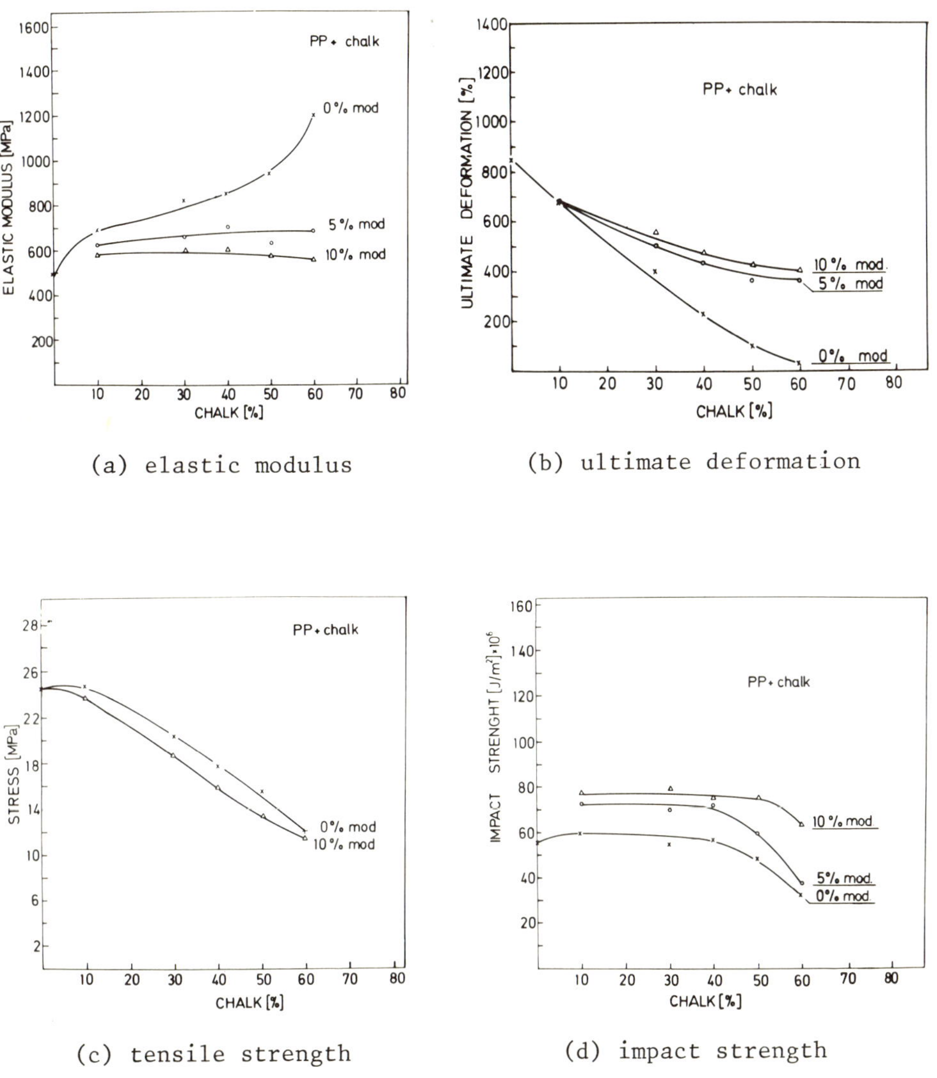

(a) elastic modulus   (b) ultimate deformation

(c) tensile strength   (d) impact strength

Figure 7.   Mechanical parameters of polypropylene filled with various
            amounts of chalk unmodified and modified with 5 and 10%
            of oligomer of ethylene oxide.

In the case of polypropylene filled with unmodified chalk an increase of elastic modulus is observed as expected for the simple physical inclusions of chalk particles into polypropylene. The samples containing chalk modified with oligomer of ethylene oxide exhibit a higher elastic modulus than the unfilled polypropylene. It remains, however, constant independently of the degree of filling. This results seems to be somohow unexpected and impossible to achieve by other methods of modification. The ultimate elongations for those samples against the amount of chalk are plotted in Figure 7b. The samples filled with unmodified chalk show a dramatic decrease of the ultimate elongation while increasing the degree of filling. The addition of a liquid modifier in amount 5-10% by weight causes a significant moderation of that decrease. The apparent effect of the modification is that the samples containing 60% by weight of modified chalk can be deformed now up to 300-400%.

The tensile strength of the composites (see Figure 7c) shows a tendency to decrease both for unmodified and modified samples with increasing amount of chalk.

The impact strength of the samples containing unmodified chalk remains constant up to 30-40% of chalk by weight (see Figure 7d) and then decreases by half for the samples containing 60% of chalk. The samples containing chalk modified with oligomer of ethylene oxide ehibit a higher impact strength, remaining constant to 30% of chalk for 5% of OEO and and 50% of chalk for 10% of OEO in the composition. For higher contents of chalk the impact strength decreases, but for the polypropylene containing 60% of chalk modified with 10% of OEO it is still higher than for unfilled polypropylene. The mechanical properties of filled polypropylene depends strongly on the condition of crystallization. Table II shows statistical mechanical parameters for samples in a form of foil for two extreme regimes of cooling of the samples from the melt:
- "fast cooling" means cooling of the samples together with brass chromium plated covers by dipping them in iced water,
- "slow cooling" means cooling of the sample together with covers in air.

Slow cooling of polypropylene from melt in the compositions causes an increase of the elastic modulus and tensile strength and a decrease of ultimate elongation.

TABLE 2

Mechanical parameters of polypropylene filled with modified chalk for different regimes of cooling.

Composition		Slow cooling	Fast cooling
1	2	3	4
PP+(chalk	E [MPa]	700±20	620±20
+10%OEO)	σ_z [MPa]	19.8±0.3	17.7±0.5
6:4	ε_z [%]	150±20	470±70
PP+(chalk	E [MPa]	700±30	570±30
+10%OEO)	σ_z [MPa]	17.5±0.2	14.5+0.4
5:5	ε_z [%]	130±40	420+70
PP+(chalk	E [MPa]	640±50	550+30
+10%OEO)	σ_z [MPa]	13.4+0.3	12.0±0.5
4:6	ε_z [%]	80±10	320±30

where E - elastic modulus for uniaxial deformation, MPa;
  σ_z - tensile strength, MPa;
  ε_z - ultimate elongation, %;

CONCLUSIONS

   All the samples containing filler covered with a liquid modifier exhibit the stress-strain behaviour, typical for termoplastic materials. After the initial elastic increase of stress the neck is formed and then the plastic deformation of the material occurs in the region of the neck formation. When all the measuring length of the sample is deformed, it fractures. No further deformation of the oriented material was observed. Such properties of highly filled polypropylene show an active role of the liquid modifier in the deformation mechanism. Thus the prediction of the analysis of the

filler particle behaviour, being surrounded by a liquid layer and
embedded in polymer matrix, was correct.

Similarly, the prediction concerning the crack propagation
through a liquid interface is confirmed by a significant increase
of the impact strength of the sample containing modified chalk.
Thus, the liquid OEO stops the crack growth. The comparison of the
respective properties of polypropylene filled with chalk industrially
produced (e.g. Mitsubishi Petrochemical Co. type Salvex JF) with the
polypropylene filled with chalk modified by a liquid layer of OEO
leads to the conclusion that the presence of a liquid decreases the
elastic modulus and increases both the impact strength and the ultimate deformation.

The independence of the elastic modulus of the compositions
containing modified chalk on the degree of filling can be explained
by an incomplete covering of the filler particles by OEO (i.e. there
are some direct contacts PP – filler increasing the modulus) or
changing the supermolecular structure of polypropylene around the
filler particles caused by the increase of the contact surface with
chalk or by the increase of the rate of cooling of the interior of
the samples. In the last case oligomer of ethylene oxide reduces
the thermal resistance on the border polypropylene-chalk and due to
the high heat conductivity of chalk the heat can be effectively
abstracted from the interior of the sample.

The tensile strength of all the samples decreases with increasing amount of chalk. The modified and unmodified samples behave
in the same way. The polypropylene does not wet the chalk particle,
so the adhesion is poor. The liquid modifier is also not able to
transmit the tensile stresses thus all the tensile stresses are concentrated in polymer matrix. Since the effective cross section of
the polymer in the samples decreases linearly increasing degree of
filling, the tensile strength decreases in the same way.

In the case of polypropylene filled with unmodified chalk the
fracture of the samples occurs at small deformation. If the liquid
modifier is introduced to the system the ultimate elongation increases several times. This leads to the conclusion that the oligomer
of ethylene oxide plays a role of inhibitor in crack formation process. We can conclude that the main function of a liquid in the
system of filled polypropylene is to influence the generation and
propagation of the cracks.

ACKNOWLEDGEMENT

The authors are indebted to the Institute of Industrial Chemistry for the financial support of the work and to Eng. I. Tomaszewska and Eng. B. Wojtowicz from that Institute for the stimulating discussions.

REFERENCES

1. W. Fisher, L. Morbitzer, Angew. Macromol. Chem., 9, 47 (1969).
2. R. Holm, Angew. Chem. int. Ed., 10,(9),591 (1971).
3. K. Kato, Japan Plastics 5, 2 (1971).
4. K. Bridger, Short Communication on 5th Discussion Conference on Phases and Interphases in Macromolecular Systems, Prague, June (1976).
5. J. B. Donnet, E. Papirer, A. Vidal, Short Communication on 5th Discussion Conference on Phases and Interphases in Macromolecular Systems, Prague, June (1976).
6. B. Vineur, in Chemistry and Physics of Carbon ed. P.L. Walker and P. A. Thrower, M. Dekker Ed. New York, p. 171 (1975).
7. Encyclopedia of Polymer Science and Technology, Vol. 6, p. 740, Interscience Publishers a division of John Wiley and Sons, Inc., Nex York, London, Sydney (1967).
8. P. V. Horton, U.S. Patent 2, 668, 134.
9. A. J. G. Allan, J. Polymer Sci. 38, 297 (1959).
10. H. E. Weeksberg, J. B. Webber, Mod. Plastics, 36,(11),101(1959).
11. R. Rossman, J. Polym. Sci., 21, 565 (1956).
12. G. Salomon, in Adhesion and Adhesive, Ed. R. Houwink and G. Salomon, Elsevier Pub. Co., Amsterdam, London, New York, p. 94 (1967).
13. R. A. Hines, Short Communication on 132 nd Am. Chem. Soc. Meeting (1957).
14. Asahi Chem. Co. Ltd., JA Patent 038320.
15. Gen. Electric Co., US Patent 374521.
16. Takuyama Soda KK,JA Patent 110136.
17. Chisso Corp., JA Patent 046933.
18. "Fillers for Plastics" Ed. W. C. Wake, Liffe Books, London, The plastic Institute (1971).
19. Encyclopedia of Polymer Science and Technology, vol. 6, p.750, Interscience Publishers a division of John Wiley and Sons, Inc. New York, London, Sydney (1967).

20. E. Heitz, Gum. Asb. Kunst, 5, 1286 (1975).

21. T. P. Bragg, M. D. Held, Plast. Eng. 9, 30 (1974).

22. Off. Plast. Caoutch., 53, 635 (1973).

23. G. Rosos, R. Kaussen, Kunst. Rdsch. 6, 245 (1972).

24. Catalog of Huls, Vestolen P 5232 T and 5272 T (1975).

25. Catalog of Ho-chst AG, Hostalen PPN 7180TV/20 (1975).

26. J. W. Axelson, Polymer Plast. Techn. Eng., 1, 93 (1975).

27. P. Arnaud, Rev. Plastics Mod., 174, 1092 (1970).

28. W. I. Buchalter, R. I. Bielova, Plast. Massy (Russian), 7, 48 (1972).

29. G. Lang, Kunstoffe, 6, 297 (1974).

30. H. Frommelt, Plaste Kautsch., 10, 721 (1973).

31. Jap. Plast. Ind. Ann., 17, 73 (1974).

32. Catalog of Freeport Kaolin Corp. (1975).

33. Catalog of Georgia Kaolin Corp. (1975).

34. M. A. Bielyj, A. F. Klimowicz, L. M. Gurinowicz, Plast. Massy (Russian) 7, 53 (1975).

35. Metal filled plastics, Composites 6, 1, 6 (1975).

36. I. R. G. Evans, D. E. Packham, J. Adhesion, 9, (4) 267-77 (1978).

37. N. I. Jegorienkov, P. G. Lin, A. I. Kuzakow, Vysokomol. Sojed. A17 (8) 1858 (1975).

38. N. I. Jegorienkov, Nowyje klei i technologija skleivanija (Russian) MDNTP, Moskawa (1973).

39. K. Okuno, R. T. Woodhams, Polym. Eng. Sci., 4, 308 (1975).

40. F. W. Maine, P.D. Shepherd, Composites, 5, 193 (1974).

41. F. Sakaguchi, Jap. Plast. Age 12 (4) 44 (1974).

42. Ju. I. Vasilenok, A. S. Diejanova, B. A. Knoplev, Plast. Massy (Russian), 8, 35 (1972).

43. C. Klason, J. Kubat, J. Appl. Polym. Sci., 19, 831 (1975).

44. E. Ja. Biejdiev, W. W. Prokopienko et al. Plast. Massy, 6, 67, (1974).

45. K. Kmitta, Kuntstoffe, 3, 116 (1975).

46. Poliplasti, 23 (207) 31 (1975).

47. Solid glass spheres cut mold cycle time, Plast. Des. Progress, 5, 18 (1975).

48. G. Lang, J. Appl. Polym. Sci., 8, 2287 (1974).

49. B. Fisa, R. H. Parchessault, J. Appl. Polym. Sci., 7, 2025 (1975).

50. T. Mureyama, Jap. Plast. Age, 1, 27 (1974).

51. L. T. Bielova, Je. G. Lubieszkina et al., Vysokomol. Sojed. 11, B5, 2485 (1973).

52. Rev. Plast. Mod. 223, (1) 118 (1975).

53. G. R. Lightsey, A. L. Hines, Plast. Engng. 5, 40 (1975).

54. Plast. Tech., 22 (8) 71 (1976).

55. Plast. Tech., 22 (4) 81 (1976).

56. S. J. Monte, P. F. Bruins, Rev. Plast. Mod., 223, 80 (1975).

57. M. Motoyoshi, Jap. Plastics Age, 33, 5 (1975).

58. E. P. Plueddemann, G. L. Stark, Rev. Plast. Mod., 34, 710 (1977).

59. Dow Corning Corp., US Patent 061505.

60. Union Carbide Corp., US Patent 862027.

61. P. A. Tierney, US Patent 3328 339 (1967).

62. P. E. Cassidy, B. J. Yager, J. Macromol. Sci. Rev. Polym.
     Technol., D1 (1), 1 (1971).

63. J. Hodgkin, D. Solomon, J. Macromol. Sci., A8 (3) 635 (1974).

64. D. Solomon, BP 1228538 (1969).

65. D. Solomon et al., J. Macromol. Sci., A5, 995 (1971).

66. D. Solomon et al., Clay Minerals, 7, 389 (1968).

67. D. Solomon, J. D. Swift, A. J. Murphy, J. Macromol. Sci. Chem.,
     A5, 587 (1971).

68. D. Solomon, J.D. Swift, J. Appl. Polym. Sci., 11, 2567 (1967).

69. B. C. Loft, D. Solomon, J. Macromol. Sci., Chem. A6, 831 (1972).

70. M. S. Akutin, Z. E. Salina, Plast. Massy, 1, 50 (1974).

71. D. Solomon, J. Hodkin, B. C. Loft, D. Hawthorne, J. Macromol.
     Sci., A8 (3) 649 (1974).

72. Asaki Chem. Ind. Co. Ltd., JA Patent 069210.

73. US Polywood Champion Papers Inc., US Patent 066107.

74. K. Thinius, B. Hosselbarth, Plaste u. Kautsch.,17, 474 (1970).

75. K. Thinius, B. Hosselbarth, Plaste u. Kautsch.,17, 567 (1970).

76. K. Thinius, B. Hosselbarth, Plaste u. Kautsch.,17, 653 (1970).

77. D. A. Novikova, Plast. Massy (Russian) 7, 56 (1975).

78. A. S. Wood, Mod. Plast. Intern., 6, 60 (1975).

79. L. M. Sharpe, J. Adhesion, 1, 51 (1972).

80. A. B. Wajnstein, E. A. Kutniev, Vysokomol. Sojed. A8, 1696
     (1974).

81. F. M. Anehev, R. G. Azrak, Plast. Engn. 7, 32 (1974).

82. T. A. Gruse, H. J. Konish, J. Polym. Plast. Techn. Eng., 1,
     41 (1975).

83. J. E. Gordon, "The New Science of Strong Materials", Penguin
     Books, London (1968).

84. Authors unpublished results.

85. Center of Molecular and Macromolecular Studies, Polish Academy
     of Sciences, Polish Patent Application P-213583.

86. Center of Molecular and Macromolecular Studies, Polish Academy
     of Sciences, Polish Patent Application P-213582, Foreign Patent
     Applications.

MILTICOMPONENT SELECTIVE MEMBRANES MADE OF IONIC/NONIONIC

POLYMERS

A. Narebska and R. Wodzki

Institute of Chemistry, N. Copernicus University

87-100 Torun ul. Gagarina 7, Poland

The multilayer models of permselective membranes
have been presented. The models relate the electric conductivity of membranes to their compositions. Applying
the models, the calculations of composition of the membranes (i.e. the volume fractions of ionic polymer in gel
state,) of inert polymer and imbibed electrolyte, were
performed.

MEMBRANES AS MULTIPHASE SYSTEMS

Ion selective membranes are composed of two kinds of polymers:
- the ion-containing polymers, like polystyrenesulphonic acid or
  polymer base with quaternary ammonium groups and
- inert polymers - often polyolefines, which ensure mechanical strength to the material.

Electrochemical properties of membranes depend on the ratio of
both components and they are better, the higher the volume fraction
of an ionic component and the concentration of functional groups
within them.

The composition and the structure of membranes vary significantly with the method of manufacturing[1]. The following main methods may be listed:

i.   preparing of block copolymers containing ionic repeat units
      in the form of thin sheets,
ii.   grafting of monomers like styrene onto a film of an inert po-
      lymer, like polyethylene, followed by sulphonation (cation se-
      lective membranes) or chloromethylation and quaternarization
      (anion selective membranes),
iii. milling an ion-exchange resin and calendering with an inert
      binder; these membranes are often supported by a reinforcing
      mesh. Some modification is the "paste method".

Due to the incompatibility of the components (method i, ii) or
to the technique of preparation (method iii) the membranes are macro-
or microheterogeneous with distinct domains of ionic and inert polymers.

When applied in electrodialysis or electrosynthesis, the membranes are in contact with an ambient solution, usually an electrolyte,
in which they swell. The swelling causes a hydration of ionic component and it is accompanied by imbibition of an external electrolyte
into the pores and those volume elements in which the concentration
of active groups and crosslinking is low. This third component influences permselectivity and cannot be neglected when the properties
of membranes are discussed. Consequently, the main components which
constitute membranes in the swollen state, are:
- gel of an ion-containing polymer (polyelectrolyte),
- an inert polymer,
- imbibed low molecular weight electrolyte of the composition and
  concentration of an external solution.
These components show the properties of immiscible or partly miscible phases. The ratio of the volume fractions of each component,
cannot be predicted quantitatively on the ground of synthesis, however it can be calculated, at least partially, on the base of the
properties of membranes and their models.

THE MODELS OF MEMBRANES

According to the classification of the heterogeneous systems
published by Barrer[2] the membranes can be considered composed of
- continuous phase of an inert polymer and the phase of nonunifor-
  mly distributed particles of gel which are separated or may form
  aggregates (membranes prepared according to methods ii and iii).

- irregular interpenetrating continuous phases of inert and ion-containing polymers (membranes produced by method i).

Except for the supporting mesh, if present, the other components are randomly distributed without any orientation. Similarly the third component, i.e. an electrolyte, is distributed through the membranes in the form of discontinuous domains. The presence of this component makes the solution of the model, a little more complex.

The theoretical models of membranes have been developed on the basis of transport, dielectric losses and the electrochemical properties. Most often the electrical conductivity is considered. All the electric models of the membranes are based on different ability of the components to conduct current and exploit the relation between the final conductivity of a material and the amount and distribution of the conducting elements. In the membranes, the gel of ion-containing polymer conducts current by counterions, an imbibed electrolyte, with usual mechanism whereas an inert polymer is not a conducting material, though it changes the specific resistance of a membrane to a degree dependent on its amount and distribution.

For the mathemathical account of the model the functional relation:

$$\varkappa_m = f\,(\varkappa_r,\,\varkappa_e,\,K) \tag{1}$$

should be known. The symbols denote:
$\varkappa_m, \varkappa_r, \varkappa_e$ - the specific conductivities of the membrane, gel phase and imbibed (external) electrolyte respectively,
K - a set of the parameters describing distribution and the shape of components.

The early models of membranes have been elaborated in the sixties. The model published by Spiegler et al.[3] was elaborated for the prediction of specific conductivity of a plug composed of ion-exchange resin beads immersed in an electrolyte solution. This model was used later for interpretation of conductivity of membranes[4]. A similar system was discussed by Arnold and Koch[5], who ascribe to the model a different electrical circuit. Both models do not make allowance for the presence of an inert nonconducting polymer, treating it as a part of polyelectrolyte. Later Guillou et al.[6] intro-

duced a nonconducting element to the model of Arnold and Koch. The
same was done by Grebieniuk et al.[7] with Spiegler's model.

In 1977-78 we examined all the models by computations of the
volume fractions of components in the cation selective membranes of
different composition and structure and found that the results fulfil the chosen experimental criteria within a limited range[8]. Sear-
ching for the source of discrepancies, the equations representing
each of the models have been analyzed. It was stated that all these equations, which relate the specific conductivity of a membrane,
$\varkappa_m$, to that of an external electrolyte, $\varkappa_e$, and to the so called
"geometric parameters" of the model may be written in a general
form:

$$\varkappa_m = \frac{A_o + A_1 \varkappa_e + A_2 \varkappa_e^2}{1 + B_1 \varkappa_e} \qquad (2)$$

It is essential that this equation is adequate for the system
in which the conductivity of a gel phase and the geometric parameters are constant at a varying concentration of an external electrolyte. The geometric parameters and $\varkappa_r$ appear in the coefficients
A_o, A_1, A_2 and B_1 as invariants. It may also be proved that equation 2 describes the conductivity of the bilayer system in which
each of the layers is composed of three components. In the layers
the volume fractions of gel and imbibed electrolyte are different
whereas that of an inert polymer is constant(Figure 1a,b). This conclusion imply why the published models present only a rough approximation of the membrane properties and structure.

With respect to the membranes the specific conductivity of a
gel phase cannot be assumed constant, at a wide range of concentrations of an ambient electrolyte, as it changes due to Donnan sorption. Similarly the magnitudes of the geometric parameters of the
models, describing the volume fraction of each component, may vary
due to the shrinkage of membranes. Finally, the preliminary calculations proved that the adequacy of the model improves, the higher
the number of layers.

For a number of layers going towards infinity n → ∞, equation 2
takes a general form:

$$\varkappa_m = \frac{\displaystyle\sum_{i=0}^{n} A_i \varkappa_e^i}{1 + \displaystyle\sum_{i=1}^{n-1} B_i \varkappa_e^i}$$

This equation appears to be quite general for the conductivity
of heterogeneous systems. At $A_o = 0$ it describes the conductivity of
a system composed of separated particles dispersed in a conducting
medium. For such a system and n=1, the equation corresponds to that
published by Nikolaev et al.[9], with n=2 – to the formula derived by
Maxwell[10], Parrish[11] and Nielsen[12]; with n=3 it takes the form of
Rayleigh's[2] and Frey's[13] equations. For. the condensed systems,
like membranes, A_o is above zero. The published 3-7 models of membranes fulfil this condition with n=2.

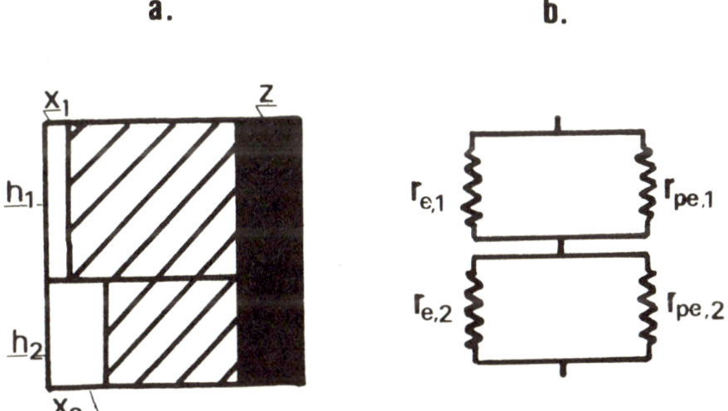

Figure 1.   Bilayer model of a membrane, a – geometric model,
b – equivalent electrical circuit, ▨ – polyelectrolyte, ▬ – inert polymer, ☐ – electrolyte.

A bilayer model

Following the conclusions drawn from the analysis, the geometric
parameters and phase composition of the membranes have been calculated according to the model presented in Figure 1. This model corresponds to the published one[5,6] with regard to the number of layers,
however, it differs in the postulated current routes and the electrical circuit. The main difference is that the imbibed electrolyte,
treated as a separate continuous phase, has been rejected. The idea
about the appearance of this phase was brought to the models from
the ion-exchange resins plugs, for which it is justified. However,
for the membranes the continuous nonselective domains would mean
open routes to the electrolyte through the membranes i.e. "holes",
which are not admissible.

An electrical circuit corresponding to the model, Figure 1b, is
composed of two elementary circuits each representing one layer.
In the circuits the parallel resistors represent the gel phase, r_{pe},
and imbibed electrolyte in voids, r_e. The geometric parameters
which describe the models are: x_1, x_2, z, h_1 and $h_2 = 1 - h_1$. They appear equation for the conductance of a layer:

$$1/r_1 = (\varkappa_e x_1 + \varkappa_r y_1)/h_1, \quad (\text{ohm}^{-1}) \tag{4}$$

$$1/r_2 = (\varkappa_e x_2 + \varkappa_r y_2)/h_2, \quad (\text{ohm}^{-1}) \tag{5}$$

whereas the equation for the specific conductivity of the membranes
takes the form:

$$\varkappa_m = \frac{y_1 y_2 \varkappa_r^2 + (x_1 y_2 + x_2 y_1)\varkappa_r\varkappa_e + x_1 x_2 \varkappa_e^2}{(y_1 h_2 + y_2 h_1)\varkappa_r + (x_1 h_2 + x_2 h_1)\varkappa_e}, \quad (\text{ohm}^{-1}\ \text{cm}^{-1}) \tag{6}$$

where: $y_1 = 1 - x_1 - z$, $y_2 = 1 - x_2 - z$

In the following calculations, the variations of the specific conductivity of the gel phase and the geometric parameters at changing
concentration of an external electrolyte have been allowed. With
the known x_1, x_2, h_1, h_2 and z, the volume fractions of the three
components have been calculated with the expressions:

$$V_e = x_1 h_1 + x_2 h_2 \tag{7}$$

$$V_i = z \tag{8}$$

$$V_{pe} = 1 - V_e - V_i \tag{9}$$

An extended model

According to the sense given to the word in this paper we call "extended" the models with n>2. In Figure 2 the scheme (a) and a corresponding electrical circuit (b) of the model of membranes composed of three layers are presented.

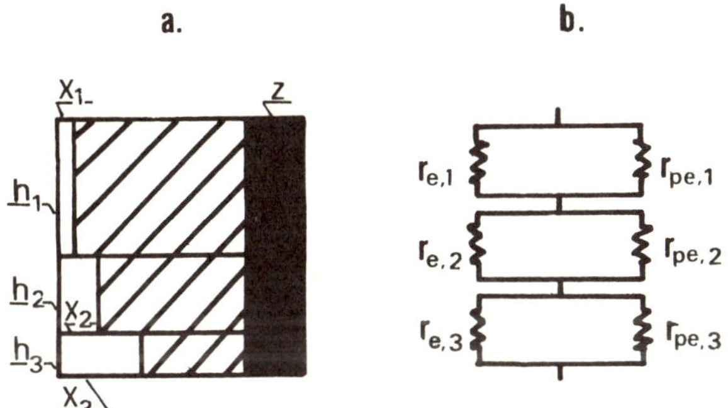

Figure 2.   Extended model of a membrane, (a) geometric model, (b) equivalent electrical circuit, polyelectrolyte, inert polymer, electrolyte.

In view of this model one can interpret the membrane as being composed of two components, gel and inert polymer and the nonuniformely distributed voids of the shape: long-narrow, middle and the wide-short. Such a model is necessarily described by an increased number of geometric parameters, namely: x_1, x_2, x_3, h_1, h_2, $h_3 = 1 - h_1 - h_2$, z and $\mathcal{H}_r$. The preliminary values of the parameters have been estimated employing the statistical sequential simplex method[14] and

the published experimental data. The conductivity at the isoconduc-
tance point has been accepted as the starting value of $\mathcal{H}_r$. With
these data further calculations have been performed as described
later on.

The equation for the conductance of one layer is:

$$1/r_i = \left[\mathcal{H}_e x_i + \mathcal{H}_r (1-x_i-z) \right] /h_i, \quad (i=1,2,3) \tag{10}$$

and for the specific conductivity of the membranes:

$$\mathcal{H}_m = 1/\sum_{i=1}^{3} h_i / \left[\mathcal{H}_e x_i + \mathcal{H}_r (1-x_i-z) \right] \tag{11}$$

The volume fractions of polyelectrolyte and inert polymer have been
calculated with expressions 9 and 8, and

$$V_e = \sum_{i=1}^{3} x_i h_i \tag{12}$$

The results presented in Tables 1, 2 and 3 prove that increasing n, the calculated volume fractions become closer to the expected
one. However the number of the geometric parameters which have to
be calculated shows that it is not, clearly, the way to further
improve the model. Instead it would be reasonable to postulate
$n \to \infty$ with some mathematical solution of the problem.

A multilayer model

As this is a general model, we shall describe it giving more
details.

A graphical representation of the mebranes prepared according
to the method iii (a) and ii (b), a geometric model (c) and an
equivalent electrical circuit (d) of the multilayer model are presented in Figure 3.

Symbols x_1, x_n and z denote the geometric parameters. Part d
represents the equivalent electrical circuit composed of n elementary
circuits arranged in series. The following assumption underlays

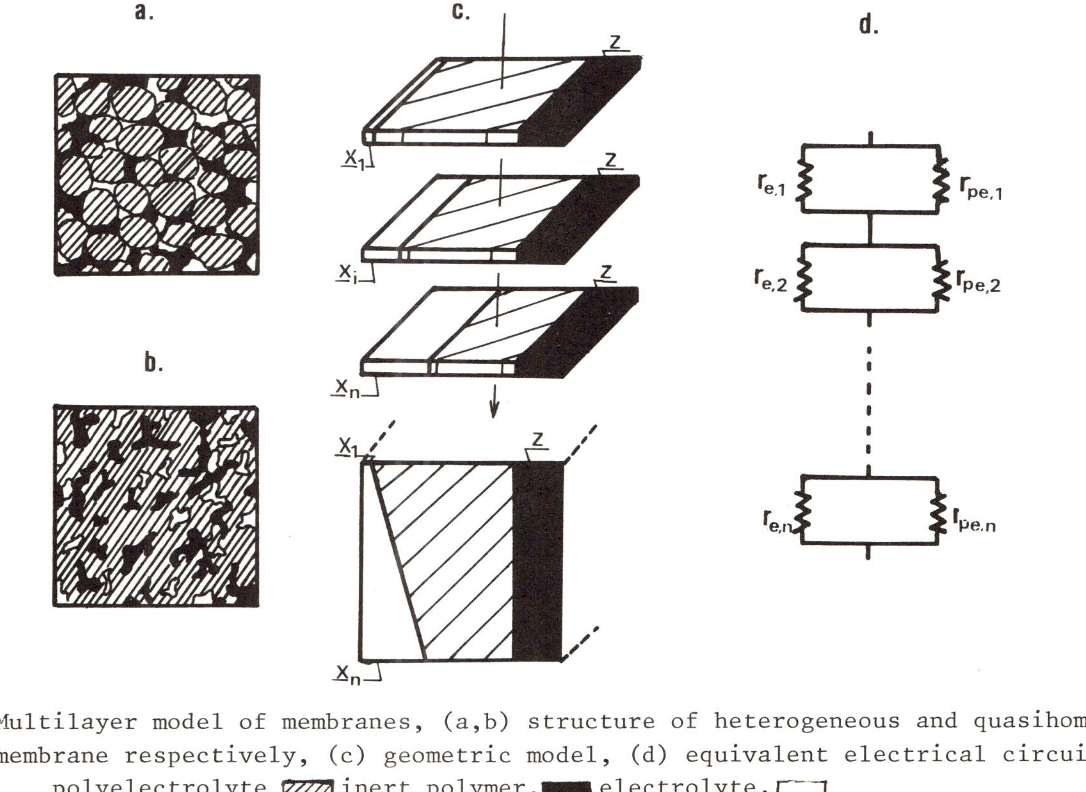

Figure 3. Multilayer model of membranes, (a,b) structure of heterogeneous and quasihomogeneous membrane respectively, (c) geometric model, (d) equivalent electrical circuit, polyelectrolyte, inert polymer, electrolyte.

the solution of the model.

i.   An unit volume element of a membrane is composed of n parallel
     layers each of hight h=1/n, n → ∞.

ii. Each layer is composed of three components: an ion containing
     polymer in a gel state, an inert polymer and some amounts of
     imbibed electrolyte in voids, as previously.

iii. The volume fraction of a gel and imbibed electrolyte in succes-
     sive layers are variables. The crossection of the volume ele-
     ment of an imbibed electrolyte in "i" layer is x_i where
     i = 1,2... n.

iv. The volume fraction of an inert polymer in each of tha layers
     is the same; its crossection area being z.

v.   Each layer can be represented by an equivalent electrical circuit
     of two parallel resistors, one representing the resistance of
     polyelectrolyte gel(r_{pe}) the other representing the resistance
     of an electrolyte (r_e).

vi. The equivalent circuit representing an unit volume element of
     a membrane is composed of n elementary circuits connected in
     series: each circuit represents one layer.

For the mathematical solution of the model it has been necessary to assume a continuous distribution of the volume fractions
of elements in successive layers and to choose the distribution
function. After some preliminary calculations the linear and hiper-
bolic functions have been chosen. In this paper we present the results with linear distribution only, whereas those with the hiper-
bolic function have already been published[15]

A model with linear distribution of nonselective domains

According to the postulated function, the crossection of the
nonselective domains (x_i) and a gel phase (y_i) in the layer i can
be written as follows:

$$x_i = x_1 \cdot \frac{i}{n} + x_n - x_1 \tag{13}$$

$$y_i = 1 - x_i - z \tag{14}$$

The conductance of the same layer i can be expressed by the specific conductivities of both components and the geometric parameters

x_1, x_n, z and $h=1/n$:

$$1/r_i = nx_i\kappa_e + ny_i\kappa_r \tag{15}$$

Summing up the resistances of n layers arranged in series, the following equation is obtained:

$$R_m = \sum_{i=n}^{n} \left[(\kappa_e-\kappa_r)(x_n-x_1)i + (\kappa_e-\kappa_r)x_1n+\kappa_r(1-z)\right]^{-1} \tag{16}$$

Since the distribution function of the volume fractions of the voids and gel has been assumed as continuous and since the number of layers tends to infinity, the integration under the same boundary conditions can be performed:

$$\sum_{i=1}^{n\to\infty} f(i) = \int_{i=1}^{n\to\infty} f(i)di \tag{17}$$

This leads to the final expression for specific conductivity of a membrane:

$$\kappa_m = (\kappa_e-\kappa_r)(x_n-x_1)/\ln \frac{(\kappa_e-\kappa_r)x_n+\kappa_r(1-z)}{(\kappa_e-\kappa_r)x_1+\kappa_r(1-z)} \tag{18}$$

The calculated parameters x_1, x_n, z and the specific conductivity of the polyelectrolyte within a membrane (κ_r) are listed in Table 1 and 2. The volume fractions of polyelectrolyte and inert polymer have been calculated using the formula 9 and 8 and V_e with 19:

$$V_e = (x_n+x_1)/2 \tag{19}$$

The results of computations have been verified by the following criteria:

i. The calculated volume fraction of nonselective domains (V_e) should be compatible with one established by an independent experimental method. For that purpose the Tye (16) treatment based on the electrolyte uptake has been applied.

ii. The calculated volume fraction of an electrolyte in nonselective domains could not exceed the amount of water in the swollen membranes determined experimentally; and so, only the result $V_e<V_{H_2O}$ is acceptable.

iii. The calculated volume fraction of an inert polymer should be
close to that resulting from the synthesis.

EXPERIMENTAL

Investigations have been performed with the following membranes:
PE-PSSA - (laboratory product of Wroclaw Technical University,
Poland) a quasi homogeneous membrane synthetized by sulphonation of
the copolymer made by grafting of styrene-divinylbenzene mixture
onto polyethylene foil; DVB content in the mixture was 10%. The
exchange capacity of the membrane is 2.1 meqv/g of dry H^+ form and
water content V_{H_2O} = 0.39. The volume fraction of an inert polymer
is V_i<0.3.

IONAC MC-3470 - (product of Ionac Chemical Corp., U.S.A.) a heterogeneous membrane made by compression of powdered mixture of an inert
polymer and polystyrenesulphonic acid resin. A supporting mesh of
inert material, poly(ethyleneterephtalate), is imbedded in the membrane. The capacity of the membrane is 1.6 meqv/g and water content
V_{H_2O}=0.37. The content of an inert polymer in the membrane V_i is
in the range 0.3 - 0.4.

The electric resistance of the membranes was measured with an
alternating current bridge (1000 Hz). The sulphuric acid solutions
of the concentrations from 0.03 up to 4 N were used as an external
electrolyte. All measurements have been performed at 298 K. The
equipment and the technique of measurements have been described
previously[17].

The curves presenting the specific conductivity of membranes
versus the concentration and conductivity of the external electrolyte can be seen in Figure 4.

The computations have been performed by fitting the experimen-
tal curves $\varkappa_m$ versus $\varkappa_e$ to the theoretical equations 6, 11 and 18.
The standard IBM Share Program N. 1428 adapted to the RIAD R-32 sy-
stem has been used. The ranges of concentrations at which the geometric parameters and $\varkappa_r$ have been assumed to be constant are listed in tables 1 and 2 together with the results of computations.

For the verification of numerical data on the volume fractions
of nonselective domains obtained from the models, the sorption of
sulphuric acid of the same concentration as in the conductivity experiments, has been measured[18]. These results were further used for
the calculations of the volume fractions of non-selective domains,
V_β, according to Tye theory[8,16].

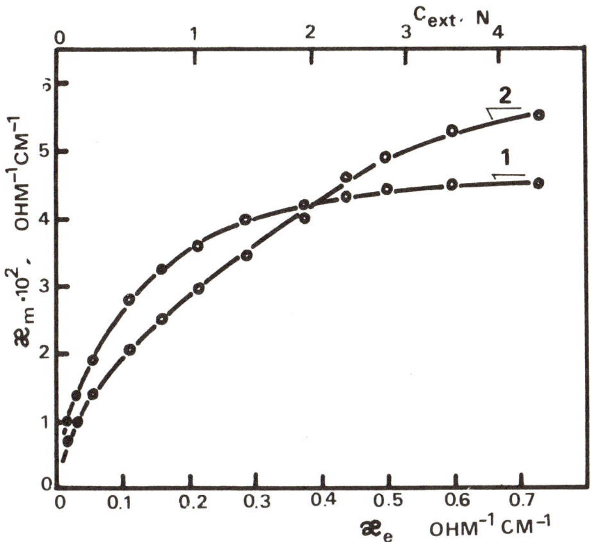

Figure 4. Specific conductivity of membranes ($æ_m$) as a function
of concentration (C_{ext}) and conductivity ($æ_e$) of external
solutions of H_2SO_4, 1 - PE-PSSA, 2 - IONAC MC-3470.

RESULTS AND DISCUSSION

In tables 1 and 2 are collected the calculated volume fractions
of the components in the swollen quasihomogeneous membrane PE-PSSA
and a heterogeneous membrane IONAC MC-347o. The mean values of V_{pe}
V_i and V_e have been compared with the measured water regain
(V_{H_2O}), with the mean volume fraction of nonselective domains calculated according to Tye and with the approximate amount of inert
polymer within the membranes (Table 3).

It should be noted here, that when the earlier models (3-7)
have been applied for the calculations of the volume fracrions of

464   A. NAREBSKA AND R. WODZKI

TABLE 1

Geometric parameters and volume fractions of components in the membrane PE–PSSA.

Model $n=2$

Model	Concentrations of external electrolyte C_{ext}, /N/	Conductivity of gel phase in membrane $æ_r \cdot 10^2$ /ohm^{-1}cm^{-1}/	x_1	x_2	h_1	h_2	V_{pe}	$V_{i=z}$	V_e
$n=2$	0.03–0.25	1.9	0.045	0.40	0.37	0.63	0.45	0.28	0.27
	0.5 – 0.8	2.5	0.044	0.41	0.61	0.39	0.56	0.25	0.19
	1.0 – 1.4	3.2	0.019	0.31	0.64	0.36	0.70	0.18	0.12
	2.2 – 2.6	3.1	0.037	0.23	0.63	0.37	0.44	0.45	0.11
	3.6 – 4.0	0.8	0.040	0.26	0.58	0.42	0.56	0.31	0.13

Model $n=3$

Model	C_{ext}, /N/	$æ_r \cdot 10^2$	x_1	x_2	x_3	h_1	h_2	h_3	V_{pe}	$V_{i=z}$	V_e
$n=3$	0.03–0.25	2.1	0.032	0.37	0.62	0.37	0.54	0.09	0.42	0.33	0.27
	0.5 – 0.8	2.7	0.032	0.41	0.45	0.54	0.39	0.07	0.47	0.32	0.20
	1.0 –1.4	3.9	0.026	0.36	0.61	0.75	0.20	0.05	0.53	0.36	0.12
	2.2 – 2.6	4.1	0.014	0.24	0.56	0.74	0.18	0.08	0.47	0.35	0.10
	3.6 – 4.0	2.7	0.029	0.13	0.12	0.57	0.41	0.13	0.40	0.52	0.08

Model $n \to \infty$

Model	C_{ext}, /N/	$æ_r \cdot 10^2$	x_1	x_n	V_{pe}	$V_{i=z}$	V_e
$n \to \infty$	0.03–0.25	2.0	0.004	0.14	0.72	0.21	0.07
	0.5 – 0.8	2.5	0.004	0.21	0.73	0.16	0.11
	1.0 – 1.4	3.5	0.009	0.09	0.77	0.18	0.05
	2.2 – 2.6	3.7	0.010	0.06	0.77	0.19	0.04
	3.6 – 4.0	4.9	0.006	0.02	0.74	0.25	0.01

TABLE 2

Geometric parameters and volume fractions of components in the membrane IONAC MC-3470.

Model	Concentrations of external electrolyte C_{ext}, /N/	Conductivity of gel phase in membrane $æ_r \cdot 10^2$ /ohm^{-1}cm^{-1}/	Geometric parameters						Volume fractions		
			x_1	x_2	h_1	h_2			V_{pe}	$V_{i=z}$	V_e
$n=2$	0.03–0.25	0.7	0.123	0.42	0.19	0.81			0.34	0.32	0.34
	0.5 – 0.8	1.6	0.052	0.23	0.46	0.54			0.49	0.36	0.15
	1.0 – 1.4	2.0	0.041	0.25	0.55	0.45			0.58	0.28	0.14
	2.2 –2.6	2.3	0.035	0.27	0.54	0.46			0.48	0.38	0.14
	3.6 –4.0	1.1	0.050	0.22	0.63	0.37			0.59	0.30	0.11
			x_1	x_2	x_3	h_1	h_2	h_3			
$n=3$	0.03–0.25	1.4	0.056	0.36	0.68	0.47	0.45	0.08	0.34	0.42	0.24
	0.5 – 0.8	1.5	0.046	0.28	0.39	0.46	0.43	0.11	0.45	0.37	0.18
	1.0 – 1.4	1.6	0.070	0.22	0.26	0.54	0.36	0.10	0.43	0.43	0.14
	2.2 – 2.6	2.5	0.033	0.29	0.27	0.61	0.31	0.08	0.52	0.35	0.13
	3.6 – 4.0	3.4	0.032	0.25	0.24	0.52	0.36	0.12	0.49	0.38	0.13
			x_1		x_n						
$n\to\infty$	0.03–0.25	0.8	0.010		0.53				0.41	0.32	0.27
	0.5 – 0.8	1.3	0.015		0.25				0.59	0.28	0.13
	1.0 – 1.4	1.3	0.035		0.23				0.52	0.35	0.13
	2.2 –2.6	1.8	0.012		0.19				0.59	0.31	0.10
	3.6 – 4.0	2.3	0.010		0.16				0.52	0.40	0.08

TABLE 3

Mean values of volume fractions of polyelectrolyte gel ($\tilde{V}_{pe}$), inert polymer ($\tilde{V}_i$) and electrolyte ($\tilde{V}_e$) in membranes.

MODEL	PE-PSSA			IONAC MC-3470		
	$\tilde{V}_{pe}$	$\tilde{V}_i$	$\tilde{V}_e$	$\tilde{V}_{pe}$	$\tilde{V}_i$	$\tilde{V}_e$
n = 2	0.54	0.29	0.16	0.49	0.33	0.18
n = 3	0.48	0.37	0.15	0.45	0.39	0.16
n → ∞	0.74	0.20	0.06	0.53	0.33	0.14
Experimental data	$V_{H_2O} = 0.35$ $\tilde{V}_\beta = 0.06$ $V_i < 0.30$			$V_{H_2O} = 0.37$ $\tilde{V}_\beta = 0.12$ $V_i = 0.3 - 0.4$		

nonselective domains, V_e has been found to vary between 0.45 to
0.15. If these data were correct the membranes would not be permselective. In this paper the results of $\tilde{V}_e$ found with a multilayer
model are close or identical with those found when apllying Tye
method. Also the variation of the conductivity of a gel phase within
a membrane, $\mathscr{H}_r$, and the composition seem to be meaningful. Increasing
C_{ext}, some amount of electrolyte enters the gel phase increasing
the conductivity. The findings confirm the expectations. At the
same time the shrinkage causes that the specific volume of membranes
decreases at a cost of the components which easily dehydrate i.e.
of the voids or the weakly crosslinked nonselective domains. Such
variations in the composition may be seen with the membrane PE-PSSA
but less with the reinforced IONAC.

Thre presented results seem to confirm that the models make it
possible to calculate the compositions of a membrane in the swollen
state and so, to comment the relation: synthesis - structure -
selectivity. They also make it possible to predict the behaviour
of a membrane under different conditions.

REFERENCES

1. M.Block, Chem.Ind., <u>16</u> , 2099 (1967).
2. R.M.Barrer,Diffusion and Permeation in Hetero-ogeneous Media,
    in "Diffusion in Polymers" ed.J.Crank and G.S.Park, Academic
    Press, New York (1968).
3. M.C.Sauer, P.F.Southwick, K.S.Spiegler, M.R.J.Wyllie,
    Ind.Eng.Chem., <u>47</u>, 2187 (1955).
4. K.Arulandan,S.S.Smith,K.S.Spiegler, Radiofrequency Properties
    of Polyelectrolyte Systems, in "Charged and Reactive Polymers,
    vol.1 Polyelectrolytes" ed.E.Selegny,D.Reidel Publishing Company,
    Dordrecht-Holland/Boston U.S.A. (1974).
5. R.Arnold, D.F.A.Koch, Aust.J.Chem., <u>19</u>, 1299 (1966).
6. M.Guillou, D.Guillou, R.Buvet, Coll.Memb.DGRST, Paris 1965,
    ed.CNRS 1967.
7. V.D.Grebieniuk, M.V.Pevnitskaya, N.P.Gnusin, Zh.Prikl. Khim.,
    <u>42</u>, 568 (1969).
8. A.Narebska,R.Wodzki,S.Koter, Angew.Makromol.Chem. in press.
9. N.I.Nikolaev, Yu.I.Kharlamov, A.M.Filimonova, Elektro-
    khimya, <u>2</u> , 465 (1966).
10. J.C.Maxwell, "A Treatise on Electricity and Magnetism" Dover
    Publication, New York (1954).

11.   J.R.Parrish,   J.Chem.Soc., 612 (1962).
12.   L.N.Nielsen,   Ind.Eng.Chem.Fund., 13,17 (1974).
13.   Son Frey,   Z.Elektrochem., 38 , 260 (1932).
14.   R.Wodzki, J.Ceynowa, Wiad.Chem., 30 , 337 (1976).
15.   R.Wodzki, A.Narebska, Angew.Makromol.Chem., in press.
16.   F.L.Tye,   J.Chem.Soc., 4784 (1961).
17.   A.Basinski, A.Narebska, J.Ceynowa, Angew.Makromol.Chem.,
      78 , 145 (1979).

POLYOLEFINIC BLENDS: RECENT APPLICATIONS AND DEVELOPMENTS

A. Mattiussi and F. Forcucci

MONTEDISON S.p.A. - Plastic Division

Largo Donegani 1/2 - 20121 Milano (Italy)

A rational development of polyolefinic blends can
take full advantage of the great versatility of the
products and these meet the most varied application
requirements. Products research, which necessarily
tends to develop specific types for each application,
works in this field to define correlations between
properties and the type and composition of the blend.

The special combination of properties offered
by polyolefinic blends opens up a number of possible
applications in such sectors, as the automotive industry, the shoe industry, sanitary ware, flexible
filter elements, flexible pipe and tubing and cable
drums.

INTRODUCTION

Elastoplastics entered upon the scene several years ago and
are now gaining the increasing favour of the market, filling the
gap existing between the traditional plastics and vulcanized elastomers as to quality. These products, in fact, possess properties
that fall mid-way between those of the previously mentioned materials; moreover, they can be converted with the equipment traditio-
nally used by the plastics industry with no need for curing proces-

ses as, instead, is the case with traditional elastomers.

Elastoplastics are heterophase materials consisting of a rubbery matrix with rigid, crystalline or glassy dispersions which at room temperature behave in a way very similar to vulcanization bonds, and in that way give the material good elastic properties, whereas at high temperatures they render the entire system fluid in the same way as thermoplastics. Elastoplastic materials have been available on the market for some time in a number of different types, which differ one from the other as to chemical structure, physical-mechanical and rheological properties. The principal types are: butadiene-styrene copolymers, copolyesters, polyolefinic polymers and polyurethanes.

POLYOLEFINIC BLENDS

Development of taylor-made grades

Intensive research work, directed towards exploiting the high potential of the existing polymers, has led to the creation of special compounds, such as polyolefins modified with EP elastomers, which take on the physionomy and a true new family of polymers.

Montedison has made a decisive contribution towards the development of these products, which are gaining increasing importance on the market, with the Dutral TP and Moplen SP grades of their own production. Most of these grades consist of mechanical blends between an elastomer, e.g. an ethylene-propylene copolymer and a plastomer such as polypropylene, low or high density polyethylene, according to specific cases.

A rational development of the blend will allow the obtainment of optimum conditions and hence the exploitation of the high versatility of these products, which then meet the most varied application requirements. The growth of consumption inevitably produces a sophistication in quality demands, and it is to satisfy such demands that Montedison has developed a wide range of materials "ad hoc", many of which are produced as standard, whilst others are available at customer request.

Montedison's development lines have been directed towards

TABLE 1 - PROPERTIES OF MODIFIED MOPLEN GRADES (*)

Property	Test method	Unit	Typical values					
			Moplen SP 150G	Moplen SP 200G	Moplen SP 32G	Moplen SP 25G	Moplen SP 21	Moplen SP 98/1
Density at 23°C	ISO-DIS 1872/A	g/cm^3	0.89	0.89	0.905	0.905	1.10	1.12
Nominal flo rate (NFR-230/2)	ASTM D 1238/L	g/10	4	5	3.5	1.5	3.5	3.5
Mould shrinkage (linear average)	Montedison	%	1.3	1.4	1.7	1.7	1.3	1.2
Flexural modulus of elasticity at 23°C	ASTM 790	Kg/cm^2	7.200	9.000	10.500	11.000	18.500	25.000
Tensile strength max(23°C)	ASTM D 638	Kg/cm^2	165	190	230	260	180	230
Elongation at break (23°C)	ASTM D 638	%	>400	>400	>400	>400	170	140
Hardness Rockwell (23°C)	ASTM D 785	alpha	−40	−13	30	32	12	22
Impact strength, Izod	ASTM D 256	Kg cm/cm						
notched (23°C)			90	75	64	90	55	25
(−20°C)			80	8	10	20	10	4
Oven softening limit (Vicat 1 Kg/mm^2)	ASTM D 1525	°C	123	137	145	148	138	144
Deflection temperature under load ($\sigma = -4.6$ Kg/cm^2)	ASTM D 648	°C	91	107	103	127	124	133

(*) Injection moulded specimens mormalized by thermal treatment in oven at 140°C for 2 hours, followed by cooling at room temperature.

TABLE 2

PROPERTIES OF DUTRAL TP

Property	Test method	Unit	Dutral TP30	Dutral TP60	Dutral TP80	Dutral TP100
Density at 23°C	–	g/cm^3	0.875	0.875	0.877	0.878
Nominal low rate	ASTM D 1238/F	g/10	20	65	85	130
Hardness at 23°C	ASTM D 676	Shore D	40	50	55	60
Tensile strength	ASTM D 412/C	Kg/cm^2	90	120	135	140
Elongation at break	ASTM D 412/C	%	350	250	200	200
Tension set at 75°	ASTM D 412	%	30	45	50	55
Flexural modulus of ela- sticity at −30 C	ASTM D 790		3000	8500	11000	15000
at 23 C			1000	2700	4000	5000
at 70 C			250	750	1300	1700
Brittle point	ASTM D 746	°C	−60	−60	−60	−60
Ozone resistance (50 pphm, 72 hours, 38°C)	– –	–	excellent(*)	excellent(*)	excellent(*)	excellent(*)

(*) – no cracking

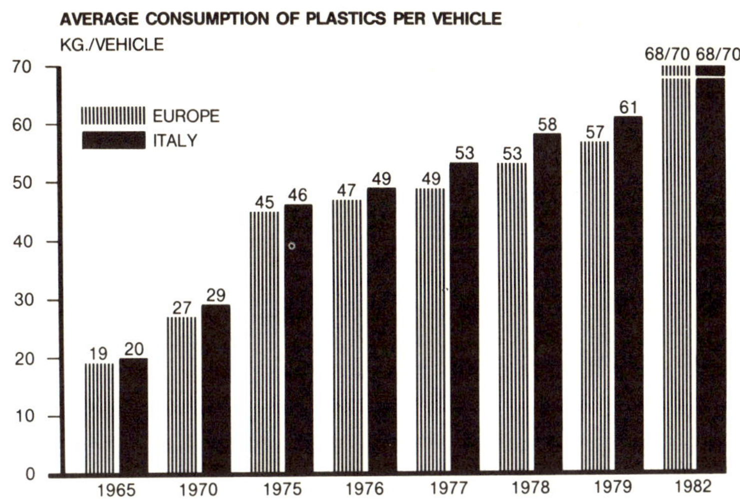

Figure 1. Average consumption of plastics per vehicle

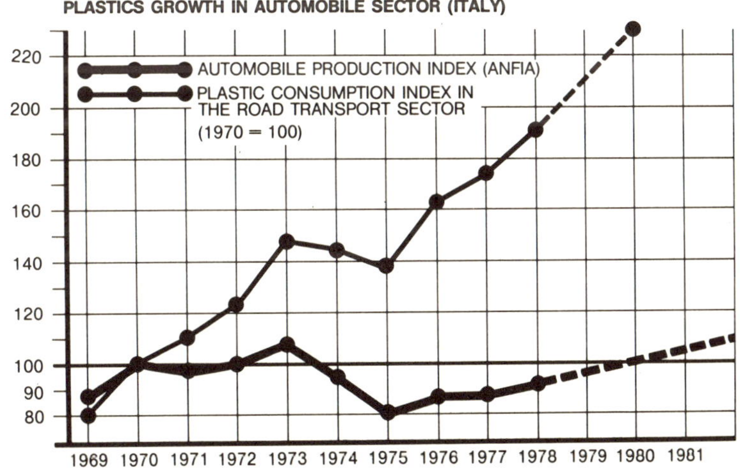

Figure 2. Plastics growth in automobile sector (Italy).

- the development of materials having high impact strength, even at
  low temperatures;
- the development of materials having a high elastic modulus and
  improved dimensional stability;
- the development of a series of low modulus grades for the application fields where this property is particularly called for;
- improvement of UV resistance and resistance to weathering in general.

To meet strict and preset service specifications of end-users, Montedison is also in a position to supply tailor-made materials for
specific applications through a definition of "identikits" in terms
of chemical-physical, mechanical and rheological properties.

The tables 1 and 2 group the main Montedison grades. A wide
range of rigidity values and a high impact strength allow the selection of a correct compromise of properties and the obtainment of
optimum article performance. All the grades indicated can be highly stabilized against hot ageing and weathering for long service
times like those required, for example, by the automotive sector.

Main applications

The automotive industry, among the various application sectors,
represents for these and for other products and exciting field of
employment and a constant stimulus for the development of new grades and the optimization of those already in existence. It acts
as a test bench for the different polymers and the experience and
know-how gained in that industry can be readily transferred to other
sectors with the obtainment of notable induced consumption.

Thanks to the dynamism of this sector and to its continual
search of the new, the average use of plastics per vehicle has been
costantly increasing (Figure 1, 2) and it may be expected that this
trend will be maintained for the future in terms of ever more sophisticated products, if not to the extent of their being created "ad
hoc" for specific applications. On the other hand, it should not
be forgotten that the constant evolution of road vehicle production
techniques, the ever tighter cost limitations, the different design trends, combine to take applications already acquired by plastics
towards obsolescence and raise the problem of their substitution.

The decisive turn of polyolefinic alloys towards market success is in fact due to the automotive sector with the introduction in 1976 of Moplen SP25 in the bumpers of the FIAT 128 car, which gave a big boost to consumption. This success has been made possible by a close cooperation between Montedison, the end-user and the converter - a cooperation that has allowed a correct identification of technical/technological and economic problems and constraints.

The plastic bumper, prompted at all times by well-defined stylistic planning, was first created as a substitution of the metal bumper and took on the basic configuration of the latter, offering a whole series of advantages over metals and/or other plastics:
- weight reduction;
- ease of processing
- elimination of finishing operations (trimming, deflasting, painting, etc.);
- high ageing resistance;
- low cost;
- reprocessability of scraps and sprues;
- less specific machinery;
- freedom from raw material storage problems;
- good design flexibility;
- less damage to bumpers and bodywork for low-speed impact;
- impact resistance with MVSS215 pendulum with central impact at 4 Km/hour, down to $-30°C$.

The versatility of the materials has favoured the evolution of the design process, which has already led to overcoming the concept of the plastic bumper as a substitute of the traditional metal bars, arriving at the creation of bumpers integrated to the bodywork or even at the creation of actual "soft-noses" (FIAT RITMO). The total weight of RITMO front grid and rear end is more than 11 Kg per car.

Also worthy of mention is the extension of the use of polyolefinic alloys to rigid instrument clusters, which are taking on an increasing importance in the European markets.

FIAT RITMO dashboard is manufactured from Moplen SP98/1, weight is about 3 Kg. Its main advantages include:
- a better cost/performance ratio;
- better acoustic properties;

- good colour uniformity (there is no flow-line problem);
- good behaviour towards ageing;
- conformity to the EEC/USA/SWEDEN standards, which require resistance to the impact of a 6.8 Kg metal ball at 24 Km/hour: deceleration must not exceed 80 g for more than 3 milliseconds, and there must be no breakages with dangerous splinter formation.

Moplen SP98/1 and Moplen SP21 have been developed for the specific application and combine properties of dimensional stability, high impact strength and rigidity.

Further examples of important applications, using the lower rigidity grades, are to be found in the following sectors:

Road transport

- cosmetic coatings for bumpers;
- gear levers;
- car side protection.

Footwear industry

- shells for ski boots, heel tips, after-ski boots, shoe soles etc.

Electrical industry and metal pole coating

- insulation of low-voltage cables, coating of metal wire and net, plugs and attachements in general.

Sport and sundry articles

- wheels for roller-skates, grips for motorcycle handle-bars, golf club grips, stoppers, sundry gasketing, multilayer films for agriculture and packing, etc.

Design evolution of high performance items, such as, for example, the bumper

In the purest sense of the word, "bumper" should only be ascribed to an element capable of absorbing or accumulating impact energy, trasferring forces of a limited energy to the body. Through a historical analysis of the solutions commonly adopted it may be said that only the introduction of plastics in this particular

field has led to the realization of such a function, as required by
the ISO and MV SS specifications.

The design of a plastic bumper has entailed the acquisition of
an appropriate designing technique and the identification of basic
parameters capable of describing and defining the behaviour of plastics under the expected dynamic conditions of service. It has
been learnt that recourse to the traditional designing techniques
and the use of the parameters by which plastics are commonly described do not leat to a reliable sizing of car bumpers.

It is indispensable to make recourse to an appropriate calculation technology, such as those based on mathematical models and
processed with the aid of a computer, by means of which the state
of stress and deformation of the bumpers can be described in every
point. It is likewise indispensable to know and to utilize all the
parameters proper to plastics under the dynamic conditions of service expected of the bumpers, which involves a well-defined approach
to the study of plastics above and beyond the traditional schemes.
Figures from n° 3 to n° 9 show some aspects of this designing dua-
lism. The behaviour of the material improves (at low energy levels)
according to time between one impact and another.

MONTEDISON has developed a know-how in the field of structural
design for bumpers and has developed new methods of analysis on the
behaviour of plastic materials. The experience in the designing
of energy absorption bumpers, acquired through a series of trials,
arrived after working on a number of functional and stylistic pro-
posal representing different degrees of design difficulties, for
the solution of which it has been necessary to gradually develop
new and more responsive finished elements and calculation procedures.

Market

Table 3 shows the high growth of consumption that has occured
following the "take-off" of the applications in the automotive sector. During 1979, total consumption is expected to reach about
11,000 tons.

Table 4 shows a similar trend to the previous, but one year
shifted. The market has already reached, in 1979, with 12,000 tons,

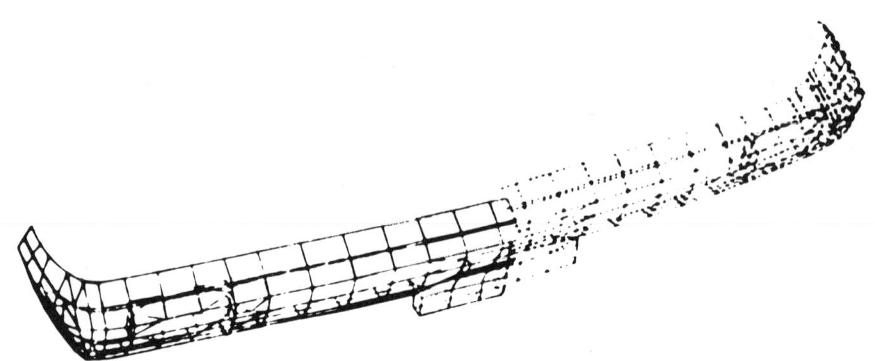

Figure 3. The mathematical model of a bumper and its deformation trend under impact conditions at a rate of 4 Km/hours in the central position.

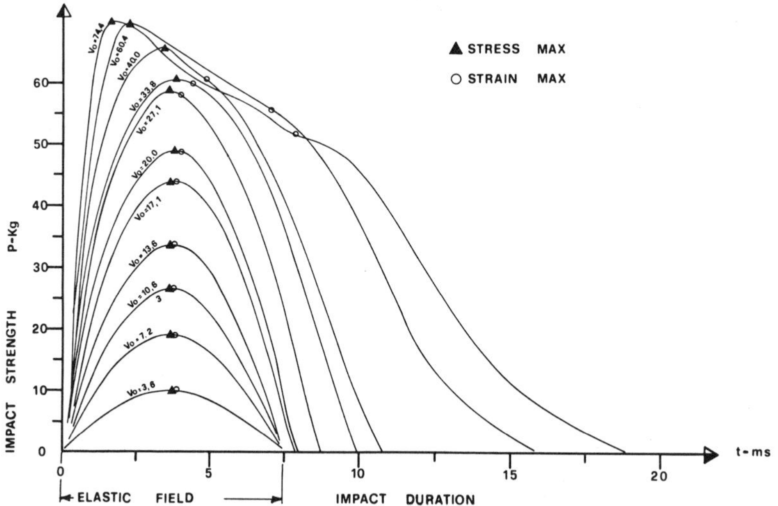

Figure 4. The trend of impact force vs. impulse time. It can be seen that when the impact occurs in the elastic field, the impulse time is constant.

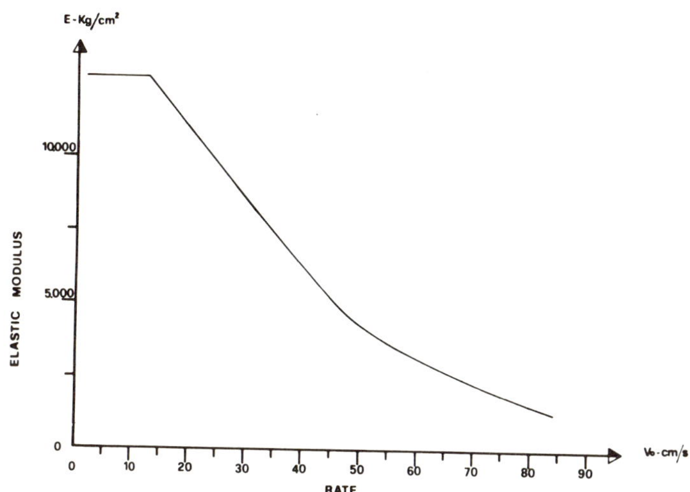

Figure 5. Modulus trend vs. impact rate (values are referred to
specimens, in practice on the bumper, under the current
condition of test the value of elastic modulus is about
14,000 Kg/cm^2).

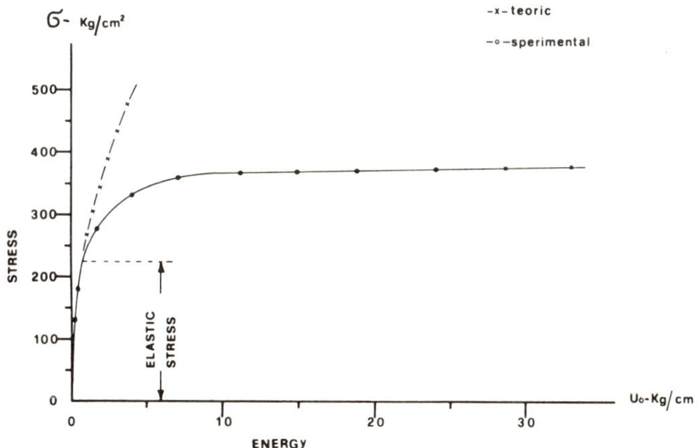

Figure 6. Unit stress vs. impact energy and impact rate.

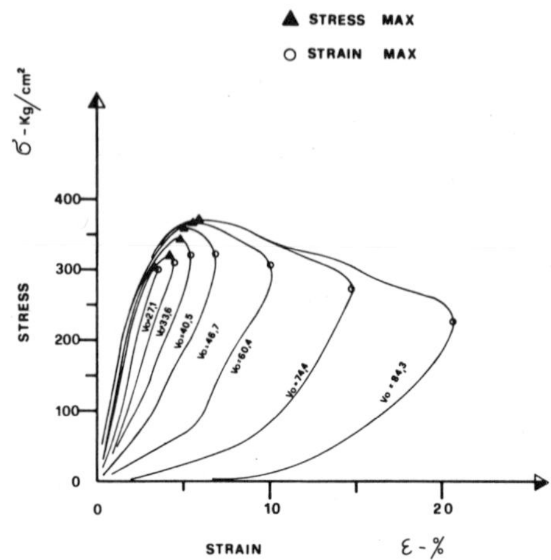

Figure 7.   The bond between stress and deformation at different
            impact speeds. It can be seen that above yield the
            residual energy remains constant even when the impact
            rate is increased.

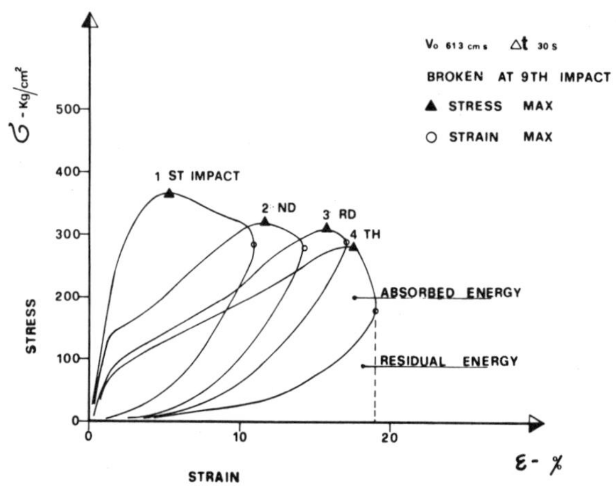

Figure 8.   The behaviour of the material, for equal imposed energy,
            vs. the number of impacts received. It can be seen that
            the absorbed energy and residual energy values remain
            unvaried, whilst there is a falloff in elastic modulus.

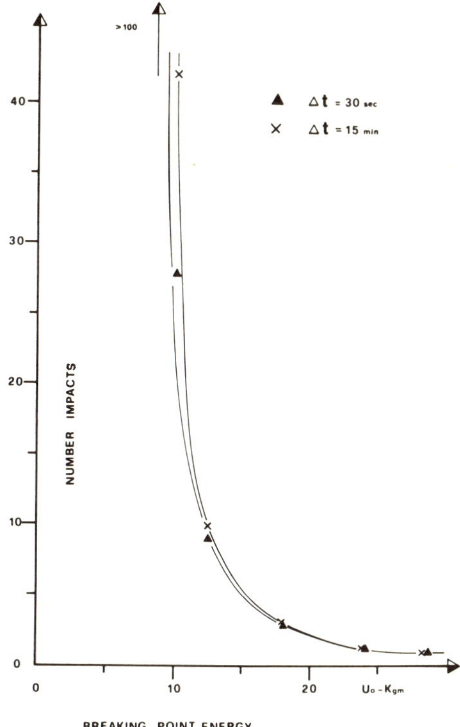

Figure 9. The number of impacts required to bring the material to rupture point vs. impact energy.

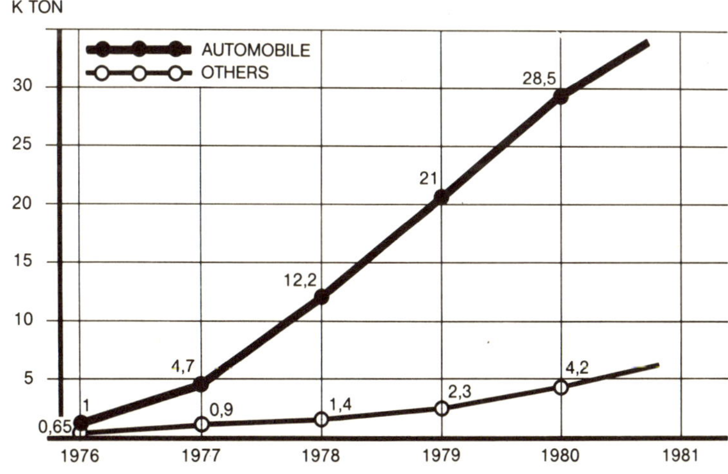

Figure 10. Total European consumption of EMP (Italy included)

TABLE 3

Consumption trend of polyolefinic alloys in Italy

	1976	1977	1978	1979	1980(*)
AUTOMOBILE SECTOR	1,000	3,500	8,000	10,000	11,500
OTHER SECTORS	500	700	1,000	1,100	1,700
TOTAL	1,500	4,200	9,000	11,100	13,200

(*) Forecast

TABLE 4

Consumption trend of polyolefin alloys in Europe (*)
(Italy not included)

	1976	1977	1978	1979	1980(**)
AUTOMOBILE SECTOR	--	1,200	4,200	11,000	17,000
OTHER SECTORS	150	200	400	1,200	2,500
TOTAL	150	1,400	4,600	12,200	19,500

(*) Spain, France, Germany, U.K., Holland, Sweden
(**) Forecast

the italian consumption, and a further heavy growth is expected for
the next 2 or 3 years.

Figures 10 and 11 visualize the overall European market trend
for EMP's, bringing to evidence the transport sector. The two curves in Figure 11 show for the 1980/1976 period, the strong average
growth rate typical of materials during the "take-off" stage. The
comparison of this trend to the one of the whole automotive sector,
that is 9-10%, or to the growth of all plastic materials in various

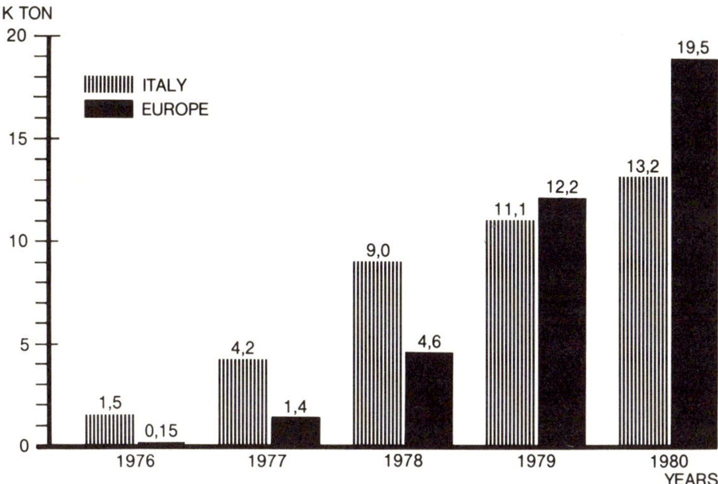

Figure 11.   Comparison between Italian and European consumption of
             EMP.

sectors, that is less than 5%, gives an idea of an importance of
these products, which have now reached, in Europe, an average weight
per vehicle of almost 3 kilos.

CONCLUSIONS

A new exciting chapter in the history of polymeric blends was
opened up a few years ago with the introduction of polyolefinic
compounds onto the market.

This group of products exploit the strong points of the individual component polymers and lend themselves in particular to processing by most of the conversion technologies in use; they withstand
the most severe weathering conditions, and when they are not filled
have density values among the lowest. A rational formulation of
the compound will put the versatility of these products to its best
advantage, and in some cases synergic effects are obtained with respect to the component materials. Optimum balance between rigidity,
flow, impact strength and high-temperature dimensional stability are
objectives that can be reached by a rational formulation of the
compound.

The blending technology, utilized in the diversification of the product grades having a polypropylene matrix, thus places it in a position of extreme versatility for the development of application sectors acquirable with the sole technology via synthesis.

The success of these materials is a highly qualified and specialistic field such as the automotive sector is proof of the high potentiality of these products to satisfy the most diverse requirements.

DISPERSIVE MIXING OF POLYMER MELTS IN THE STATIC MIXER

J. Swietoslawski, J. Morawiec and T. Pakula

Center of Molecular and Macromolecular Studies

Polish Academy of Sciences, 90-362 Łódź, Poland

INTRODUCTION

The multicomponent polymer systems obtained by mixing polymer
melts are very important and interesting, owing to the technical
ease of mixing, which consists mostly of extruding the preliminarly
mixed components. Mixing technology has considerably improved in
recent years due to the development of static mixers[1] in which the
blending and the dispersion of components proceeds by multiple division of the flowing stream of molten polymers on a relatively short
length of the mixer. The efficiency of the mixer depends on the
construction of static mixer units. The geometric design of the
unit results in a unique pattern of flow division and radial and tangential displacement of individual substreams.

This paper will present the results of efficiency studies of
dispersive mixing in the static mixer, originally constructed in
our Laboratory[2].

The term dispersive mixing is applied to those mixing processes
that reduce sizes of minor component particles as well as randomize
their positions in the matrix. In the case of polymer melts only
the laminar mixing is possible because of the too high viscosity of
melts. Usually the blending results from two processes: forced mechanical mixing and spontaneous mixing. The former one is realized
by external forces, developing flow rate gradients for which the
initial configurations of components undergo some deformations.

This usually leads to the formation of structures thermodinamically
unstable. The instability of the deformed structure causes a spontaneous mixing that in the case of incompatible polymers is usually
observed as spontaneous breaking-up of threads of polymer melts
suspended in the molten matrix[3]. Both processes are interrelated.
Higher flow gradients, for example, involve a higher-deformation of
suspended component and, consequently, higher rates of spontaneous
changes. The breaking-up of elongated elements results in the formation of smaller size particles that can be deformed again. The
process looks like a cyclic one, as long as the flow rate gradients
are high enough to involve a considerable deformation of the molten
particles. This leads to better and better dispersion in the blend.
The flow causes changes in particle positions which can lead to increased randomness of the structure. It can result, however, in
counter-active process, i.e. the aglomeration of particles and their
coalescence. Structural changes involving the melt flow of two polymers are schematically illustrated in Figure 1. All the individual
processes can contribute, with various intensities, to any mixing
process depending on the conditions and the type of mixing equipment.
The forced mixing is extremely intensive in static mixers due to the
multiple division of the flowing stream of molten polymers into
substreams which follow different paths inside the mixer. The number
of divisions increases along the mixer and characterizes the mixer
construction.

In all mixing operations one is concerned with the efficiency
of mixing. A given mixer can be characterized by the determination
of parameters describing uniformity and dispersion of the mixture
at various stages of the mixing process.

CHARACTERIZATION OF UNIFORMITY AND DISPERSION OF BLENDS

A complete description of the state of a blend would require
the determination of forms, sizes and positions of all dispersed
particles. Obviously, this is not possible to do and one must be
satisfied with the description based on some mean parameters which
could characterize adequately uniformity, texture and local structure
of the blend.

The uniformity of the blend can be characterized by the distribution of compositions in small volume elements chosen randomly from
the blend as a whole. The sampling volume v_s should satisfy the

relation

$$V \gg v_s \gg v_e \tag{1}$$

where V is the volume of the sample under investigation and v_e is
the volume of the smallest particles in the blend. If the blends
were uniformed then the concentrations would follow the binomial
distribution. To determine whether or not the mixture is uniformed
one must withdraw samples, analyse them and then determine if the
sample compositions are distributed according to the binomial distribution. An extremely large number of samples would be required to
generate enough data, to make a direct comparison against the binomial distribution. Therefore, in most cases, only the deviation
of certain parameters of the observed distribution from the theoretical one, can be used as a measure of the degree of uniformity.
Each distribution can be characterized by the mean and the variance,
the determination of which demands analysis of a relatively small
number of samples. The mean composition $\bar{x}$ of N samples is defined
by the equation

$$\bar{x} = \frac{1}{N} \sum_{i=1}^{N} x_i \tag{2}$$

where x_i is the composition in the i-th sample. If the blend consists of two components A and B the composition is defined as

$$x = v_A = 1 - v_B \tag{3}$$

where v_A and v_B are volumetric contributions of components A and B,
respectively.

The mean $\bar{x}$ should not significantly differ from the nominal or
expected composition p (the fraction of component A in the blend as
a whole). The differences in these two quantities can indicate that
the sampling technique is false or that the mixing process does not
proceed properly. It could become an important characteristic especially for static mixers in which cumulation of one component is
possible. The normal proportion test can be used to determine whether the observed difference between $\bar{x}$ and p is significant. The
test is applied by calculating the quantity

$$z = \frac{(\bar{x} - p)\,\sqrt{N}}{s} \tag{4}$$

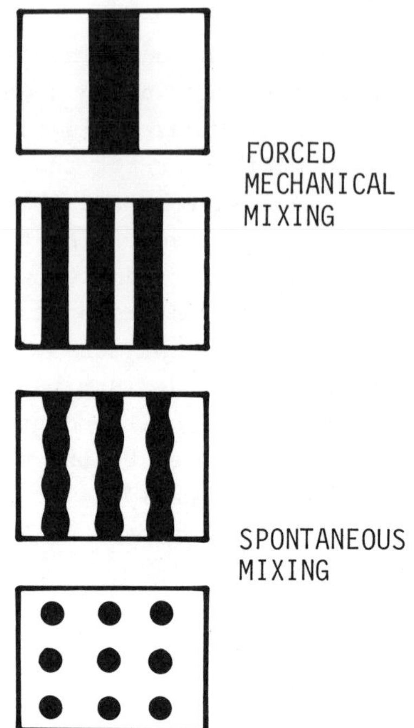

Figure 1. Scheme of structural changes involved by flow of melts
           of two polymers.

and with reference to the tables of the stadarized normal distribution, one can determine the probability that the observed difference
occurred by chance.

The variance s^2 of measured composition is defined by the equation

$$s^2 = \frac{1}{N-1} \sum_{i=1}^{N} (x_i - \bar{x})^2 \tag{5}$$

Many methods have been proposed to specify the degree uniformity on
the basis of discrepances between the measured variance and that
predicted by the binomial distribution of compositions[4].

A useful index has been defined by Lacey[5]

$$M = \frac{s_o^2 - s^2}{s_o^2 - \sigma^2} \qquad (6)$$

where σ^2 represents the variance of the perfectly mixed state, the value of which, according to binomial distribution, is given by

$$\sigma^2 = \frac{p\,(1-p)\,v_e}{V} \qquad (7)$$

and s_o^2 represents the variance of the perfectly unmixed state. It can be calculated by the equation

$$s_o^2 = p\,(1-p) \qquad (8)$$

The index defined by (6) has the advantage of ranging from zero for perfectly unmixed state to unity for the perfectly mixed state and is further called the degree of uniformity. When the value of v_e is very small, the variance σ^2 can be assumed to be zero and eq. 6 reduces to simpler form.

The determination of compositions of samples withdrawn from the blend can be made by measurements of any physical property R related to the composition. If the functional dependence $R = f(p)$ is known and the function $f(p)$ is exactly monotonic, then the monotonic antifunction $p = \varphi(R)$ exists also. In this way, from the measurements of the physycal property R in the i-th sample, the value of composition can be determined uniquely. A further analysis of blend uniformity can be performed as described above. It shows the possibility to perform a comparable analysis of blend uniformity, using different experimental methods.

Local structure of the blend should be characterized by shapes and sizes of particles of a dispersed component. In this paper we consider the case when the particles have the spherical form. It involves considerable limitation of the analysis. It cannot be directly applied to the blend with composition close to the phase inversion point, at which the structure is frequently of the mosaic type. For compositions significantly different from the phase inversion point the commonly observed structures are the emulsion type: the spherical particles of the minor component are dispersed in the matrix.

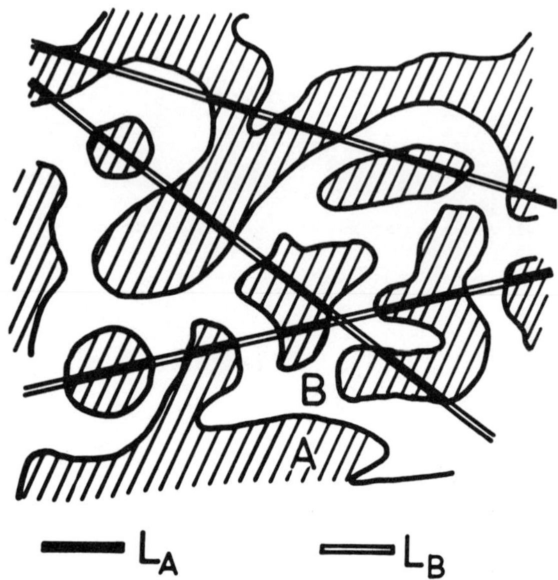

Figure 2.   Illustration of transversal lengths through the components along random lines drawn through the system (redrawn
according to ref. 4)

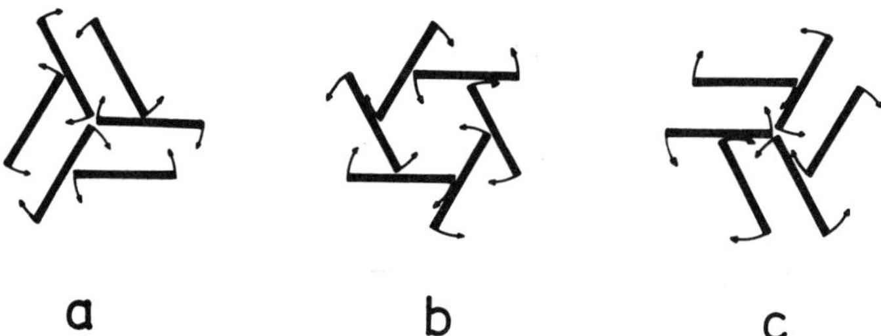

Figure 3.   Cross-sections of six element mixer at various distances
from its entrance: (a) at the entrance, (b) after 1/3
of helix pitch, (c) after 2/3 of helix pitch.

For such blends the local structure is well characterized by the determination of the particle size distribution. Various experimental methods can be applied depending on the range of size of particles in the blend. Direct measurements of sizes from microscopical pictures, light scattering or X-ray scattering techniques are used for this purpose. To simplify the introduction of experimental data, the mean and the distribution variance of sizes can be considered as parameters characterizing the particle size distribution

$$\bar{R} = \int_{0}^{R_{max}} R \cdot n(R)dR \qquad (9)$$

$$S_R^2 = \int_{0}^{R_{max}} (R - \bar{R})^2 \cdot n(R)dR \qquad (10)$$

where $n(R)$ is the size distribution function.

The value of

$$D_R = \frac{\bar{R}_o - \bar{R}}{\bar{R}_o} \qquad (11)$$

will be defined as the dispersion degree and the value

$$D_{SR} = \frac{S_R^2}{\bar{R}} \qquad (12)$$

will be treated as a measure of the uniformity of dispersion, ($\bar{R}_o$ is the average particle size of the initial, reference state of the blend).

For blends with composition close to the phase inversion point and consequently with the mosaic structure, another description of the dispersion may be applied. Useful parameters can be obtained in such a case from distribution of transversal lengths through the phases[6]. The transversal lengths through phases are measured along random lines drawn through the system as illustrated in Figure 2. The average transversal lengths through both phases and the variances of transversal length distributions will characterize, in such a case, the degree and the uniformity of dispersion of components, respectively. The average transversal lengths of components of two-phase system are related to the composition of the system by the relations

$$\overline{L}_A = L\,p \tag{13}$$

$$\overline{L}_B = L(1-p) \tag{14}$$

where $\overline{L}_A$ and $\overline{L}_B$ are the average transversal lengths of components A and B respectively, and

$$\overline{L} = \overline{L}_A + \overline{L}_B \tag{15}$$

is a characteristic length for a given degree of dispersion of the whole system.

As for eqs. 11 and 12 the degree of dispersion will be

$$D_L = \frac{L_o - \overline{L}}{L_o} \tag{16}$$

where L_o characterizes the initial state of the blend.

The uniformity of dispersion for both components independently will be given by

$$D_{SA} = \frac{S_{LA}^2}{\overline{L}_A} \qquad\qquad D_{SB} = \frac{S_{LB}^2}{\overline{L}_B} \tag{17}$$

where S_{LA}^2 and S_{LB}^2 are the variances of the transversal lengths of components A and B respectively. If the local composition or particle size changes in the blend from place to place and this change is correlated then the blend is regarded as textured. The best characterization of the texture can be obtained by determination of the composition correlation function and size correlation function. If the composition is determined for a number of point pairs separated by the distance r and when the concentrations are x' and x", the correlation function of concentration is defined as

$$K_x(r) = \frac{1}{n\,S^2} \sum_{i=1}^{n} (x_i' - p)\,(x_i'' - p) \tag{18}$$

where n is the number of pairs of points.

Analogously the correlation function of particle size can be defined by

$$K_R(r) = \frac{1}{n\,S_R^2} \sum_{i=1}^{n} (R_i' - \overline{R})(R_i'' - \overline{R}) \qquad (19)$$

where R_i' and R_i'' are mean particle sizes, determined at points separated by distance r, and $\overline{R}$ is the mean value of sizes of all points. In general, function K(r) may assume values close to zero (indicating random correlation of point pairs) and unity in the case of perfect correlation. On the basis of correlation functions the following parameters characterizing the texture of the blend can be defined: the linear scale of correlation

$$S_L = \int_O^{r_m} K(r)\ dr \qquad (20)$$

and the volumetric scale of correlation

$$S_V = 2\pi \int_O^{r_m} r^2\ K(r)\ dr \qquad (21)$$

where r_m is the distance at which K(r) = 0. These parameters can describe the scales of correlation for both compositions and particle sizes when determined from compositions or particle size correlation functions respectively.

All parameters which could characterize the uniformity, the dispersion and the texture of blends are summarized in Table I.

EXPERIMENTAL

The static mixer used in these studies (Pakula mixer) has been constituted by number of right-and-left-hand helices packed closed together in a tube[2]. The helices, having the length of the mixer unit and the axes parallel to each other, form a characteristic internal structure of the mixer which determines the number of divisions of the flowing stream. The number of divisions I increases

TABLE I

Parameters characterizing the state of blend

		Parameters determined experimentally	Related characteristics of the blend
Composition uniformity		mean composition $\bar{x}$ (Eq. 2)	probability that the deviation from the nominal composition occurred by chance $p(z)$ (Eq.4)
		variance of composition distribution S^2 (Eq.5)	degree of uniformity M (Eq.6)
Local structure (dispersion)	particles dispersed in the matrix	mean particle size $\bar{R}$ (Eq.9)	degree of dispersion D_R (Eq.11)
		variance of particle size distribution S^2_R (Eq.10)	uniformity of dispersion D_{SR} (Eq.12)
	All structures	mean transversal length $\bar{L}$ (Eq.15)	degree of dispersion D_L (Eq.16)
		variances of transversal length S^2_{LA} and S^2_{LB}	uniformity of dispersion D_{SA} and D_{SB} (Eq.17)
Texture		spatial distribution of composition x', x''	composition correlation function $K_x(r)$ (Eq.18)
		spatial distribution of particle sizes or transversal lengths	size correlation function $K_R(r)$ or $K_L(r)$ (Eq.19)

among the mixer and for the mixer used, it can be estimated by the
following expression

$$I = k \, 6^{3n-1}$$

where k is the number of helices in the cross-section of the mixer,
and the number of helix pitches along the mixer is n. The construction of the mixer is schematically illustrated in Figure 3 by plotting cross-sections at different distances from the entrance of the
mixer unit with six helices. At the distance of one helix pitch
the cross-section is like that at the entrance and changes again
according to the presented scheme. The arrows in the Figure show
the directions of forced displacements of the mixed material after
passing a given cross-section. This indicates a complex picture of
the movement of elemental streams inside the mixer. In Figure 4
a photograph of the mixer unit with six helices is shown. It is a
basic unit of the described mixer. In general, the mixer can be
built up of larger number of helices. The mixer with 48 helices and
diameter of 30 mm was used in our experiments. The length of the
mixer was 300 mm and was divided into 10 units separated by distances
of 2 mm each from the other. The samples, later analysed for the
structure of blended material were taken from these spaces.

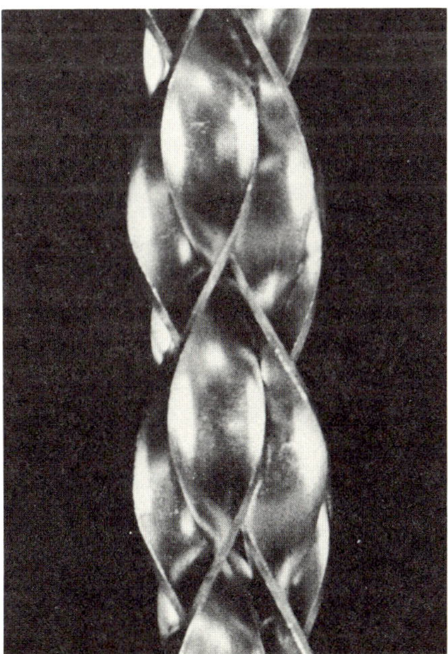

Figure 4. A photograph of six element mixer unit.

As a reference mixing device, Kenics mixer[1] was used with
external dimensions, identical with dimensions of the mixer described
above. Both mixers have been used for blending of two incompatible
polymers: polystyrene and polypropylene. The components to be mixed
were coextruded with mass velocity 25 kg/h into the mixer. A schematic diagram of the equipment used for the extrusion is shown in
Figure 5. The Figure shows also the structure of the extrudate at
the entrance of the mixer. The temperature of the mixer during
extrusion was 520°K.

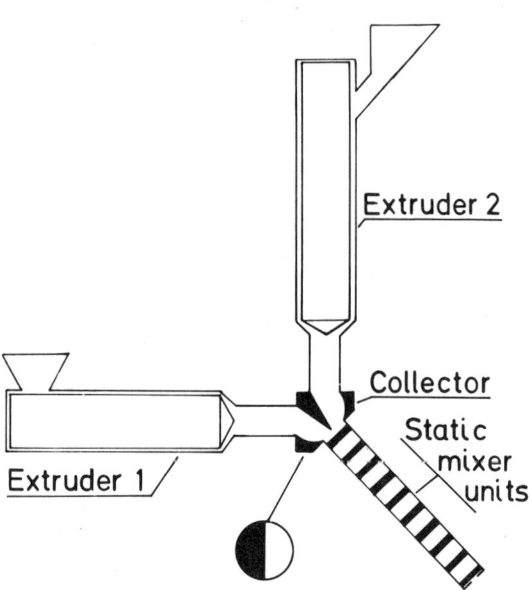

Figure 5. Schematic diagram of the equipment used for extrusion.

To analyse the structure of the blend inside the mixer the
extrusion was suddenly stopped and the mixer, with the material inside, was quenched in water bath. We believe that the actual structure of the blend was fixed in this way. The solidified material
was taken out from the mixer tube and 1mm thick layers perpendicular
to mixer axis were cut out from spaces between mixer units. Each
layer was cut into 40 microvolume elements for which the composition
was determined as a function of the position of the subsample in the
cross-section of the mixer.

RESULTS

Changes of blend homogeneity along static mixers will be characterized according to the analysis described in the first section of this paper.

Figure 6 shows photographs of cross-sections of blended material at various distances from the entrance of the mixers (m is the distance from the entrance expressed by mixer diameters). The photographs show qualitative changes of the spacial distribution of components along both mixers.

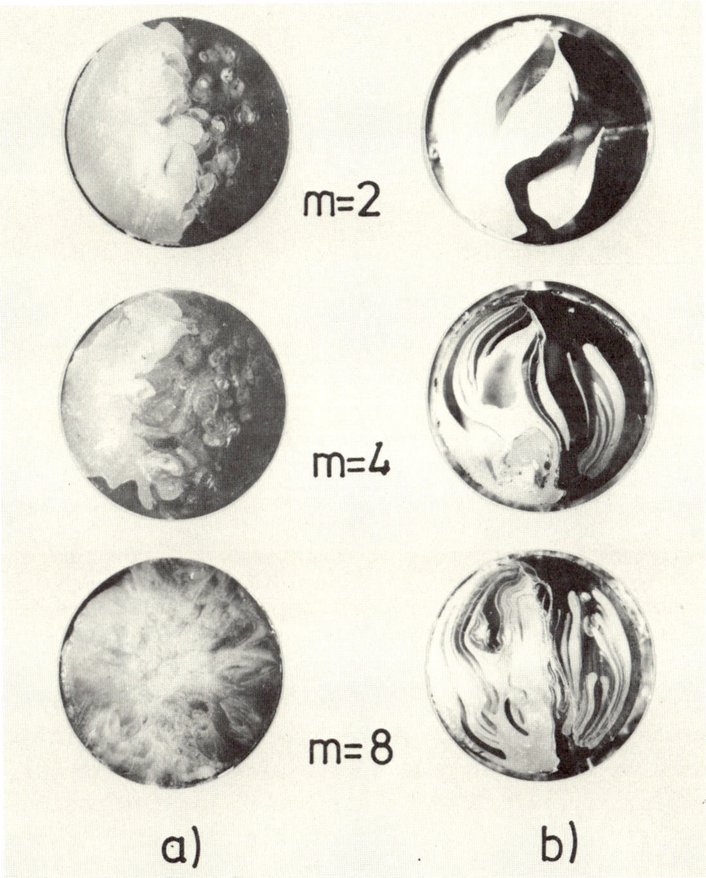

Figure 6. Photographs of cross-sections of blended material at
various distances from the entrance of mixers.
(a) Pakula mixers and (b) Kenics mixer.

From the analysis of compositions in microvolume elements at
a given cross-section of the mixer, the composition distribution can
be determined. Figure 7 shows examples of these distributions
in the form of histograms for both mixers. The distributions n(x)
are normalized so that Σn(x)=1. Each value of n(x) describes the
relative number of microsamples found in a given cross-section of
the mixer which have the composition ranging from x−Δx/2 to x+Δx/2
(where Δx is the elemental composition interval of the histogram).
For the presented distributions the composition is defined as a concentration of polystyrene in the microvolume sample.

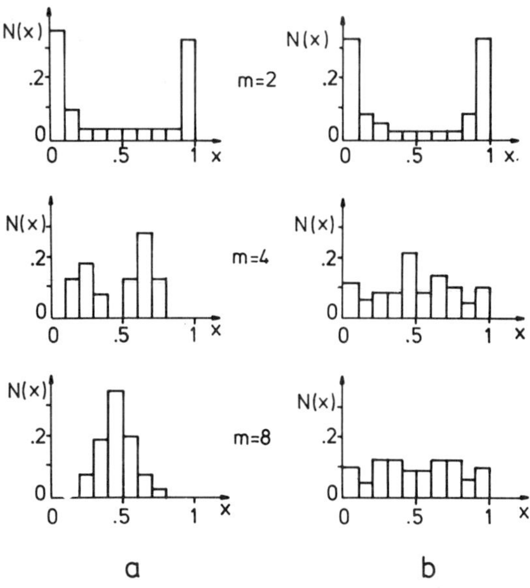

Figure 7.   Composition distributions in samples from two mixers.
             (a) Pakula mixer and (b) Kenics mixer.

As the mixing proceeds along the mixer, the composition distribution in the cross-section of the mixer changes specifically. At
the entrance of the mixer most of the microvolume samples contain
pure components therefore the distribution has two maxima at edges
of the composition range. Increasing distance from the entrance
more and more volume elements will contain both components which
lead to broadening of the maxima of the initial distribution as it
is, for example, well observed in the Kenics mixer. If a good dispersion and homogeneity of the blend are reached, the microvolume
samples will have compositions close to the nominal one and the

composition distribution should show in such a case a maximum at the
nominal composition as for example at the end of the Pakula mixer.
The zero values of the composition distribution at the edges of the
composition range indicate the lack of volume elements containing
only one component. It should be observed in the case of well mixed
material.

It was observed that for both mixers the values of the mean
composition were only slightly deviated from the nominal composition
and it was established by normal distribution test that these deviations occurred by chance with the probability higher than 0.8. To
characterize the degree of uniformity of blended material in both
mixers the parameter M was determined as a function of mixer length.
The results are shown in Figure 8. It is observed that in our mixer
the value of M close to one is reached at relatively short length
of the mixer.

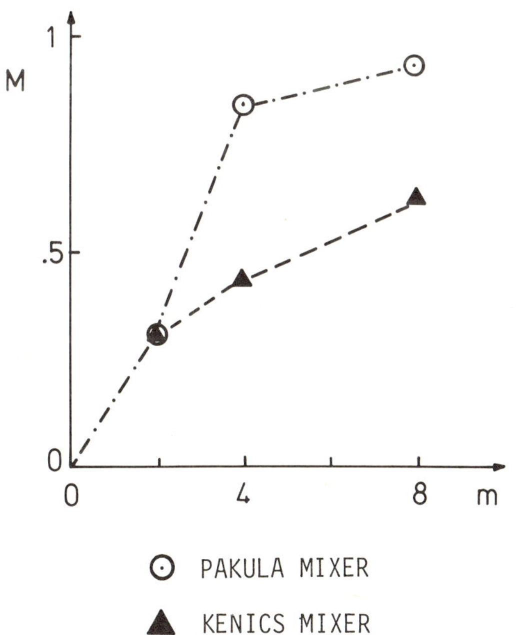

Figure 8. The degree of uniformity as a function of mixer length.

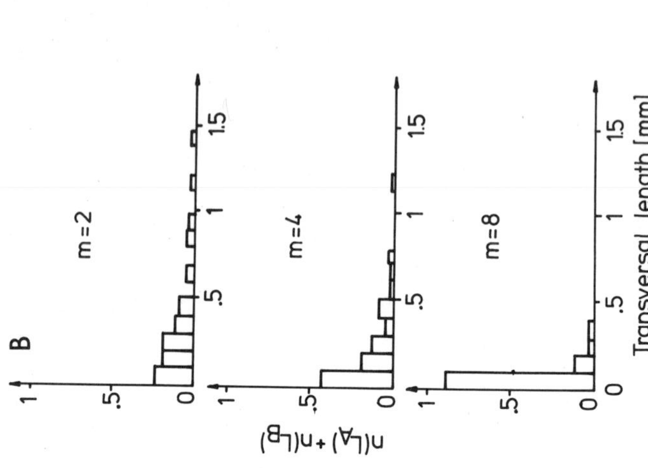

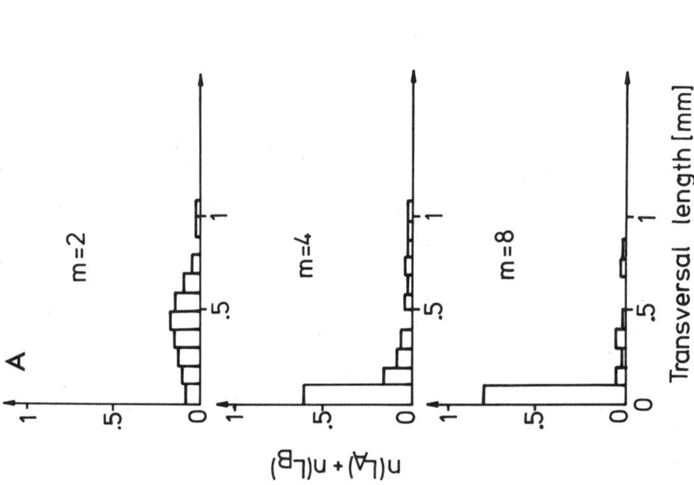

Figure 9. Distributions of transversal lengths in cross-sections of Kenics mixer (a) and Pakula mixer (b).

The dispersion of blends at various distances from the entrance of the mixer was characterized by the distribution of transversal lengths in both components. Distributions of n (L_A) + n (L_B) for both mixers are shown in Figure 9. The transversal lengths were measured from magnified photographs of cross-sections of material blended in the mixer (Figure 6). The smallest sizes which were able to be distinguished by these measurements were 50 μm. In fact the dispersion at the end of our mixer was better than the one noticed from the presented distributions. The similarity of distributions of transversal lengths for both mixers is a result of a low resolution of the measurements. The degree and uniformity of dispersion were, however, calculated from the distributions and respective values for both mixers are collected in Table II.

TABLE II

The degree of dispersion and the homogeneity of dispersion at various distances from the entrance of mixers

m (mixer length)	Kenics mixer			Pakula mixer		
	D_L	D_{SA}	D_{SB}	D_L	D_{SA}	D_{SB}
2	.55	.90	.75	.66	.88	.71
4	.86	1.25	1.41	.82	1.00	1.30
8	.95	1.30	1.60	.98	.74	.82

D_L - the degree of dispersion (Eq. 16).

D_{SA} and D_{SB} - the homogeneity of dispersion for components A and B respectively (Eq. 17).

To characterize the changes of the texture along mixers the composition correlation functions were calculated according to eq.18 for the initial texture (Figure 5) determined by the geometrical configuration of components at the entrance of mixers and for the final state, assuming that, in this case, the blend has homogeneously the nominal composition. The correlation functions for these extremal states of the blend are compared with the correlation functions

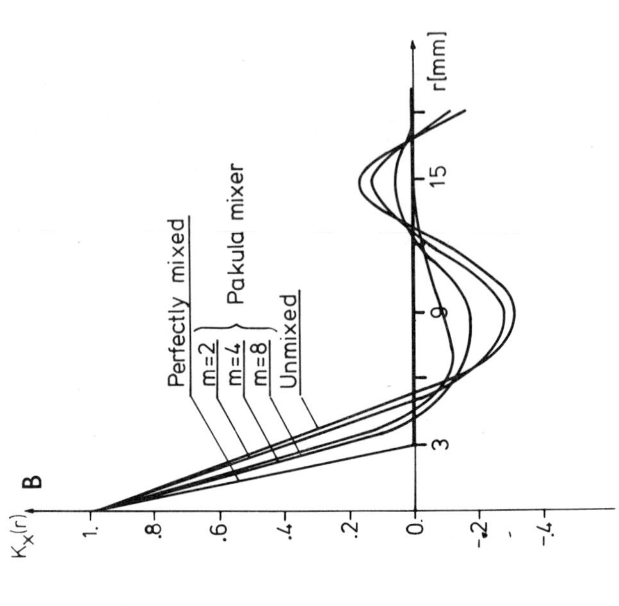

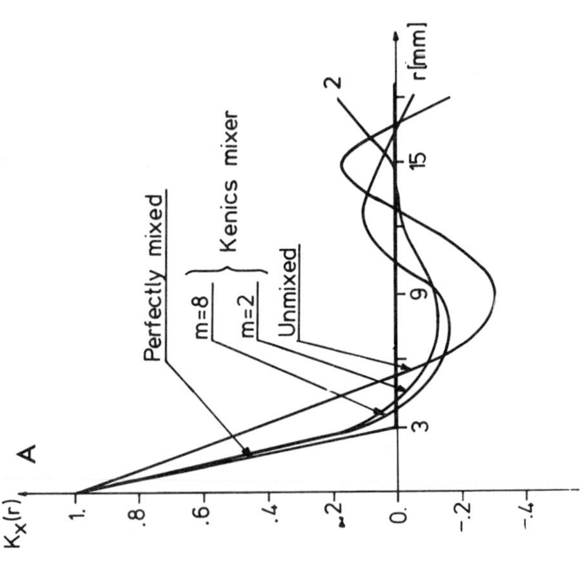

Figure 10. Composition correlation functions (a) for Kenics mixer and (b) for Pakula mixer.

determined from experimental data of composition at various distances
from the entrance of mixers. It is seen from Figure 10a that for
Pakula mixer the correlation function changes regularly along the
mixer from the shape characteristic for the initial texture to the
shape close to that of perfectly homogeneous blend. The changes in
texture are more rapid in the Kenics mixer (Figure 10b) but they do
not lead to the nontextured state, as in our mixer. Characteristics
of texture changes along both mixers obtained in this way, are in
good agreement with qualitative observations (see photographs in
Figure 6).

CONCLUSIONS

The efficiency of two static mixers have been studied by characterization of changes of homogeneity and structure of blends along
the mixers. Three main characteristics of blends have been obtained:
the composition uniformity, the dispersion of components and the
texture. All parameters introduced changes along the mixer dependently on its construction and the determination of all of them is
necessary to complete the characterization of the mixer efficiency.

REFERENCES

1. S.M. Skoblar, Plastics Technology, 20-11, 37 (1974).
2. T. Pakula, Polish Patent, N. 187 386 (1976).
3. T. Pakula, J. Grebowicz, M. Kryszewski, prepared for publication
4. S.S. Weidenbaurn, "Mixing of Solids" in Advances in Chemical
    Engineering, vol. II Ed. T.B. Drew and J.W. Hoopes, Academic
    Press (1958).
5. P.M.C. Lacey, J. Appl. Chem., 4, 257 (1954).
6. O. Kratky, Pure and Appl. Chem., 12, 483 (1966).